Micro/Nanophotonic Devices in Europe

Micro/Nanophotonic Devices in Europe

Editors

Luigi Sirleto

Giancarlo C. Righini

MDPI • Basel • Beijing • Wuhan • Barcelona • Belgrade • Manchester • Tokyo • Cluj • Tianjin

Editors

Luigi Sirleto
Institute of Applied Sciences
and Intelligent Systems (ISASI)
National Research Council
of Italy (CNR)
Napoli
Italy

Giancarlo C. Righini
Nello Carrara Institute of
Applied Physics (IFAC)
National Research Council
of Italy (CNR)
Firenze
Italy

Editorial Office
MDPI
St. Alban-Anlage 66
4052 Basel, Switzerland

This is a reprint of articles from the Special Issue published online in the open access journal *Micromachines* (ISSN 2072-666X) (available at: www.mdpi.com/journal/micromachines/special_issues/micro_nanophotonic_devices_in_europe).

For citation purposes, cite each article independently as indicated on the article page online and as indicated below:

LastName, A.A.; LastName, B.B.; LastName, C.C. Article Title. *Journal Name* **Year**, *Volume Number*, Page Range.

ISBN 978-3-0365-7525-4 (Hbk)
ISBN 978-3-0365-7524-7 (PDF)

Contents

About the Editors

Luigi Sirleto

Luigi Sirleto has been a researcher at National Research Council (CNR) in Naples, Italy, since 2001. As visiting scientist, from 2001 to 2004 he spent several months at the Institute of Nanotechnology, the University of Twente (Netherlands) and at the Electrical Engineering Department of UCLA (University of California Los Angeles), USA. In 2006 he founded the Non-linear and Ultrafast Optics Lab at CNR in Naples and has kept leading the research activities of the same. In April 2018, he received his qualification as Full Professor of Experimental Matter Physics.

He co-edited with Giancarlo Righini the book *Non-Linear Photonics Devices* (MDPI, 2021) and the book *Advances in Non-Linear Photonics* (Elsevier, 2023). Since 2021, he has been a section editor of Optik and an associate editor of the Journal of Non-Linear Optical Physics and Materials. He has published about 200 research papers, mostly on non-linear optics and photonics devices. He has served as a scientific committee member at many national and international conferences.

Giancarlo C. Righini

Giancarlo C. Righini is a former director of the Enrico Fermi Center in Rome, Italy, and of the National Department of Materials and Devices, National Research Council of Italy (CNR). He was for many years research director at the Nello Carrara Institute of Applied Physics (IFAC-CNR) in Florence. In 1991, he was one of the founders of the *Italian Society of Optics and Photonics* (SIOF), and later one of its presidents. He also was vice-president of the *International Commission for Optics* and of the *International Union of Pure and Applied Physics*. He is Honorary Chair of the series of international workshops on Photoluminescence in Rare Earths (PRE).

His research interests have concerned fiber and integrated optics, glass materials, and microresonators, with reference to both linear and non-linear phenomena. He has published more than 500 research papers (ORCID: 0000-0002-6081-6971) and co-edited various books, the most recent being *Advances in Nonlinear Photonics*, co-edited with Luigi Sirleto (Elsevier, 2023). He collaborated with several international journals, both as reviewer and editor; since January 2023, he is associate editor of Ceramics International. He is a fellow of Optica (formerly OSA), SPIE, SIF (Italian Physical Society), EOS (European Optical Society) and SIOF.

micromachines

MDPI

Editorial

Editorial for the Special Issue on Micro/Nanophotonic Devices in Europe

Luigi Sirleto [1,*] and Giancarlo C. Righini [2,*]

1 Institute of Applied Sciences and Intelligent Systems (ISASI), CNR, 80131 Napoli, Italy
2 "Nello Carrara" Institute of Applied Physics (IFAC), CNR, Sesto Fiorentino, 50019 Florence, Italy
* Correspondence: luigi.sirleto@cnr.it (L.S); righini@ifac.cnr.it (G.C.R.)

Citation: Sirleto, L.; Righini, G.C. Editorial for the Special Issue on Micro/Nanophotonic Devices in Europe. *Micromachines* **2023**, *14*, 861. https://doi.org/10.3390/mi14040861

Received: 11 April 2023
Accepted: 14 April 2023
Published: 16 April 2023

Photonics has often been defined as the key technology of the 21st century. In recent decades, the overall trend of research and development (R&D) towards the miniaturization of devices and systems, pushed by the tremendous development of microelectronics, has led to continuous development of micro/nanophotonic devices. They have huge potential for low-cost, scalable production, as well as the integration of high-density components. Their evolution would enable advanced applications to emerge not only in optical processing and computing, but also in metrology, single-molecule sensing, imaging, microscopy, mid-infrared photonics, terahertz generation, microwave photonics, and biomedicine.

Many research centers worldwide are working on projects related to advances in micro/nanophotonic devices. Although it is challenging to define the borders of such a research area, it is certainly true that a significant number of scientific projects have been or are being carried out in Europe, often in collaboration with colleagues from abroad. The growth of such projects has certainly been facilitated by the multi-year programs funded by the European Commission; among the successful examples of cooperation projects at the European level, one can mention the regional photonics clusters, corresponding to national technology platforms, and the European Technology Platform Photonics21.

By examining the scientific production in the areas of microphotonics and nanophotonics, one can see (perhaps with some surprise) that the publications listed under the topic "microphotonics" are 20 times fewer (≈550) than those appearing under the topic "nanophotonics" (≈11,050), according to the Clarivate Web of Science™. If we refer, for the sake of simplicity, to the sole area of nanophotonics, and we filter the results on the basis of the authors' country of residence (as indicated in each publication), the result is that 3778 publications, i.e., over 34% of the total, are authored by at least one scientist from one of the 28 countries of the European Community, plus Switzerland and the UK. The image in Figure 1 gives a graphical representation of the larger geographical contributions; it also makes the importance of the collaboration with abroad countries, such as USA and China, evident.

The present Special Issue (SI) of *Micromachines*, entitled "Micro/Nanophotonic Devices in Europe", represents a weak effort to shed some light on the ongoing researches and the achieved results in the European laboratories. Although it is far from an exhaustive overview, this Special Issue presents a few examples of the potentiality of micro/nanophotonics devices in a relatively wide spectrum of applications.

In the first group of papers, three reviews are proposed, aiming to highlight the state of the art of integrated photonics and its applications in optical communications and optical processing.

Figure 1. Chart presenting some of the countries of authors of publications relevant to the "nanophotonics" topic, according to a search in Clarivate Web of Science™; the search was refined by including only the 28 countries of European Community, plus Switzerland and the UK, in the "Countries/Regions" field. Here, only the countries with greater numbers of publications are shown. Since co-authors of a same publication may be affiliated with institutions in different countries (even outside Europe), the total of the numbers in the chart is greater than the number obtained from the refined search. Note: The areas on the chart are not strictly proportional to the values of each entry.

Silicon photonics is currently a commercially established and yet fast-growing technology for communication systems. To monitor the status of fabrication processes and characterize the performance of photonic circuits after production, wafer-scale testing is an essential technology in a semiconductor production line. Various techniques can be used to couple the optical signal from optical fibers to photonic chips, such as prism coupling, butt coupling, end-fire coupling, and grating coupling. However, these techniques allow for the performance of the entire circuit to be measured during wafer-scale testing, although it is not possible to test any subsystems or individual devices on the photonic chips. Therefore, in order to routinely test some key sensitive subsystems and individual devices on the photonic chips at the wafer scale, erasable grating couplers and directional couplers are highly desirable. Germanium (Ge) ion implantation into silicon waveguides induces lattice defects in the silicon, which can eventually change the crystal silicon into amorphous silicon and increase the refractive index from 3.48 to 3.96. A subsequent annealing process, using either an external laser or integrated thermal heaters, can partially or completely remove these lattice defects and gradually change the amorphous silicon back into its crystalline form, therefore reducing the material's refractive index. Keeping in mind this change in optical properties, various erasable photonic devices which can be used to implement a flexible and commercially viable wafer-scale testing method for a silicon photonics fabrication line were reviewed in the paper by Yu and others [1]. In addition, Ge ion implantation and annealing were also demonstrated to enable post-fabrication trimming of ring resonators and Mach–Zehnder interferometers (MZIs). Finally, a proof of principle of nonvolatile programmable photonic circuits with the Ge implantation technology was also implemented. It is certain that, based on the refractive index change associated with the aforementioned Ge ion implantation and annealing, the flow of an optical signal within the photonic circuit can be switched or rerouted using erasable directional couplers or trimmable MZIs [1].

The combination of integrated optics technologies with nonlinear photonics, which has led to the growth of nonlinear integrated photonics, has also opened the way for groundbreaking new devices and applications. In this Special Issue, two companion review papers, dedicated to nonlinear photonics and aimed at offering a broad overview, are proposed [2,3]. In the former, the main physical processes involved in nonlinear photonics applications are introduced, starting from traditional second-order and third-order phenomena and progressing to ultrafast ones (self- and cross-phase modulation, supercontinuum generation, and optical solitons). The applications, on the other hand, were made possible by the availability of suitable materials with high nonlinear coefficients, and/or by the design of guided-wave structures, which can enhance the materials' nonlinear properties. A summary of the most common nonlinear materials was also presented, together with a discussion of the innovative ones. Silicon and related materials, such as SiN, a-Si, and SiC; glasses, such as silica, high-index glass, and chalcogenide ones; III–V semiconductors; lithium niobate; and, more recently, investigated materials such as tantalum pentoxide and vanadium dioxide were discussed, and their pros and cons were pointed out. Lastly, two-dimensional layered materials and zero-index media were also introduced [2]. The latter review aimed at describing the development of integrated photonics, which began with centimeter-long circuits, progressed through microphotonics, and finally arrived at nanophotonics. The first step, from the centimeter to micrometer scale, was motivated by the wish to investigate light behavior and its interaction with micro-objects on the microscale. The key challenges were reducing the sizes of optical devices and improving their performances. The main goal was to develop a reliable platform for dense integration. The second step, from micro to nano, was motivated by the demand that, in the near future, it is expected for devices to control light in very thin nanoscale layers or in nanoparticles of nonlinear material. Various waveguide structures, fabrication processes, and integration platforms were discussed. Finally, a brief, and certainly not exhaustive, overview of nonlinear photonics devices was given; for simplicity, the examples were classified into three groups, each characterized by a common goal, even if quite broad: all-optical digital devices, devices for all-optical processing, and nonlinear light sources [3].

Nanophotonics is an attractive field of recent development. In the second group of papers, two contributions were proposed: the first related to metamaterials and the second related to dielectric nanostructures. These are among the most promising approaches for nanophotonics devices.

Toroidal multipoles are a class of fundamental electromagnetic excitations that complement the more familiar electric and magnetic multipole families. Recently, metamaterials have provided fertile ground for the observation of toroidal dipole and higher-order multipoles through the ability to judiciously shape the meta-atom/meta-molecule geometry in the unit cell. In the referenced paper [4], a conductive meta-atom of toroidal topology was studied both theoretically and experimentally, demonstrating a sharp and highly controllable resonant response. Simulations were performed both for a free-space periodic metasurface and a pair of meta-atoms inserted within a rectangular metallic waveguide. A quasi-dark state (almost non-radiating) with controllable radiative coupling was supported, allowing for the linewidth (quality factor) and lineshape of the supported resonance to be tuned via the appropriate geometric parameters. The structure was fabricated with a 3D printer and coated with silver paste. Importantly, the structure was planar, consisted of a single metallization layer, and did not require a substrate when neighboring meta-atoms were touching, resulting in a practical, thin, and potentially low-loss system. Measurements were performed in the 5 GHz regime with a vector network analyzer, and a good agreement with the simulations was demonstrated.

Plasmon-based devices operating at an optical and near-infrared frequency have been demonstrated to reach extraordinary field confinement capabilities, with localized mode volumes of only down to a few nanometers. Although such levels of energy localization are substantially unattainable with dielectrics, it is possible to operate subwavelength field confinement by employing high-refractive index materials with proper patterning, e.g., photonic crystals and metasurfaces. In the last 20 years, optical surface modes, i.e., optical modes strongly confined at the truncation interface of planar dielectric multilayers (called Bloch surface waves (BSWs)), on flat and patterned dielectric multilayers have been investigated in many frameworks. In another paper [5], BSWs, as a viable alternative to plasmonic platforms allowing for easy wavelength and polarization manipulation and reduced absorption losses, are investigated. A computational study on the transverse localization of BSWs by means of quasi-flat Fabry–Perot microcavities, which have the advantage of being fully exposed toward the outer environment, was proposed. These structures are constituted by defected periodic corrugations of a dielectric multilayer top surface. The dispersion and spatial distribution of the cavity mode of BSWs were studied. In addition, the hybridization of BSWs with an A exciton in a 2D flake of tungsten disulfide (WS2) was also addressed. WS2 monolayers display a narrow and intense excitonic resonance at 2.03 eV that is well suited to promoting mode hybridization with BSWs. A strong coupling, involving not only propagating BSWs, but also localized BSWs, namely, band-edge and cavity modes, was pointed out.

The recent developments in biophotonics have pushed the application of photonic devices towards microscopy and imaging. The third group of papers [6–8] can be placed in this frame.

Long-focusing range beams (LFRB), also known as quasi-Bessel beams, are appealing due to their non-diffractive properties. Their use in various important applications, such as in optical manipulation, high-resolution lithography, and microscopy, can be enhanced if the focal spot has a submicron full width at half maximum (FWHM) diameter. Another desirable feature when working with LFRBs is the possibility to tune the light distribution without moving parts in the experimental set-up. A possible solution can be obtained through forming LFRBs by exploiting bilayers composed of weakly absorbing film and thicker layers possessing thermo-optical properties. The working principle is based on the phase change of an input Gaussian beam induced in the latter film via the thermo-optical effect, while the heating power is produced by the optical absorption of the former film. The phase-modified beam is transformed by an objective into a long focus with a submicron diameter. A difficult challenge is the fabrication of ultrathin thin absorbers with (1) high flat spectral transmission, (2) sufficient absorption in the Vis–NIR spectral region, (3) high and uniform heat conduction, and (4) high chemical and structural stability. In this context, particular attention is focused on the emerging 2D materials, thanks to their intrinsic characteristic of low effective thickness and interesting optical properties for photonics applications. The theoretical and experimental characterizations of an elastomeric polydimethylsiloxane/single-layer graphene (PDMS/SLG) axicon-like tunable device, which is able to generate diffraction-resistant submicrometric spots, were reported in [6]. The results demonstrated that in spite of its low absorption, the SLG can significantly raise its temperature and even deliver moderate optical pump powers by heating the PDMS sufficiently to push its temperature close to the tolerance limit (~470 K) and generating a remarkable refractive index gradient. The proposed thermo-optical device can, thus, be useful for the simple and cheap improvement of the spatial resolution on long focus lines.

Traditionally, in carrying out macro tensile tests, the strain components are measured by means of a strain gauge attached to the sample. Of course, for miniaturized specimens, these instrumentation devices may not be suitable due to the reduced sample size. A non-contact optical technique can be more suitable and more conveniently used for accessing full-field strains over the target region. For this purpose, several white-light and interferometric optical methods have been proposed in the literature. Among them, the Digital Image Correlation (DIC) has the merit to allow one to measure full-field kinematic quantities by tracking unique features of images taken at different deformation stages. The DIC technique uses a camera-lens optical setup, and it allows for variable sensitivity and resolution in strain measurements. Reference [7] described how the digital image correlation (DIC) technique may be coupled with mechanical tests to quantify the deformation of a material across the central region of interest for both macro and miniaturized specimens. The development of equipment dedicated to performing uniaxial, tensile–compression tests on miniaturized metallic specimens was presented. The experimental equipment, called MSTD (miniaturized specimen tester device), permitted the testing of specimens on a miniaturized scale, thus opening up a broad field of applications, one of them being compressive loading tests, as the equipment has the advantage of overcoming buckling effects. In addition to uniaxial tensile, compression, and cyclic tests, it is also possible and expected that equipment will adapt to other fundamental tests for metallic sheets, such as shear and ductile damage tests.

Non-contact current surface texture analysis techniques include coherence scanning interferometry, confocal microscopy, and scanning laser microscopy. A drawback of the currently employed surface texture analysis techniques is their inability to capture the full surface of a sample, thus leaving users restricted to being able to analyze only a limited area. A unique solution to the aforementioned drawback may be found in electrically tunable lenses (ETL), which possess the ability to alter their optical power in response to an electric signal. This feature allows such systems not only to image the areas of interest, but also to obtain spatial depth perception. The mechanism of ETL is achieved via two main approaches: local variations in the refractive index, which include liquid crystals, and controlling the shape of the lens with two main techniques that use electrowetting and shape-changing polymers. A low-cost, non-destructive, ETL-based imaging system for the 3D surface characterization of materials was described in reference [8]. The system was based on a polymer-based liquid ETL. The tuning of the voltage and feeding of the ETL enables the adjustment of the ETL focal power. Images were captured via a camera attached to the ETL system. The system as then employed to measure the amplitude, spatial, hybrid, and volume surface texture parameters of a model material (pharmaceutical dosage form), and these were validated against the parameters obtained using optical profilometry.

One of the most important and significant area of applications for photonics devices is sensors and actuators. The fourth group of papers [9–11] is related to this field of applications.

Optical fiber sensors have provided effective sensing devices for decades. A wide variety of configurations have been successfully reported, according to the development of optical fiber fabrication techniques. Single-mode–multimode–single-mode (SMS) fiber structures have received significant attention in the last years due to their unique spectral features, high sensitivity, easy fabrication, and potential low cost. The general sensing principle of an SMS fiber sensor is based on the self-imaging effect in a multimode waveguide, i.e., multimode interference. In this framework, a new and attractive approach is to consider specialty fibers. An interesting example is given by square-core fibers, which produce an optical beam with uniform intensity over the core area, because the shape of the core promotes mode mixing as light propagates through the fiber. In one of the papers included in this Special Issue [9], the sensing performance of a simple multimode-interference-based fiber sensor, containing a square-core fiber 30 cm in length, was investigated for temperature and strain measurement. In addition, a comparison to other specialty fibers, which were mostly circular core shapes, was also presented. The experimental results demonstrated that this fiber sensor was able to exhibit great potential for applications in strain-insensitive temperature measurement.

The conversion of light into mechanical work has been a hot topic for many years. This approach is of great interest because actuators can be temporally, spatially, and remotely controlled by light, which can be beneficial in various applications. One of the most investigated photo-responsive liquid crystalline molecules is Azobenzene, which can be used as a molecular engine in photo-mobile materials. The use of azobenzene molecules embedded as moieties in liquid crystal elastomers yields photo-mobile materials that efficiently convert light into mechanical output. In reference [10], the thermo/mechanical phenomena, regulating the actuation of three LC mixtures consisting of 100% azobenzene moieties, were investigated. The nematic temperature of the LC elastomers was measured and the PMPs were carefully characterized for their bending and speed capabilities. The results demonstrated that the nematic temperature of the LC mixture was greatly influenced by the ratio of the linear and cross-linker monomer; when their ratio was in favor of the linear monomer, the PMPs bent more and the process was faster compared to the other mixtures. Finally, it was shown that the thermal effects did not play a predominant role in the movement of the film, and that such material could be useful for designing polarization-selective switches.

Quartz-enhanced photoacoustic spectroscopy (QEPAS) is a well-known technique used for the detection of specific trace gases in complex mixtures. The high performances in terms of selectivity and sensitivity allow for the exploitation of this technique in a wide range of applications. In QEPAS, acoustic waves are generated between the prongs of a quartz tuning fork (QTF) by the absorption of modulated light from the gas molecules via non-radiative relaxation processes. QTFs are employed as piezoelectric-sensitive elements to transduce pressure waves in an electric signal. These quartz resonators are characterized by a good immunity to environmental acoustic noise due to their high quality factors (Q) and compact dimensions. Due to the sharp resonance, external noise sources outside of the resonator's small bandwidth (~4 Hz at atmospheric pressure) do not influence the QTF signal. Suitable front-end electronics must be designed to read out the signal generated by the QTF and to optimize the signal-to-noise ratio (SNR), as well as the minimum detection limit of the gas concentration. A theoretical study of the SNR trend in a voltage-mode read-out of QTFs was presented [11], mainly focusing on the effects of (i) the noise contributions of both the QTF-equivalent resistor and the input bias resistor RL of the preamplifier; (ii) the operating frequency; and (iii) the bandwidth (BW) of the lock-in amplifier low-pass filter. As a result of the study, general guidelines for the choice of resistor RL and the most suitable operating frequency for the QEPAS system implementing the voltage-mode read-out of QTFs was derived. Furthermore, the proposed analysis allows for the study of the noise contributions at different bandwidths to optimize the acquisition time of QEPAS measurements.

Finally, we would like to thank all the authors for their submissions to this Special Issue; we very much appreciate their contributions. We also thank all the reviewers for dedicating their time and helping to ensure the quality of the submitted papers. Last but not least, we are grateful to the staff at the editorial office of *Micromachines*, and in particular, to Mr. Dikies Zhang, for their efficient assistance.

Conflicts of Interest: The authors declare no conflict of interest.

References

1. Yu, X.; Chen, X.; Milosevic, M.M.; Shen, W.; Topley, R.; Chen, B.; Yan, X.; Cao, W.; Thomson, D.J.; Saito, S.; et al. Ge Ion Implanted Photonic Devices and Annealing for Emerging Applications. *Micromachines* **2022**, *13*, 291. [CrossRef] [PubMed]
2. Sirleto, L.; Righini, G.C. An Introduction to Nonlinear Integrated Photonics Devices: Nonlinear Effects and Materials. *Micromachines* **2023**, *14*, 604. [CrossRef] [PubMed]
3. Sirleto, L.; Righini, G.C. An Introduction to Nonlinear Integrated Photonics: Structures and Devices. *Micromachines* **2023**, *14*, 614. [CrossRef] [PubMed]
4. Tsilipakos, O.; Viskadourakis, Z.; Tasolamprou, A.C.; Zografopoulos, D.C.; Kafesaki, M.; Kenanakis, G.; Economou, E.N. Meta-Atoms with Toroidal Topology for Strongly Resonant Responses. *Micromachines* **2023**, *14*, 468. [CrossRef] [PubMed]
5. Marcucci, N.; Guo, T.-L.; Pélisset, S.; Roussey, M.; Grosjean, T.; Descrovi, E. Bloch Surface Waves in Open Fabry–Perot Microcavities. *Micromachines* **2023**, *14*, 509. [CrossRef] [PubMed]
6. Margheri, G.; Barbosa, A.N.; Freire, F.L., Jr.; Del Rosso, T. Thermooptical PDMS-Single-Layer Graphene Axicon-like Device for Tunable Submicron Long Focus Beams. *Micromachines* **2022**, *13*, 2083. [CrossRef] [PubMed]
7. Cruz, D.J.; Xavier, J.; Amaral, R.L.; Santos, A.D. A Miniaturized Device Coupled with Digital Image Correlation for Mechanical Testing. *Micromachines* **2022**, *13*, 2027. [CrossRef] [PubMed]
8. Nirwan, J.S.; Lou, S.; Hussain, S.; Nauman, M.; Hussain, T.; Conway, B.R.; Ghori, M.U. Electrically Tunable Lens (ETL)-Based Variable Focus Imaging System for Parametric Surface Texture Analysis of Materials. *Micromachines* **2022**, *13*, 17. [CrossRef] [PubMed]
9. Wang, K.; Dong, X.; Kienle, P.; Fink, M.; Kurz, W.; Köhler, M.H.; Jakobi, M.; Koch, A.W. Optical Fiber Sensor for Temperature and Strain Measurement Based on Multimode Interference and Square-Core Fiber. *Micromachines* **2021**, *12*, 1239. [CrossRef] [PubMed]
10. Sagnelli, D.; Rippa, M.; D'Avino, A.; Vestri, A.; Marchesano, V.; Petti, L. Development of LCEs with 100% Azobenzene Moieties: Thermo-Mechanical Phenomena and Behaviors. *Micromachines* **2022**, *13*, 1665. [CrossRef] [PubMed]
11. Di Gioia, M.; Lombardi, L.; Marzocca, C.; Matarrese, G.; Menduni, G.; Patimisco, P.; Spagnolo, V. Signal-to-Noise Ratio Analysis for the Voltage-Mode Read-Out of Quartz Tuning Forks in QEPAS Applications. *Micromachines* **2023**, *14*, 619. [CrossRef] [PubMed]

micromachines

Review

Ge Ion Implanted Photonic Devices and Annealing for Emerging Applications

Xingshi Yu [1], Xia Chen [1], Milan M. Milosevic [1], Weihong Shen [1,2], Rob Topley [1], Bigeng Chen [3], Xingzhao Yan [1], Wei Cao [1], David J. Thomson [1], Shinichi Saito [4], Anna C. Peacock [1], Otto L. Muskens [1] and Graham T. Reed [1,*]

1 Optoelectronics Research Centre, University of Southampton, Southampton SO17 1BJ, UK; xy1m13@soton.ac.uk (X.Y.); xia.chen@soton.ac.uk (X.C.); M.Milosevic@soton.ac.uk (M.M.M.); shenweihong@sjtu.edu.cn (W.S.); rob.topley@gmail.com (R.T.); Xingzhao.Yan@soton.ac.uk (X.Y.); wei.cao@soton.ac.uk (W.C.); d.thomson@soton.ac.uk (D.J.T.); acp@orc.soton.ac.uk (A.C.P.); o.muskens@soton.ac.uk (O.L.M.)
2 State Key Laboratory of Advanced Optical Communication Systems and Networks, Shanghai Jiao Tong University, Shanghai 200240, China
3 Zhejiang Lab, Hangzhou 311100, China; chenbg@zhejianglab.com
4 Electronics and Computer Science, University of Southampton, Southampton SO17 1BJ, UK; ss1a11@ecs.soton.ac.uk
* Correspondence: G.Reed@soton.ac.uk

Abstract: Germanium (Ge) ion implantation into silicon waveguides will induce lattice defects in the silicon, which can eventually change the crystal silicon into amorphous silicon and increase the refractive index from 3.48 to 3.96. A subsequent annealing process, either by using an external laser or integrated thermal heaters can partially or completely remove those lattice defects and gradually change the amorphous silicon back into the crystalline form and, therefore, reduce the material's refractive index. Utilising this change in optical properties, we successfully demonstrated various erasable photonic devices. Those devices can be used to implement a flexible and commercially viable wafer-scale testing method for a silicon photonics fabrication line, which is a key technology to reduce the cost and increase the yield in production. In addition, Ge ion implantation and annealing are also demonstrated to enable post-fabrication trimming of ring resonators and Mach–Zehnder interferometers and to implement nonvolatile programmable photonic circuits.

Keywords: silicon photonics; optical waveguide; Ge ion implantation; annealing; wafer-scale testing; post-fabrication trimming; programmable photonic circuits

Citation: Yu, X.; Chen, X.; Milosevic, M.M.; Shen, W.; Topley, R.; Chen, B.; Yan, X.; Cao, W.; Thomson, D.J.; Saito, S.; et al. Ge Ion Implanted Photonic Devices and Annealing for Emerging Applications. *Micromachines* **2022**, *13*, 291. https://doi.org/10.3390/mi13020291

Academic Editor: Hugo Aguas

Received: 15 January 2022
Accepted: 7 February 2022
Published: 12 February 2022

Publisher's Note: MDPI stays neutral with regard to jurisdictional claims in published maps and institutional affiliations.

1. Introduction

Silicon photonics is currently a commercially established and yet fast-growing technology for communication systems. It has also started to play a significant role in many other applications, including chemical and biomedical sensing, LiDAR application in driving assistance systems, and in applications for building quantum networks. The integration with III–V light source materials such as indium phosphide through bonding has also been reported to solve the lack of intrinsic light emission within the silicon platform [1–9].

To monitor the status of fabrication processes and characterise the performance of photonic circuits after production, wafer-scale testing is an essential technology in a semiconductor production line [10,11]. However, testing the silicon photonics devices optically at wafer scales is nontrivial. Various techniques can be used to couple the optical signal from optical fibres to photonic chips, such as prism coupling, butt coupling, end-fire coupling, and grating coupling [12–15]. Among them, grating couplers provide an effective coupling technique that can couple between integrated waveguides and the out-of-plane fibres at the wafer scale without dicing the photonic chips [16,17]. However, grating couplers are normally only used as the input/output for the whole photonic circuit. Therefore, apart from measuring the performance of the entire circuit during wafer-scale testing, it is not possible to test any subsystems or individual devices on the photonic chips. Once a problem within the circuits, such as a fabrication error, occurs, early identification and analysis can bring huge cost savings. Therefore, it is desirable to routinely test some key sensitive subsystems or individual devices on the photonic chips, at a wafer scale. Erasable grating couplers and directional couplers are, therefore, proposed for such applications, based on germanium (Ge) ion implantation and annealing technologies [18,19].

With current semiconductor fabrication technology, the uniformity of the fabrication processes for some silicon photonics devices, such as ring resonators and Mach–Zehnder interferometers (MZIs), exhibits variability. The optical phase error induced by the variation in waveguide dimensions across the chip will significantly change the device performance. This is one of the major factors affecting the yield of final silicon photonic modules in production [20,21]. For example, the resonant wavelength and Q factor of ring resonators will be affected by variations in the fabrication process [22,23]. The transmission of MZI mainly depends on the phase difference between the two arms, and this will also suffer from the variations in the fabrication process [24,25].

Silicon photonic modulators and filters have benefited from the resonance properties of ring resonators. However, this structure is extremely susceptible to fabrication variations [26]. Several methods have been investigated to tune the resonant wavelength of optical ring resonators. Electron beam and laser irradiation are two techniques that have been used for trimming, with each utilising a different process to realise a refractive index change [27]. Inducing strain and compaction into an oxide cladding by e-beam can change the effective index of the optical mode in the devices [28,29]. The polymethyl methacrylate cladding allows the resonant wavelength of ring resonators to be trimmed after irradiation by ultraviolet light [30]. Post-fabrication trimming can also be realised by introducing photosensitive materials onto the waveguide surface and subsequently exposing them to lasers [31]. The optical properties of the waveguide can also be shifted by depositing and partial etching of a thin SiN film on top of devices [32]. Laser-induced lattice damage can also change the refractive index of silicon waveguides and can provide a shift in resonant wavelength [33]. In this study, we demonstrated that Ge ion implantation and annealing can be applied for post-fabrication trimming [34–37], with some unique advantages over other techniques, such as a large change in refractive index change and easy implementation.

In addition to the applications of wafer-scale testing and post-fabrication device trimming, we also implemented a proof of principle of nonvolatile programmable photonic circuits with the Ge implantation technology [38]. Based on the refractive index change associated with Ge ion implantation and annealing mentioned above, the flow of an optical signal within the photonic circuit can be switched or rerouted using erasable directional couplers or trimmable MZIs. Such devices were used as the basic building blocks for large-scale programmable photonics circuits. After fine-tuning the transmission, the working states of the directional couplers or MZIs are fixed and do not need a continuous electrical power supply to retain the operating point [34]. This will greatly reduce the power consumption, compared with more traditional programmable photonic circuits, implemented with MZI arrays controlled by integrated thermal heaters [39].

2. Ion Implantation and Annealing

2.1. Ge Ion Implantation

Ion implantation is a widely used fabrication process to achieve impurity doping in a solid target [40]. Typically, in the CMOS industry, ions of elements from groups III and V are implanted to change the free carrier concentration. The process to alter the charge density present in the sample via doping is different from the ion implantation process used in this study to change the refractive index of the silicon waveguide via crystal damage. For crystalline silicon waveguides, lattice disorder can be induced during the implantation process. Therefore, the effective index of propagating mode in optical devices can be altered through this process via a damage-induced change in the refractive index. During the implantation process, the ions are accelerated in an electric field and injected into the sample, causing lattice damage in the crystalline silicon. The alteration in the refractive index is attributed directly to this lattice damage [41,42]. Therefore, a variety of implanted species can be utilised to cause lattice damage.

In this project, Ge was chosen as the implanted ion in the ion implantation process to create amorphous sections in a crystalline silicon waveguide. Firstly, Ge is a group IV element, which means that the free carrier concentration of the silicon waveguide will not be changed. Secondly, it is a CMOS-compatible element, and compared with carbon and silicon, Ge has a higher relative mass and can cause greater lattice damage than other group IV elements for the same implanted dose [41,43]. Consequently, amorphisation can be achieved for a lower dose than for lighter ions. Ion implantation is also a temperature-dependent process. Low ion implantation temperature can reduce the required ion implantation dose and implant time [42]. However, it also increases the utilisation time of the implantation equipment and production costs because self-heating via the ion beam must be avoided. The lattice disorder in silicon crystal has been reported with a very mild temperature dependency below 323 K [44]. Therefore, room temperature can be used to reduce fabrication costs.

Figure 1 shows the lattice damage profile in a 220 nm thick silicon waveguide [45]. The King and King 3D software packages were used to simulate the damage profile of silicon waveguides after Ge ion implantation [46–48]. The energy and implantation dose in this simulation are 100 keV and 1×10^{15} ions/cm^2, respectively. The real part of the refractive index of the implanted silicon waveguide regions was increased by approximately 0.5 due to 80% lattice disorder or more, which was taken to be the definition of amorphisation for the purposes of this research. The induced lattice defects in silicon will also result in a small increase (0.04) of the imaginary part of the refractive index [49].

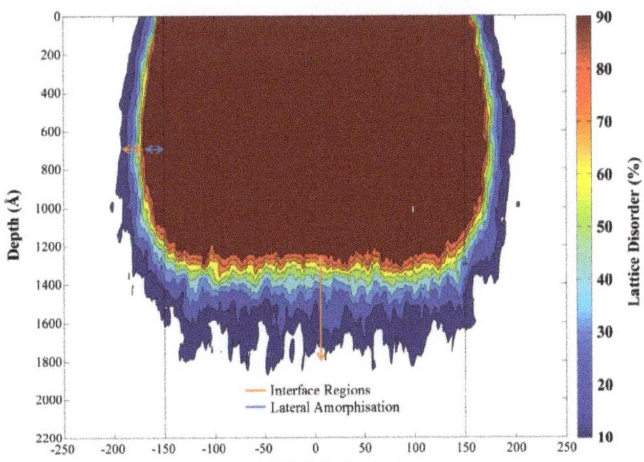

Figure 1. Lattice damage profile at the cross-section of waveguide [45–48].

The mask patterns to define the impanated area (i.e., implantation masks) were fabricated using a DUV lithography scanner. Since shrinkage of a photoresist mask layer during the ion implantation can occur, a SiO_2 layer was used as a hard mask, to eliminate this uncertainty in the fabrication process. The ion implantation process was carried out at the Ion Beam Centre at the University of Surrey. An ion energy value of 130 keV and a dose of 1×10^{15} ions/cm^2 were used in order to efficiently implant Ge into silicon and create deeper implantation in the silicon waveguides, based on our previous studies [50].

2.2. Annealing

The lattice damage in silicon, created by implantation in the waveguide, can be repaired by a thermal heating process. The refractive index will, therefore, decrease during this recrystallisation process [51]. Several annealing methods were used in this study—namely, rapid thermal annealing (RTA), laser annealing, and heating via an integrated electrical heater.

Rapid thermal annealing (RTA) is a widely used process in standard CMOS fabrication lines. It can heat the whole silicon wafers/chips, in order to improve the electrical properties, and it activates dopants, changes the properties of films, and anneals the implantation damage. The temperature of the annealing process can be precisely controlled [35,52]. Therefore, it is a useful tool to characterise refractive index variations at specific temperatures but is not useful for local device heating.

In our study, continuous wave-laser exposure and pulse-laser exposure were used to achieve the annealing of implanted sections in waveguides. More details of laser annealing setups can be found in our previously published studies [34–38]. Compared with RTA, this method can locally anneal the implanted regions of single devices. As the energy of the annealing laser photons is directly transferred into the lattice of the sample, the annealing process is accomplished by scanning the laser spot over the implanted regions, a process that can be completed in a few seconds [53,54]. Although laser annealing is quicker than RTA, it is still not an ideal annealing method in large-scale production. Achieving the correct power density on the sample surface can be time-consuming, and the devices cannot be processed by laser irradiation after packaging.

Integrated electrical heaters have been used in photonic devices to correct the operating points and realign the resonant wavelength of ring resonators via the thermo-optic effect [55]. However, we can also achieve annealing via integrated microheaters on top of optical devices. The target area on the chip can be heated to the annealing temperature by applying the correct voltage to the heater. This approach can enable localised and rapid annealing, and facilitate multiple simultaneous annealing points in large-scale photonic integrated circuits, even after packaging.

Results from our previous study revealed that the difference in residual insertion loss between the crystalline silicon and annealed silicon devices is negligible. The resultant peak concentration of Ge ions in the silicon waveguides is less than 0.3%, compared with a silicon concentration in the same volume [18,34,45,50], indicating that the creation of a SiGe alloy is not a concern.

3. Erasable Grating Couplers

A typical structure of a conventional grating coupler is based on periodically etched waveguides to cause a periodic variation in the effective refractive index. By means of the Ge ion implantation, erasable grating couplers can be fabricated using the lattice damage induced by Ge ion implantation in silicon waveguide to cause a similar periodic change in the index, thus diffracting the light and maintaining the surface planarity at the same time. After annealing, the implanted grating coupler can be totally erased, become a transparent waveguide and make no difference to the entire circuit, which are significant advantages for wafer-scale testing.

For proof of principle of implanted gratings for wafer-scale testing, a bow-tie-shaped design was fabricated, as shown in Figure 2. In the implanted grating, the etched part of a conventional grating was replaced by the amorphous regions induced by Ge ion implantation. After sweeping the parameter values of grating period and implantation depth at a fixed duty cycle of 0.5, we chose a period of 600 nm and 130 nm implanted depth, to obtain an optimal coupling efficiency of 44%. The resulting implanted depth is almost twice the depth of the etched grating depth. This is because the lower Δn of the amorphous to the crystalline interface, compared with that of silicon-to-air or -dioxide interface, results in a lower coupling strength for gratings of similar dimensions. Therefore, to achieve similar coupling efficiency to etched gratings, implanted gratings require deeper amorphous regions, to compensate for the smaller index contrast.

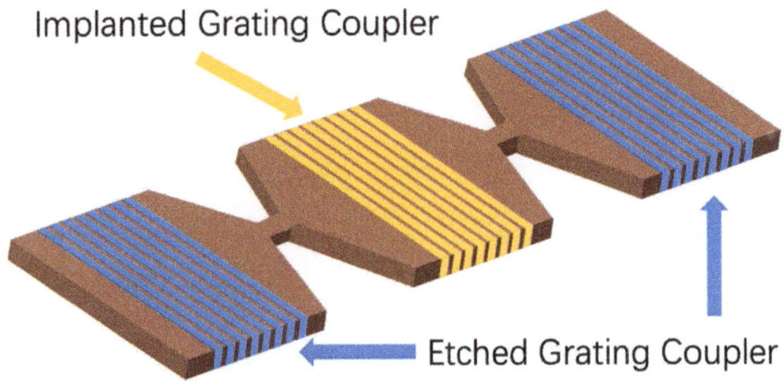

Figure 2. Structure of implanted grating coupler in photonic circuits.

An implanted grating coupler such as this can be positioned anywhere within the optical circuit to evaluate the performance of individual components or to monitor any device or subsystem. Furthermore, after testing, it can be erased by annealing to eliminate any additional insertion loss introduced to the circuit.

A continuous-wave (CW) laser system was initially used for annealing [18]. To guarantee the regrowth of crystal lattice in silicon, the laser spot was scanned over the pattern of the implanted gratings. After full exposure, the implanted grating coupler can be completely erased. Material quality before and after annealing was measured by Raman spectroscopy, to determine whether amorphous regions were successfully annealed and recrystallised [18].

The coupling efficiency of the erasable grating coupler before and after annealing was measured. The implanted grating coupler shows a peak coupling efficiency of ~5.5 dB per coupler [45]. After laser annealing, light passes through the erased grating coupler with almost no additional loss and couples out via the conventional output grating coupler. The residual propagation loss through the erased grating coupler is suppressed to under −25 dB.

4. Implanted Mach–Zehnder Interferometer

The Mach–Zehnder interferometer (MZI) is one of the most popular components in photonics design. The working performance of a single MZI is controlled by the phase difference between the two arms. In the conventional design of an MZI, the phase difference can be induced by a length difference of the two arms and the thermo-optic effect, applied via an integrated heater. In our design, the variation of the phase difference is achieved by Ge ion implantation and subsequent annealing techniques, to choose a specific operating point.

Figure 3 illustrates a basic structure of an implanted MZI. The device was formed of rib waveguides with a 100 nm slab layer. Implanted sections were induced into each arm of the MZI with different lengths of 7 μm and 2 μm. The longer one was primarily responsible for the phase difference and, therefore, controlled the power distribution at the output ports. The shorter one was used to balance any unwanted reflection at the interface of the crystalline Si and implanted Si regions.

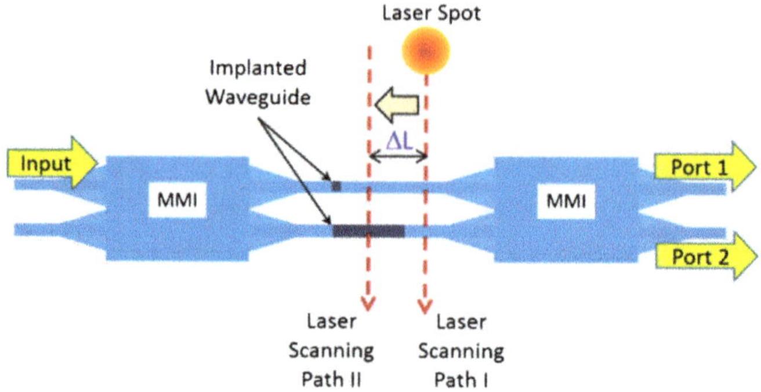

Figure 3. Structure of implanted MZI and scanning route of laser spot [34].

4.1. *Post-Fabrication Trimming of MZI*

For optical modulators, a specific operating point is usually desirable, such as the quadrature point. Due to the fabrication tolerances and environmental factors, the operating point of any MZI normally will be shifted, compared with the original design. Post-fabrication trimming of MZIs has been studied by several groups, and several effective methods were reported. A similar method was proposed by Samarao et al. [56], but a refractive index change of only 0.02 was achieved. In our study, Ge ion implantation can bring much higher refractive index changes, up to one order of magnitude larger.

To experimentally demonstrate this, a CW laser spot, with a diameter of 2.5 μm and power of 45 mW, was used to scan across the implanted sections in the two arms of an MZI (Figure 3). The measurement results of its transmission during the annealing process are shown in Figure 4.

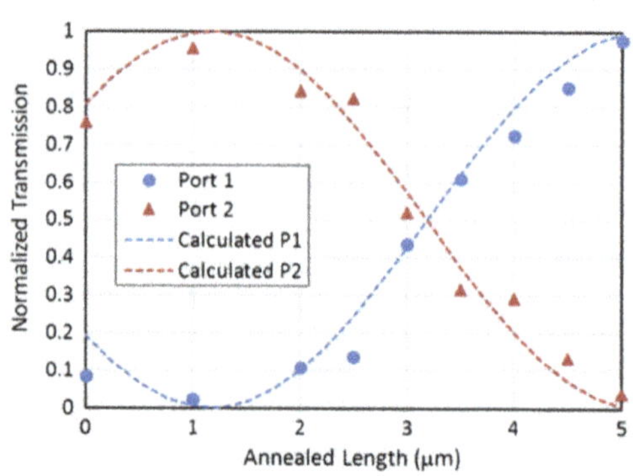

Figure 4. Normalised transmission of implanted MZI during the annealing process [34].

Before annealing, the 5 μm difference of the implanted sections in both arms results in approximately 87% of the signal coupled into the output port 2. After 1 μm of the implanted section was annealed by a CW laser, over 95% of the signal was detected in port 2. Annealing the remaining 4 μm length can switch the output from port 2 to port 1, demonstrating full control of the operating point of the MZI.

4.2. *Real-Time Phase Tuning of MZI*

To implement trimming in commercial applications, real-time measurements are necessary. Therefore, we studied real-time trimming as proof of principle. The implanted MZI was annealed with a real-time optical measurement setup, which is shown in Figure 5. Optical signal T (mV) was collected by the photodetector, coupled to output 1. In this setup, a lock-in amplifier was used to extract the optical signal even within a high-noise environment.

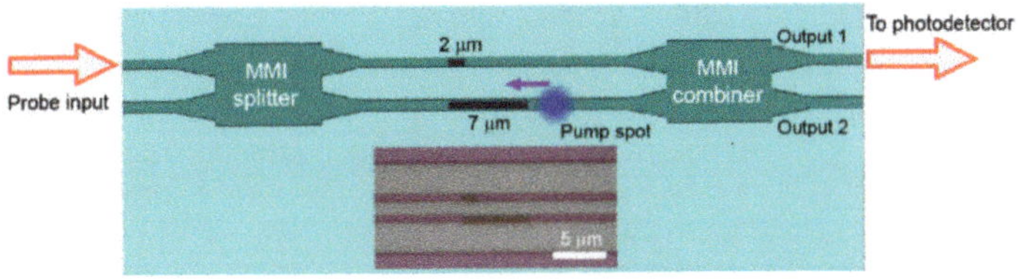

Figure 5. Schematic of real-time pulse laser annealing for implanted MZI [37].

During the annealing process (with reference to Figure 5), the annealing laser spot, with a power of 2.4 mW, was scanned linearly from the right-hand side of the 7 μm long implanted section to its left. As the phase difference between the two arms changes with the annealing process, the resultant transmission shift in the implanted MZI output is shown in Figure 6a. Taking the right-hand side of 7 μm long implanted section as the reference position (0 μm), the change in T starts from position 0 μm to 6.6 μm. Figure 6b shows the curve of ΔT, which is collected by the lock-in amplifier. ΔT is the differential of transmission T. In Figure 6b, ΔT starts at −1 mV. Then, this curve follows a downward trend and reaches the lowest point of the curve at 3 μm, which indicates the location of the quadrature point with the largest derivative of the MZI output. There is a small discontinuity of the ΔT curve, which is caused by the shaking of the piezo stage during the scanning. The phase-dependent transmission of the signal T should follow the trend of $(1 + \cos\theta)$ where θ represents the phase difference between the two arms of MZI. Therefore, the derivative of this transmission should follow the trend of $-\sin\theta$. The negative maximum value of this curve is located at approximately −1.8 mV, corresponding to a phase difference of 0.5π between the arms. Based on the former discussion, it can be estimated that the whole phase shift corresponds to approximately 0.72π, after the pump laser annealing.

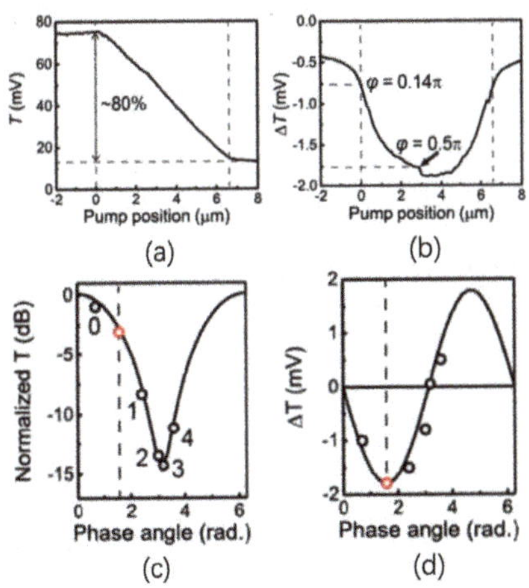

Figure 6. (**a**) Measured transmission (T) signal shifts as the function of the pump spot position during the scanning along the whole 7 μm implanted section; (**b**) measured ΔT signal dependent on the pump spot position. Measured values of (**c**) T and (**d**) ΔT for four subsequent annealing cycles, labelled 1–4 in (**d**), with red dot corresponding to a balanced working point as determined from minimum in ΔT of (**c**) [37].

The average power of the annealing laser is 2.4 mW, which is insufficient to achieve full annealing for the implanted sections in one annealing cycle. Figure 6c,d show the phase-dependent shift, using θ as a free parameter, of T and ΔT in four annealing cycles. This result also shows consistency with former results [34].

4.3. Electrical Annealing of MZI

Electrical annealing of implanted MZIs with integrated TiN heaters was also demonstrated. The top and cross-sectional views of an implanted MZI are shown in Figure 7. A 600 nm thick SiO_2 layer was deposited to separate the optical mode from the heater. The top SiO_2 cladding layer was deposited to protect the heater filament from the air at high temperatures.

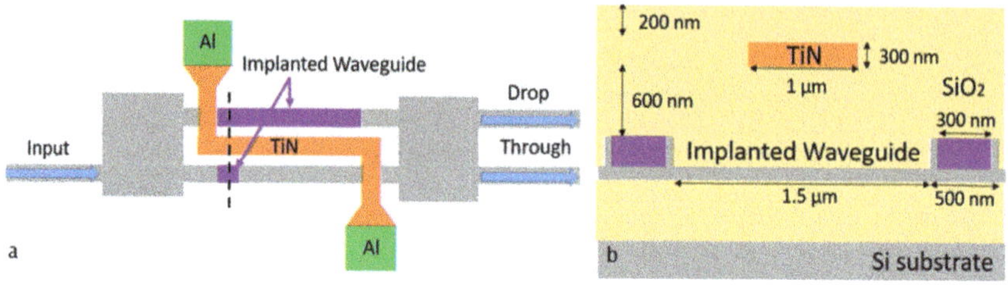

Figure 7. Top view (**a**) and cross-section (**b**) of impanated MZI with the integrated TiN heater on top.

The optical transmission spectrum of an implanted MZI is shown in Figure 8. Over 90% of the optical signal is coupled into the drop port before annealing. We applied 7 V in this experiment, corresponding to electrical thermal power of 150 mW. After 10 s annealing time, approximately 90% of the input optical signal was detected at the through port of the MZI. Due to the phase error between the two arms of MZI, 10% insertion loss can be further improved by precisely controlling the annealing temperature and annealing length. After 10 s annealing time, the implanted sections were fully annealed. No further change in optical transmission was observed. The simulated temperature distribution (cross-section) of this device is shown in Figure 9. The temperature of the implanted waveguides was predicted to be around 600 °C to 700 °C in the annealing process.

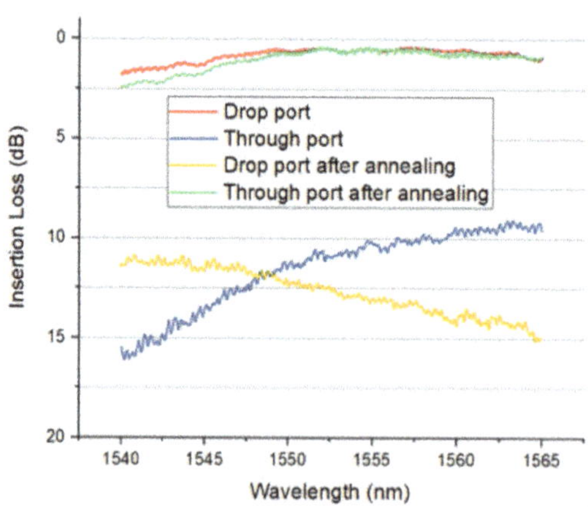

Figure 8. Measured transmission of implanted MZI before and after electrical annealing [19].

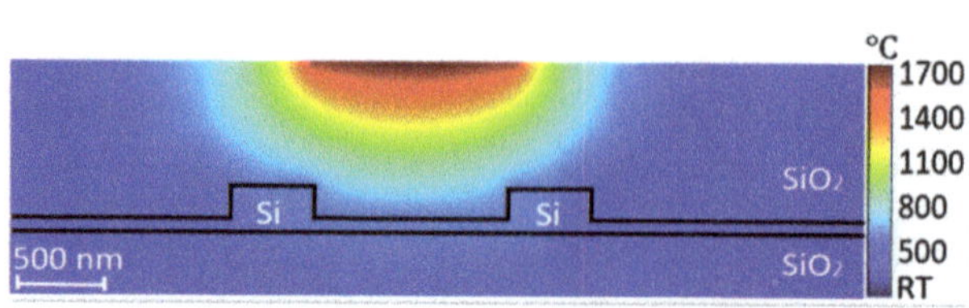

Figure 9. Temperature distribution of implanted MZI at 7 V [19].

5. Implanted Ring Resonators

Ring resonators represent the basic building component of many photonics circuits. They are widely used for optical switching, modulation, sensing, and optical signal filtering applications. However, these devices are also very susceptible to fabrication tolerances and environmental conditions; even a small variation in waveguide width or ambient temperature may result in a significant resonant wavelength shift; therefore, thermal heaters are typically required which will add to the overall power consumption of the photonic integrated circuit. The resonant wavelength shift (of the order of nanometres) caused by dimension variations are orders of magnitude larger than the shift (order of picometers) caused by ambient temperature variations [57]. Therefore, by utilising ion implantation technology and localised annealing, it is possible to induce localised refractive index changes and, therefore, trim the extinction ratio and resonant wavelength of fabricated optical filters. This process saves a significant amount of power required for accurate repositioning of the resonant wavelength due to a shift caused by fabrication tolerances.

Figure 10 shows a schematic of a Ge ion implanted silicon ring resonator. We used a 220 nm silicon-on-insulator (SOI) platform, to fabricate 500 nm wide rib waveguides (100 nm thick silicon slab layer; 120 nm etch depth) and ring resonators with different bend radii. By implanting a 6 μm long section of germanium into the silicon ring resonators of 10 μm in radius, we were able to locally increase the effective refractive index of silicon and, therefore, shift the resonant wavelength to a longer wavelength. We then performed localised laser annealing using the same setup used for trimming MZIs, to accurately position the resonant wavelength at any wavelength within the full free spectral range (FSR) of the ring resonator. Example transmission spectra before and after annealing are shown in Figure 11.

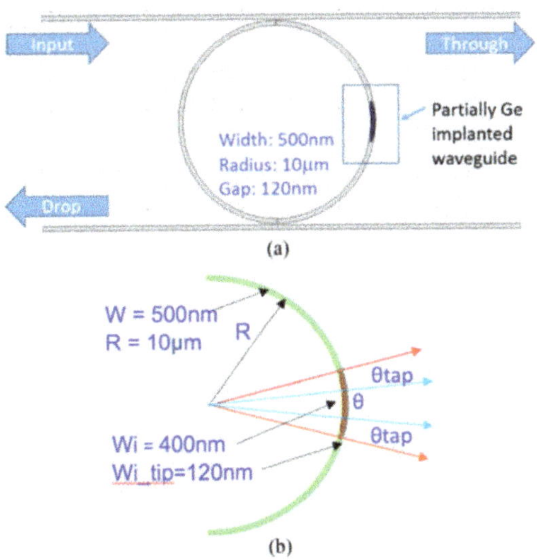

Figure 10. Structure of implanted ring resonator: (**a**) top view of the implanted ring resonator. (**b**) schematic of implanted part in rings [35].

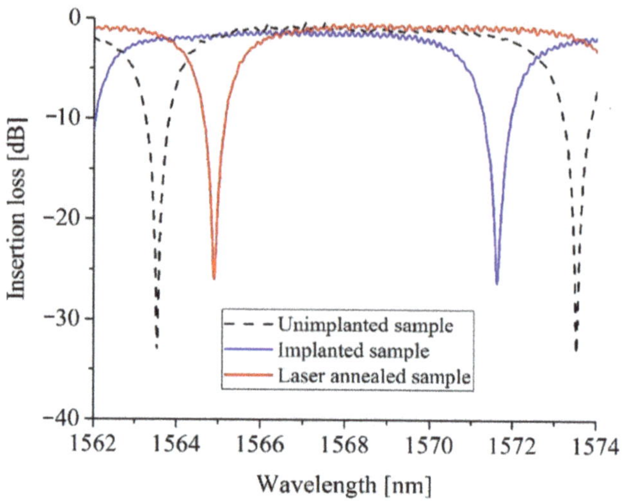

Figure 11. Transmission spectra of implanted ring resonator (with θ = 18°) before and after annealing [35].

Ring resonator trimming was also performed in real time, and as can be seen from Figure 12, we were able to manually achieve a precision of 3% within the targeted wavelength position of 1550 nm for a variety of ring resonators, even when manually controlling the trimming time.

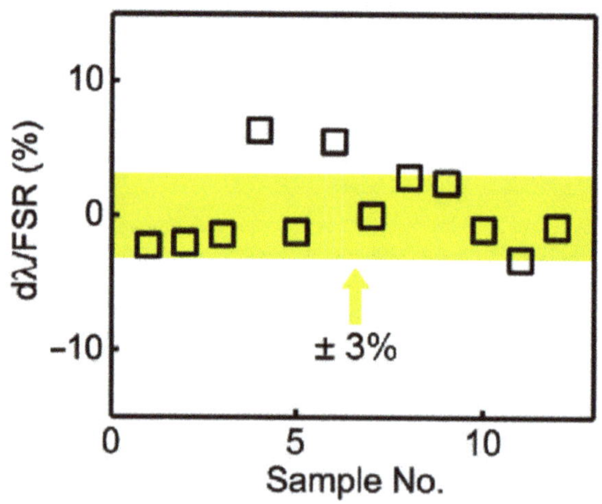

Figure 12. Variation between actual, trimmed resonances and target wavelength 1550 nm normalised to the FSR [37].

Although ion implantation and localised laser annealing represent very accurate procedures for trimming resonant wavelengths and do not require any additional fabrication steps for electrical heaters fabrication, electrical annealing is popular and is expected to become mainstream in silicon photonics. With silicon photonics applied to more industry applications, electrical heater annealing is more achievable in photonic circuits with high integration levels.

For racetrack resonators with 25 μm radii, Figure 13 shows the shift of the resonance peak around 1550 nm during the electrical annealing process. The applied voltage is 1.6 V, and the electrical power is approximately 43 mW, which can bring a wavelength shift of 0.5 nm. Figure 14 shows the wider shift of resonant wavelength under higher electrical powers, (59 mW, 120 mW, and 160 mW). Compared with laser annealing, the applied power and annealing time can be directly controlled, to adjust the resonant wavelength with high accuracy, resulting in a simpler operational process.

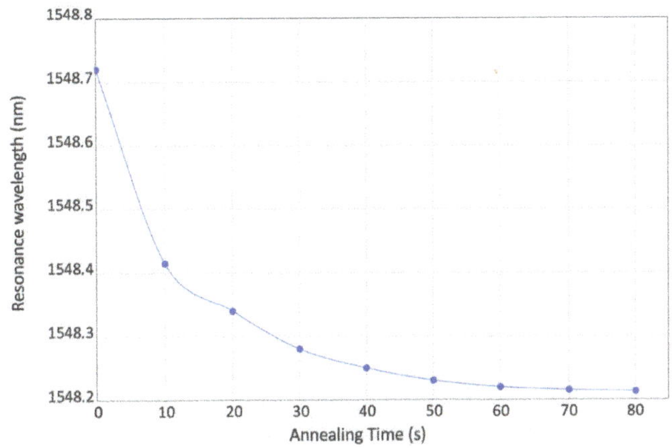

Figure 13. Resonant wavelength shift under 43 mW applied electrical power.

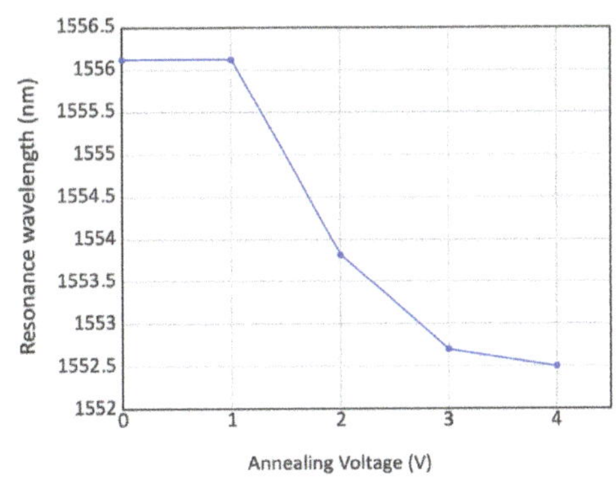

Figure 14. Resonant wavelength shift under applied electrical power of 59 mW, 120 mW, and 160 mW.

6. Erasable Directional Couplers

6.1. Erasable Directional Couplers

Directional couplers (DCs) have been reported as the basic building blocks for configurable photonic circuits. Compared with other devices that can also realise similar switching functions, such as MZIs and ring resonators, they have smaller dimensions. In the semiconductor and integrated photonics industries, the footprint has been an essential consideration in production cost. Furthermore, the performance of DCs is relatively insensitive to temperature fluctuations [58]. A cross-section of a typical device is shown in Figure 15a, and the simulated profile of lattice damage within the implanted waveguide is shown in Figure 15b. Implanted waveguides are successfully formed in the slab region of the conventional rib waveguide to couple light in/out, which can be erased by a laser annealing process.

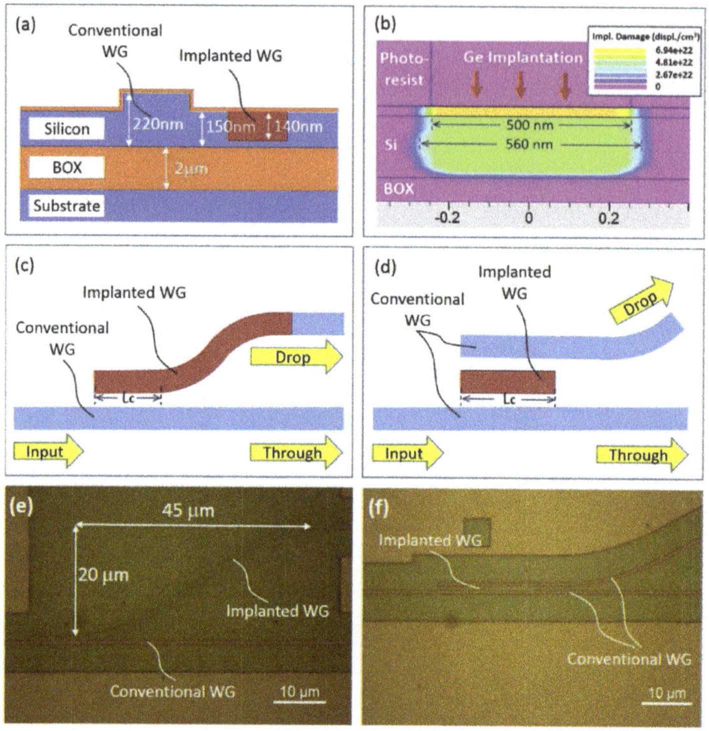

Figure 15. (**a**) Cross-section of implanted waveguide and conventional rib waveguide; (**b**) simulation result for the Ge implanted waveguide using the Silvaco software; (**c**) structure of single-stage DC; (**d**) structure of two-stage DC. Optical microscope images of (**e**) a fabricated single-stage DC and (**f**) a two-stage DC on an SOI wafer [38].

In applications discussed so far in this paper, the optical signals propagate in traditional silicon waveguides fabricated by a dry etching process. However, in erasable DCs, the coupling waveguides were formed solely by an ion implantation process, which was first demonstrated in silicon photonics. Two types of DCs were designed and fabricated, as shown in Figure 15c (single-stage DC) and Figure 15d (two-stage DC). Optical microscope images of a typical fabricated single-stage DC and two-stage DC are shown in Figure 15e,f, respectively.

The devices were designed and optimised using the variational FDTD solver from Lumerical MODE solutions and fabricated in the cleanroom facilities in The University of Southampton, UK. The propagation loss of the germanium-ion-implanted waveguides with various widths was first characterised using the cut-back method. A loss of 32.6 dB/mm was measured for implanted waveguides with 560 nm width. Whilst this loss is very high, only very short lengths of a few micrometres are required in DCs. Figure 16a,b show the measured transmission of single-stage DCs and two-stage DCs, respectively. For single-stage DCs with a coupling length of around 6 µm, over 80% coupling efficiency was experimentally achieved. For two-stage DCs with a coupling length of around 12 µm, over 90% coupling efficiency was measured. The simulated results are also provided, shown as dotted lines in both figures.

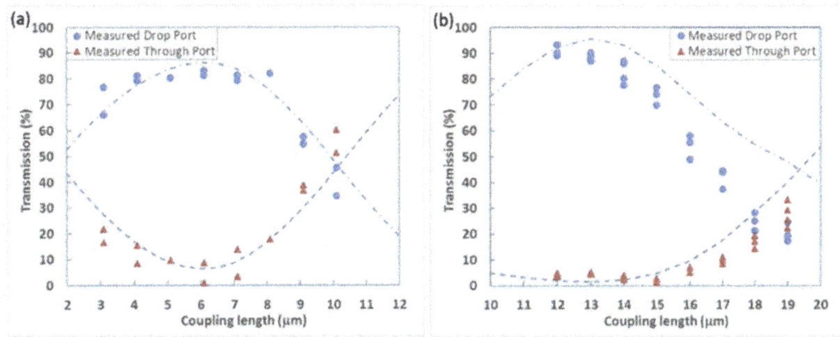

Figure 16. Measured transmission of erasable directional couplers: (**a**) single-stage DCs; (**b**) two-stage DCs [38].

6.2. *One-Time Programmable Photonic Circuits*

The previous devices were used to build 1 × 4 and 2 × 2 OTP (one-time programmable) switching circuits. No subsequent continuous electrical power supply is needed to maintain the operating point because the silicon lattice is permanently changed with our technology. Microscopic images of the switching circuits are shown in Figure 17.

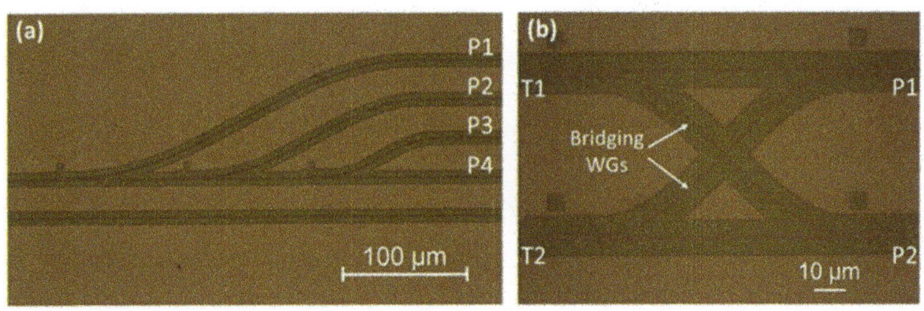

Figure 17. Microscopic image of switching circuits: (**a**) 1 × 4 switch, (**b**) 2 × 2 switch [38].

The 1 × 4 switching circuit comprises one through port (P_4) and three drop ports (P_1, P_2, and P_3). These drop port waveguides are connected with the through port bus waveguide by three two-stage DCs in series. The input optical signal can be directed to any of these four outputs by suitable programming of the circuit. Less than −11.6 dB crosstalk is achieved between each channel.

Figure 17b shows a 2 × 2 OTP switching circuit comprising four two-stage DC structures. This 2 × 2 transmission array allows cross-coupling between two optical paths. Compared with the simple 2 × 2 switch, using two outputs and inputs of a single DC, our design offers lower crosstalk [59]. By using the annealing technique, the 2 × 2 OTP switching circuit can be permanently programmed into bar and cross-operating modes. According to the measurement results, crosstalk of −18.3 dB is achieved at 1550 nm for both channels.

6.3. *Electrical Annealing of Directional Couplers*

As laser annealing cannot usually be applied after the photonic devices are packaged, electrical annealing using integrated TiN heaters was also demonstrated for DCs. The structure of the implanted two-stage directional coupler with an integrated heater is shown in Figure 18. The design of the integrated heaters is similar to the ones used for trimming the MZIs introduced previously.

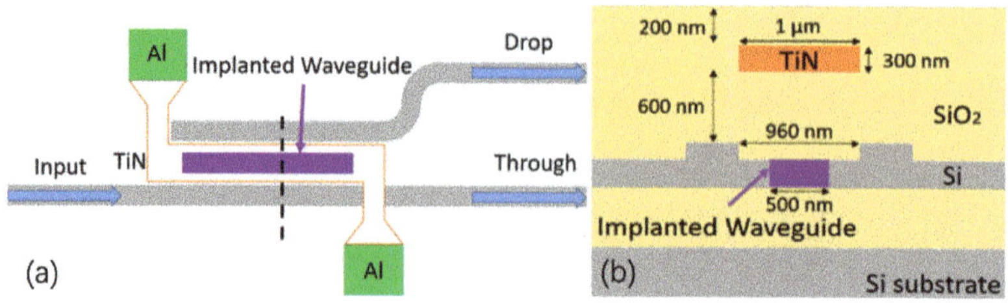

Figure 18. Top view (**a**) and cross-section (**b**) of the ion-implanted two-stage DC [19].

Figure 19 shows the transmission spectra of the implanted two-stage directional coupler before and after annealing. Before annealing, the directional coupler output exhibited a random level of approximately 65% of the optical signal at the drop port. The remaining 35% was detected at the through port. After electrical annealing with an applied voltage of 8 V, the signal at the drop port decreased to around 1.5%. Meanwhile, over 95% signal was detected at the through port. This indicates that most of the optical power remains in the input waveguide and is coupled out at the through port because the implanted waveguide was successfully annealed using the integrated heater.

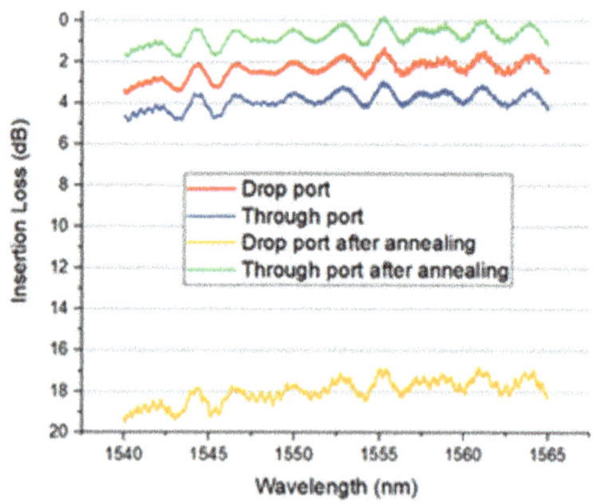

Figure 19. Optical transmission of DC before and after annealing with 19 μm long TiN filament [19].

7. Applications and Discussions

The ion implantation technique has been widely reported for modification and doping of materials. The performance of devices can, therefore, be improved after this treatment [60–64]. In previous experiments, Ge ion implantation was used to induce lattice damage into crystalline silicon waveguides, and the real part of the refractive index can be increased to approximately 3.96. The implantation process will not generate free carriers in silicon waveguides because, similar to silicon, Ge is a group IV material. The transmission of silicon optical devices can be shifted after the implantation process. This alteration of the performance of optical devices is larger than the typical variation brought about by fabrication errors, therefore allowing for correction or 'trimming' of such devices. The annealing process, also widely reported, is the mechanism utilised in removing the lattice defects created by implantation [65–68]. We used the annealing technique here to tune the performance of implanted optical devices for different applications, including wafer-scale testing, post-fabrication trimming, and programmable photonic circuits.

7.1. *Wafer-Scale Testing*

The traditional testing methods, such as butt coupling, are not suitable for large-scale comprehensive testing. They offer limited opportunities for accessing intermediate testing points in photonic circuits for monitoring individual optical devices within a photonic circuit. One possible solution is to use another device, such as an MZI or directional coupler, to tap a small portion of the optical signal from the testing point in a photonic circuit for monitoring and testing purposes. However, these couplers are not removable and will induce permanent optical loss throughout the operational life of photonic circuits [59,69].

On the other hand, our erasable grating couplers and directional couplers were demonstrated to enable flexible comprehensive wafer-scale testing for large-scale integrated photonic circuits without inducing additional loss during subsequent operation of the photonic circuits. The optical signal used for monitoring can be coupled in/out through implanted devices, which can be erased after testing. After the annealing process, the measured residual insertion loss is typically the order of −25 dB for the erasable grating couplers and −18 dB for the erasable directional couplers [18,19].

The implanted sections of these devices can be annealed by laser or electrical heaters, with the latter being a more feasible choice for commercial silicon photonic chips with high volume production and large-scale integration, as thermal heaters can be activated even in packaged devices. Tapers with a length of at least 700 μm are required for each erasable grating coupler, to minimise residual loss after annealing and to facilitate straightforward alignment within a commercial wafer prober, whereas erasable directional couplers can be much smaller [50]. A smaller insertion loss should be potentially achieved if a shorter implanted waveguide is used. Furthermore, there is typically no requirement to launch 90% of an optical signal for testing purposes.

7.2. *Post-Fabrication Trimming of Optical Devices*

Post-fabrication trimming techniques were investigated to fine-tune the optical phase in waveguide-based devices. In this study, the refractive index of the silicon crystalline waveguide was controlled by introducing implanted sections and annealing technologies. The effective index of a propagating mode and the working performance of devices, including MZIs, ring resonators, and directional couplers could, therefore, be trimmed.

For ring resonators, fabrication tolerances will influence the resonant wavelength, the accuracy of which is essential for applications such as sensing and modulation. In our study, the effective index change of the propagating mode in the ring waveguide depends on the length of the induced implantation section. This induced shift can also be controlled by annealing methods such as laser and integrated thermal heating. Our research also successfully demonstrated post-fabrication trimming of the critical coupling condition for racetrack resonators. To achieve this, the same technique was used to implant the slab layer between and bus waveguide and the ring waveguide. Then, the coupling efficiency could be trimmed through partial annealing of implanted section.

We also studied post-fabrication trimming of MZIs using similar techniques. Implanted sections were created in the two arms of MZIs. Then, an annealing process could gradually anneal the implanted waveguide sections and trim the operating point, hence the transmission performance. One example application is for optical modulation, where MZIs can offer the highest modulation efficiency working at the quadrature point.

7.3. *Programmable Photonic Circuits*

Based on the mature fabrication and integration technologies, photonic circuits can support complex applications. Programmable photonic circuits with application flexibility can be designed and developed to realise a variety of functions with the same chip when programmed in different ways. These circuits are built with waveguide meshes, phase shifters, and tuneable couplers, which can be programmed by software [39]. The flow of light in the programmable PICs can be electrically manipulated to realise different circuit functionality from the same chip. Conventional programmable circuits use an electrical signal to control integrated heaters to induce thermo-optic effects in thermal phase shifters. However, hundreds of heaters are typically required for such a system, which has also brought new issues into programmable photonics circuits, such as high power consumption, heat dissipation, and crosstalk between each phase shifter.

The erasable and tuneable optical devices investigated in this study can be used to fabricate nonvolatile actuators in programmable photonics circuits. By introducing a section of an implanted waveguide and then controlling annealing, the routing of an optical signal in an implanted MZI or directional coupler can be tuned. There are also other techniques proposed for building nonvolatile programmable circuits, such as phase-change materials [70,71] and mechanically latched MEMS [72,73]. However, our proposed programmable photonic circuits based on ion implantation and annealing techniques are more CMOS compatible.

8. Conclusions

We used Ge ion implantation and annealing technologies to realise erasable and trimmable silicon optical devices, including erasable grating couplers, erasable directional couplers, tuneable MZIs, and ring resonators. The applications of these devices were also discussed.

Erasable grating couplers and directional couplers were demonstrated as flexible testing points used for wafer-scale testing of photonic integrated circuits, which facilitates comprehensive circuit testing. These testing points can then be erased permanently after testing, with negligible residual insertion loss. Our proposed technology was also used in the post-fabrication trimming of silicon photonic devices with many unique advantages. For trimming of ring resonators, a large trimming range, up to 10 nm (across the entire free-spectral range) of 10 μm rings, was achieved. Table 1 illustrates a comparison of the current trimming methods of ring resonators.

Table 1. Comparison of current trimming of ring resonators.

	Ref. [29]	Ref. [30]	Refs. [74,75]	Ref. [36]
Methods	E-beam	UV light	Sb_2Se_3	Ion implantation and annealing
CMOS compatibility of materials	Yes	No	No	Yes
Throughput (minutes/device)	Low (>17)	Low (>10)	Low (<1)	High (<1)
Effective index change	0.06	<0.1	0.017	0.19

For trimming of MZIs, a trimming accuracy of 0.078 rad and a trimming range of 1.2π were achieved with active feedback control. These trimmable and tuneable devices, such as MZIs and directional couplers, were also demonstrated for rerouting the optical signals of a photonic circuit after fabrication. This enabled us to design a multipurpose photonic integrated circuit with the capability to be programmed for various specific applications after fabrication or packaging. This one-time programmable photonic circuit can potentially reduce the overall production costs, with an increase in production volume, and can speed up the development or prototyping cycles for new photonic circuits.

Author Contributions: Devices design and measurements, X.Y. (Xingshi Yu), X.C., M.M.M., B.C., W.C. and R.T.; devices fabrication, X.C., X.Y. (Xingzhao Yan) and X.Y. (Xingshi Yu); writing the paper and editing: X.Y. (Xingshi Yu), X.C., M.M.M., W.S. and G.T.R.; supervision, G.T.R., D.J.T., S.S., A.C.P. and O.L.M. All authors have read and agreed to the published version of the manuscript.

Funding: This research was supported by the UK Engineering and Physical Sciences Research Council (EPSRC) under the projects 'Silicon Photonics for Future Systems' (EP/L00044X/1), 'Electronic–Photonic convergence' (EP/N013247/1), and 'CORNERSTONE' (EP/L021129/1). 'Laser Engineered Silicon' (EP/M022757/1).

Acknowledgments: Weihong Shen acknowledges the support from China Scholarship Council. D. J. Thomson acknowledges funding from the Royal Society for his University Research Fellowship.

Conflicts of Interest: The authors declare no conflict of interest.

References

1. Richard, J.; Doussiere, P.; Driscoll, J.B.; Lin, W.; Yu, H.; Akulova, Y.; Komljenovic, T.; Bowers, J.E. Heterogeneously integrated InP/silicon photonics: Fabricating fully functional transceivers. *IEEE Nanotechnol. Mag.* **2019**, *13*, 17–26. [CrossRef]
2. Minkenberg, C.; Krishnaswamy, R.; Zilkie, A.; David, N. Co-packaged datacenter optics: Opportunities and challenges. *IET Optoelectron.* **2021**, *15*, 77–91. [CrossRef]
3. Westerveld, W.J.; Mahmud-Ul-Hasan, M.; Shnaiderman, R.; Ntziachristos, V.; Rottenberg, X.; Severi, S.; Rochus, V. Sensitive, small, broad-band and scalable optomechanical ultrasound sensor in silicon photonics. *Nat. Photonics* **2021**, *15*, 341–345. [CrossRef]
4. Baets, R. Silicon-Photonics-Based Spectroscopic Sensing for Environmental Monitoring and Health Care. In Proceedings of the Optical Fiber Communications Conference and Exhibition (OFC), San Francisco, CA, USA, 6–10 June 2021; pp. 1–42. Available online: https://www.photonics.intec.ugent.be/download/pub_4760.pdf (accessed on 8 February 2022).
5. Chen, X.; Meng, F.; Fortune, S.A.; Compston, A.J.; Ebert, M.; Yan, X.; Du, H.; Banakar, M.; Tran, D.T.; Littlejohns, C.G.; et al. Silicon photonic beam steering module with backside coupling elements toward dense heterogeneous integration with drive electronics. *APL Photonics* **2021**, *6*, 116106. [CrossRef]
6. Shen, Y.; Harris, N.C.; Skirlo, S.; Prabhu, M.; Baehr-Jones, T.; Hochberg, M.; Sun, X.; Zhao, S.; Larochelle, H.; Englund, D.; et al. Deep learning with coherent nanophotonic circuits. *Nat. Photonics* **2017**, *11*, 441–446. [CrossRef]
7. Harris, N.C.; Carolan, J.; Bunandar, D.; Mihika, P.; Michael, H.; Baehr-Jones, T.; Michael, L.F.; Smith, A.M.; Tison, C.C.; Alsing, P.M.; et al. Linear programmable nanophotonic processors. *Optica* **2018**, *5*, 1623–1631. [CrossRef]
8. Apple is Entering Silicon Photonics. Available online: http://www.yole.fr/Silicon_Photonics_Market_Update_2021.aspx (accessed on 8 February 2022).
9. Chen, X.; Milosevic, M.M.; Stanković, S.; Reynolds, S.; Bucio, T.D.; Li, K.; Thomson, D.J.; Gardes, F.; Reed, G.T. The emergence of silicon photonics as a flexible technology platform. *Proc. IEEE* **2018**, *106*, 2101–2116. [CrossRef]
10. Silicon Photonics Reaches Tipping Point, with Transceivers Shipping in Volume. Available online: http://www.semiconductor-today.com/features/PDF/semiconductor-today-april-may-2018-Silicon-photonics.pdf (accessed on 8 February 2022).
11. Traverso, M.; Mazzini, M.; Lakshmikumar, K.; Sunder, S.; Kurylak, A.; Appel, C.; Muzio, C.; Tummidi, R.; Cervasio, A.; Nadeau, M.; et al. Integrated Silicon Photonics Transceiver Module for 100Gbit/s 20km Transmission. In Proceedings of the 2021 Optical Fiber Communications Conference and Exhibition (OFC), San Francisco, CA, USA, 6–10 June 2021; pp. 1–3. Available online: https://ieeexplore.ieee.org/stamp/stamp.jsp?tp=&arnumber=9489560 (accessed on 8 February 2022).
12. Jaturaphagorn, P.; Chattham, N.; Limsuwan, P.; Chaisakul, P. Optimization of end-fire coupling between an LED mid-IR light source and SiNx optical waveguides for spectroscopic sensing. *Results Opt.* **2021**, *5*, 100174. [CrossRef]
13. Han, Y.-T.; Yun, S.-J.; Jung, H.-D.; Kim, S.-T.; Shin, J.-U.; Park, S.-H.; Lee, S.-Y.; Baek, Y. A hybrid-integrated 400G TROSA module using chip-to-chip optical butt-coupling. In Proceedings of the Optical Fiber Communication Conference, San Diego, CA, USA, 8–12 March 2020; Optical Society of America: Washington, DC, USA, 2020; p. 1. [CrossRef]
14. Zhang, W.-Q.; Gao, M.-X.; Guo, B. Tunable modulation of photonic spin Hall effect by using a prism-coupling waveguide with hyperbolic metamaterials. *JOSA B* **2020**, *37*, 3777–3783. [CrossRef]
15. Selvaraja, S.K.; Vermeulen, D.; Schaekers, M.; Sleeckx, E.; Bogaerts, W.; Roelkens, G.; Dumon, P.; Thourhout, D.V.; Baets, R. Highly efficient grating coupler between optical fiber and silicon photonic circuit. In Proceedings of the Conference on Lasers and Electro-Optics and 2009 Conference on Quantum electronics and Laser Science Conference, Baltimore, MD, USA, 2–4 June 2009; pp. 1–2.
16. Verenbergh, V.T.; Sun, P.; Hooten, S.; Jain, M.; Wilmart, Q.; Seyedi, A.; Huang, Z.; Fiorentino, M.; Beausoleil, R. Wafer-level testing of inverse-designed and adjoint-inspired vertical grating coupler designs compatible with DUV lithography. *Opt. Express* **2021**, *29*, 37021–37036. [CrossRef]
17. Cheng, L.; Mao, S.; Li, Z.; Han, Y.; Fu, H.Y. Grating couplers on silicon photonics: Design principles, emerging trends and practical issues. *Micromachines* **2020**, *11*, 666. [CrossRef] [PubMed]
18. Topley, R.; Martinez-Jimenez, G.; O'Faolain, L.; Healy, H.; Mailis, S.; Thomson, D.J.; Gardes, F.Y.; Peacock, A.C.; Payne, D.; Mashanovich, G.Z.; et al. Locally erasable couplers for optical device testing in silicon on insulator. *J. Lightwave Technol.* **2014**, *32*, 2248–2253. [CrossRef]
19. Yu, X.; Chen, X.; Milosevic, M.M.; Yan, X.; Saito, S.; Reed, G.T. Electrically Erasable Optical I/O for Wafer Scale Testing of Silicon Photonic Integrated Circuits. *IEEE Photonics J.* **2020**, *12*, 1–8. [CrossRef]
20. Liu, J.; Huang, G.; Wang, R.N.; He, J.; Raja, A.S.; Liu, T.; Engelsen, N.J.; Kippenberg, T.J. High-yield, wafer-scale fabrication of ultralow-loss, dispersion-engineered silicon nitride photonic circuits. *Nat. Commun.* **2021**, *12*, 1–9. [CrossRef]

21. Park, J.-W.; Sim, E.-D.; Beak, Y.-S. Improvement of fabrication yield and loss uniformity of waveguide mirror. *IEEE Photonics Technol. Lett.* **2005**, *17*, 807–809. [CrossRef]
22. Weinstein, D.; Bhave, S.A.; Tada, M.; Mitarai, S.; Morita, S.; Ikeda, K. Mechanical coupling of 2D resonator arrays for MEMS filter applications. In Proceedings of the IEEE International Frequency Control Symposium Joint with the 21st European Frequency and Time Forum, Geneva, Switzerland, 29 May–1 June 2007; IEEE: Piscataway, NJ, USA, 2007; pp. 1362–1365. [CrossRef]
23. Horikawa, T.; Shimura, D.; Jeong, S.-H.; Tokushima, M.; Kinoshita, K.; Mogami, T. The impacts of fabrication error in Si wire-waveguides on spectral variation of coupled resonator optical waveguides. *Microelectron. Eng.* **2016**, *156*, 46–49. [CrossRef]
24. Lee, T.; Lee, D.; Chung, Y. Design and simulation of fabrication-error-tolerant triplexer based on cascaded Mach–Zehnder inteferometers. *IEEE Photonics Technol. Lett.* **2007**, *20*, 33–35. [CrossRef]
25. Yen, T.-H.; Hung, Y.J. Fabrication-tolerant CWDM (de) multiplexer based on cascaded Mach-Zehnder interferometers on silicon-on-insulator. *J. Lightwave Technol.* **2020**, *39*, 146–153. [CrossRef]
26. Jayatilleka, H.; Frish, H.; Kumar, R.; Heck, J.; Ma, C.; Sakib, M.N.; Huang, D.; Rong, H. Post-Fabrication Trimming of Silicon Photonic Ring Resonators at Wafer-Scale. *J. Lightwave Technol.* **2021**, *39*, 5083–5088. [CrossRef]
27. Lushchik, A.; Kuzovkov, V.N.; Kotomin, E.A.; Prieditis, G.; Seeman, V.; Shablonin, E.; Vasil'chenko, E.; Popov, A.I. Evidence for the formation of two types of oxygen interstitials in neutron-irradiated α-Al_2O_3 single crystals. *Sci. Rep.* **2021**, *11*, 1–10. [CrossRef]
28. Schrauwen, J.; Van, T.D.; Baets, R. Trimming of silicon ring resonator by electron beam induced compaction and strain. *Opt. Express* **2008**, *16*, 3738–3743. [CrossRef] [PubMed]
29. Prorok, S.; Petrov, A.Y.; Eich, M.; Luo, J.; Jen, A. Trimming of high-Q-factor silicon ring resonators by electron beam bleaching. *Opt. Lett.* **2012**, *37*, 3114–3116. [CrossRef] [PubMed]
30. Zhou, L.; Okamoto, K.; Yoo, S.J.B. Athermalizing and trimming of slotted silicon microring resonators with UV-sensitive PMMA upper-cladding. *IEEE Photonics Technol. Lett.* **2009**, *21*, 1175–1177. [CrossRef]
31. Canciamilla, A.; Grillanda, S.; Morichetti, F.; Ferrari, C.; Hu, L.; Musgraves, J.D.; Richardson, K.; Agarwal, A.; Kimerling, L.C.; Melloni, A. Photo-induced trimming of coupled ring-resonator filters and delay lines in As_2S_3 chalcogenide glass. *Opt. Lett.* **2011**, *36*, 4002–4004. [CrossRef]
32. Atabaki, A.H.; Eftekhar, A.A.; Askari, M.; Adibi, A. Accurate post-fabrication trimming of ultra-compact resonators on silicon. *Opt. Express* **2013**, *21*, 14139–14145. [CrossRef]
33. Bachman, D.; Chen, Z.; Fedosejevs, R.; Tsui, Y.Y.; Van, V. Permanent fine tuning of silicon microring devices by femtosecond laser surface amorphization and ablation. *Opt. Express* **2013**, *21*, 11048–11056. [CrossRef] [PubMed]
34. Chen, X.; Milosevic, M.M.; Thomson, D.J.; Khokhar, A.Z.; Franz, Y.; Runge, A.; Mailis, S.; Peacock, A.C.; Reed, G.T. Post-fabrication phase trimming of Mach-Zehnder interferometers by laser annealing of germanium im-planted waveguides. *Photonics Res.* **2017**, *5*, 578–582. [CrossRef]
35. Milosevic, M.M.; Chen, X.; Cao, W.; Runge, A.; Franz, Y.; Littlejohns, C.G.; Mailis, S.; Peacock, A.C.; Thomson, D.J.; Reed, G.T. Ion implantation in silicon for trimming the operating wavelength of ring resonators. *IEEE J. Sel. Top. Quantum Electron.* **2018**, *24*, 1–7. [CrossRef]
36. Milosevic, M.M.; Chen, X.; Yu, X.; Dinsdale, N.J.; Aktas, O.; Oo, S.Z.; Khokhar, A.Z.; Thomson, D.J.; Muskens, O.L.; Chong, H.M.H.; et al. Ion implantation of germanium into silicon for critical coupling control of racetrack resonators. *J. Lightwave Technol.* **2020**, *38*, 1865–1873. [CrossRef]
37. Chen, B.; Yu, X.; Chen, X.; Milosevic, M.M.; Thomson, D.J.; Khokhar, A.Z.; Saito, S.; Muskens, O.L.; Reed, G.T. Real-time monitoring and gradient feedback enable accurate trimming of ion-implanted silicon photonic devices. *Opt. Express* **2018**, *26*, 24953–24963. [CrossRef]
38. Chen, X.; Milosevic, M.M.; Runge, A.F.J.; Yu, X.; Khokhar, A.Z.; Mailis, S.; Thomson, D.J.; Peacock, A.C.; Saito, S.; Reed, G.T. Silicon erasable waveguides and directional couplers by germanium ion implantation for configurable photonic circuits. *Opt. Express* **2020**, *28*, 17630–17642. [CrossRef] [PubMed]
39. Pérez-López, D.; López, A.; DasMahapatra, P.; Capmany, J. Multipurpose self-configuration of programmable photonic circuits. *Nat. Commun.* **2020**, *11*, 1–11. [CrossRef]
40. Kotomin, E.; Kuzovkov, V.; Popov, A.I.; Maier, J.; Vila, R. Anomalous kinetics of diffusion-controlled defect annealing in irradiated ionic solids. *J. Phys. Chem. A* **2018**, *122*, 28–32. [CrossRef] [PubMed]
41. Baranova, E.C.; Gusev, V.M.; Martynenko, Y.V.; Starinin, C.V.; Hailbullin, I.B. On silicon amorphisation during different mass ions implantation. *Rediation Eff.* **1973**, *18*, 21–26. [CrossRef]
42. Townsend, P.D.; Chandler, P.J.; Zhang, L. *Optical Effects of Ion Implantation;* Cambridge University Press: Cambridge, UK, 1994.
43. Heidemann, K.F. Complex-refractive-index profiles of 4 MeV Ge ion-irradiation damage in silicon. *Philos. Mag. B* **1981**, *44*, 465–485. [CrossRef]
44. Picraux, S.T.; Westmoreland, J.E.; Mayer, J.W.; Hart, R.R.; Marsh, O.J. Temperature dependence of lattice disorder created in Si by 40 keV Sb ions. *Appl. Phys. Lett.* **1969**, *14*, 7–9. [CrossRef]
45. Topley, R.; O'Faolain, L.; Thomson, D.J.; Gardes, F.Y.; Mashanovich, G.Z.; Reed, G.T. Planar surface implanted diffractive grating couplers in SOI. *Opt. Express* **2014**, *22*, 1077–1084. [CrossRef]
46. Lulli, G.; Bianconi, M.; Parisini, A.; Napolitani, E. Structural characterization and modeling of damage accumulation in In implanted Si. *J. Appl. Phys.* **2004**, *95*, 150–155. [CrossRef]

47. Albertazzi, E.; Bianconi, M.; Lulli, G.; Nipoti, R.; Carnera, A.; Cellini, C. Dynamic Monte Carlo simulation of nonlinear damage growth during ion implantation of crystalline silicon. *Nucl. Instrum. Methods Phys. Res. Sect. B Beam Interact. Mater. At.* **1996**, *112*, 152–155. [CrossRef]
48. Lulli, G.; Albertazzi, E. King Software; CNR-IMM Bologna, via Gobetti 101, 40129 Bologna. 2007. Available online: https://publications.cnr.it/doc/186001 (accessed on 8 February 2022).
49. Kang, H. Crystalline Silicon vs. Amorphous Silicon: The Significance of Structural Differences in Photovoltaic Applications. *IOP Conf. Ser. Earth Environ. Sci.* **2021**, *726*, 012001. [CrossRef]
50. Topley, R. Erasable Diffractive Grating Couplers in Silicon on Insulator. Doctoral Thesis, University of Southampton, Southampton, UK, 2014. Available online: https://eprints.soton.ac.uk/371751/ (accessed on 8 February 2022).
51. Heidemann, K.F. The propagation of light waves through oxygen irradiation induced depth profiles of the complex refractive index in silicon. *Phys. Status Solidi A* **1981**, *68*, 607–617. [CrossRef]
52. Lee, B.H.; Kang, L.; Nieh, R.; Qi, W.J.; Lee, J.C. Thermal stability and electrical characteristics of ultrathin hafnium oxide gate dielectric reoxidized with rapid thermal annealing. *Appl. Phys. Lett.* **2000**, *76*, 1926–1928. [CrossRef]
53. Bell, A.E. Review and analysis of laser annealing. *Rca Rev.* **1979**, *40*, 295.
54. Phillips, K.C.; Gandhi, H.H.; Mazur, E.; Sundaram, S.K. Ultrafast laser processing of materials: A review. *Adv. Opt. Photonics* **2015**, *7*, 684–712. [CrossRef]
55. Sharma, J.; Li, H.; Xuan, Z.; Kumar, R.; Hsu, C.M.; Sakib, M.; Liao, P.; Rong, H.; Jaussi, J.; Balamurugan, G. Silicon Photonic Micro-Ring Modulator-based 4 × 112 Gb/s O-band WDM Transmitter with Ring Photocurrent-based Thermal Control in 28 nm CMOS. In Proceedings of the 2021 Symposium on VLSI Circuits, Kyoto, Japan, 13–19 June 2021; IEEE: Piscataway, NJ, USA, 2021; pp. 1–2. [CrossRef]
56. Samarao, A.K.; Ayazi, F. Postfabrication electrical trimming of silicon micromechanical resonators via joule heating. *J. Microelectromechanical Syst.* **2011**, *20*, 1081–1088. [CrossRef]
57. Kim, G.-D.; Lee, H.-S.; Park, C.-H.; Lee, S.S.; Lim, B.T.; Bae, H.K.; Lee, W.G. Silicon photonic temperature sensor employing a ring resonator manufactured using a standard CMOS process. *Opt. Express* **2010**, *18*, 22215–22221. [CrossRef]
58. Morino, H.; Maruyama, T.; Iiyama, K. Reduction of wavelength dependence of coupling characteristics using Si optical waveguide curved directional coupler. *J. Lightwave Technol.* **2014**, *32*, 2188–2192. [CrossRef]
59. Shoji, Y.; Kintaka, K.; Suda, S.; Kawashima, H.; Hasama, T.; Ishikawa, H. Low-crosstalk 2 × 2 thermo-optic switch with silicon wire waveguides. *Opt. Express* **2010**, *18*, 9071–9075. [CrossRef]
60. Kumar, V.; Maan, A.S.; Akhtar, J. Defect levels in high energy heavy ion implanted 4H-SiC. *Mater. Lett.* **2022**, *308*, 131150. [CrossRef]
61. Kumar, V.; Maan, A.S.; Akhtar, J. Selective SHI irradiation for mesa type edge termination in semiconductor planar junction. *J. Phys. Conf. Ser.* **2013**, *423*, 012057. [CrossRef]
62. Sourani, F.; Enayati, M.H.; Ashrafizadeh, F.; Sayyedan, F.S.; Chu, P.K. Enhancing surface properties of (Fe, Cr) Al-Al_2O_3 nanocomposite by oxygen ion implantation. *J. Alloy. Compd.* **2021**, *853*, 156892. [CrossRef]
63. Xu, F.; Tan, Y.; Xie, Z.; Zhang, B. Implantation energy-and size-dependent light output of enhanced-efficiency micro-LED arrays fabricated by ion implantation. *Opt. Express* **2021**, *29*, 7757–7766. [CrossRef] [PubMed]
64. Pazniak, H.; Benchakar, M.; Bilyk, T.; Liedl, A.; Busby, Y.; Noël, C.; Chartier, P.; Hurand, S.; Marteau, M.; Houssiau, L.; et al. Ion Implantation as an Approach for Structural Modifications and Functionalization of $Ti_3C_2T_x$ MXenes. *ACS Nano* **2021**, *15*, 4245–4255. [CrossRef]
65. Yan, C.; Zeng, Q.; He, W.; Zhu, J. Enhanced surface hardness and tribocorrosion performance of 60NiTi by boron ion implantation and post-annealing. *Tribol. Int.* **2021**, *155*, 106816. [CrossRef]
66. Castaldini, A.; Cavallini, A.; Rigutti, L.; Nava, F. Low temperature annealing of electron irradiation induced defects in 4H-SiC. *Appl. Phys. Lett.* **2004**, *85*, 3780–3782. [CrossRef]
67. Wu, J.; He, Z.; Guo, Z.; Tian, R.; Wang, F.; Liu, M.; Yang, X.; Fan, Z.; Yang, F. Pulsed Laser Annealing of Phosphorous-Implanted 4H-SiC: Electrical and Structural Characteristics. *J. Electron. Mater.* **2022**, *51*, 172–178. [CrossRef]
68. Roccaforte, F.; Giannazzo, F.; Greco, G. Ion Implantation Doping in Silicon Carbide and Gallium Nitride Electronic Devices. *Micro* **2022**, *2*, 23–53. [CrossRef]
69. Miller, D.A.B. Self-aligning universal beam coupler. *Opt. Express* **2013**, *21*, 6360–6370. [CrossRef]
70. Ríos, C.; Stegmaier, M.; Hosseini, P.; Wang, D.; Scherer, T.; Wright, C.D.; Bhaskaran, H.; Pernice, W.H. Integrated all-photonic non-volatile multi-level memory. *Nat. Photonics* **2015**, *9*, 725–732. [CrossRef]
71. Wuttig, M.; Bhaskaran, H.; Taubner, T. Phase-change materials for non-volatile photonic applications. *Nat. Photonics* **2017**, *11*, 465–476. [CrossRef]
72. Errando-Herranz, C.; Takabayashi, A.Y.; Edinger, P.; Sattari, H.; Gylfason, K.B.; Quack, N. MEMS for photonic integrated circuits. *IEEE J. Sel. Top. Quantum Electron.* **2019**, *26*, 1–16. [CrossRef]
73. Quack, N.; Sattari, H.; Takabayashi, A.Y.; Zhang, Y.; Verheyen, P.; Bogaerts, W.; Edinger, P.; Errando-Herranz, C.; Gylfason, K.B. MEMS-enabled silicon photonic integrated devices and circuits. *IEEE J. Quantum Electron.* **2019**, *56*, 1–10. [CrossRef]

74. Ríos, C.; Du, Q.; Zhang, Y.; Popescu, C.C.; Shalaginov, M.Y.; Miller, P.; Roberts, C.; Kang, M.; Richardson, K.A.; Gu, T.; et al. Ultra-compact nonvolatile photonics based on electrically reprogrammable transparent phase change materials. *arXiv* **2021**, arXiv:2105.06010.
75. Delaney, M.; Zeimpekis, I.; Du, H.; Yan, X.; Banakar, M.; Thomson, D.J.; Hewak, D.W.; Muskens, O.L. Nonvolatile programmable silicon photonics using an ultralow-loss Sb_2Se_3 phase change material. *Sci. Adv.* **2021**, *7*, 3500. [CrossRef] [PubMed]

micromachines

MDPI

Review

An Introduction to Nonlinear Integrated Photonics Devices: Nonlinear Effects and Materials

Luigi Sirleto [1,*] and Giancarlo C. Righini [2,*]

1 National Research Council (CNR), Institute of Applied Sciences and Intelligent Systems (ISASI), Via Pietro Castellino 111, 80131 Napoli, Italy
2 National Research Council (CNR), Institute of Applied Physics (IFAC) "Nello Carrara", Via Madonna del Piano 10, 50019 Sesto Fiorentino, Florence, Italy
* Correspondence: luigi.sirleto@cnr.it (L.S.); righini@ifac.cnr.it (G.C.R.)

Abstract: The combination of integrated optics technologies with nonlinear photonics, which has led to the growth of nonlinear integrated photonics, has also opened the way to groundbreaking new devices and applications. Here we introduce the main physical processes involved in nonlinear photonics applications, and we discuss the fundaments of this research area, starting from traditional second-order and third-order phenomena and going to ultrafast phenomena. The applications, on the other hand, have been made possible by the availability of suitable materials, with high nonlinear coefficients, and/or by the design of guided-wave structures, which can enhance the material's nonlinear properties. A summary of the most common nonlinear materials is presented, together with a discussion of the innovative ones. The discussion of fabrication processes and integration platforms is the subject of a companion article, also submitted for publication in this journal. There, several examples of nonlinear photonic integrated devices to be employed in optical communications, all-optical signal processing and computing, or quantum optics are shown, too. We aimed at offering a broad overview, even if, certainly, not exhaustive. We hope that the overall work could provide guidance for those who are newcomers to this field and some hints to the interested researchers for a more detailed investigation of the present and future development of this hot and rapidly growing field.

Keywords: photonics devices; nonlinear photonics; integrated photonics; photonic structures; optical materials; all-optical signal processing; all-optical computing; nonlinear optical sources

Citation: Sirleto, L.; Righini, G.C. An Introduction to Nonlinear Integrated Photonics Devices: Nonlinear Effects and Materials. *Micromachines* **2023**, *14*, 604. https://doi.org/10.3390/mi14030604

Academic Editor: Aaron Hawkins

Received: 31 January 2023
Revised: 24 February 2023
Accepted: 3 March 2023
Published: 6 March 2023

1. Introduction

Photonics has often been defined as the key technology of the 21st century. The term "photonics" has certainly been coined in the 20th century, even if there is some incertitude on the precise creation date of this word. Some ambiguity also remains about its frontiers and the differences with respect to optoelectronics and electro-optics. One of the claims is that the first appearance of the word was in 1952 [1]. However, many authors consider the French scientist Pierre Agrain as the "father" of photonics, in 1967, but likely there was an almost simultaneous invention of the word by a group of French physicists working in lasers and fiber optics and by a Dutch group of high-speed photography specialists. A very interesting analysis of the use of the term photonics, embracing history, philosophy, and sociology of science, was published some years ago [2].

The word photonics, however, started to be broadly used only in the 1980s, when the operators of telecommunications networks switched from electrical to fiber-optic data transmission. Optical fiber systems are typically composed of discrete elements, such as lasers, modulators, and detectors often packaged in a rack scale module. In 1969, inspired by the rapid development of integrated electronics, the concept of integrated optical circuits (IOCs) was proposed by researchers at Bell Labs [3]. Some twenty years later, the growth in complexity of the optical architectures led to the introduction of the term photonic integrated circuits (PICs), and in 1991 a paper, again by researchers at AT&T Bell Labs, presented an early review of InP-based PICs [4]. Then, it took more than a decade to have the announcement of a breakthrough in photonic integration with the industry's first large-scale PIC: a 2005 article described a 100 Gb/s dense wavelength division multiplexed (DWDM) transmitter and receiver PIC, fabricated through the integration of over 50 discrete functions onto a single monolithic InP chip [5].

In most cases, the response of a material to an applied optical field is linear (i.e., the strength of the response is proportional to the strength of the optical field), but all the way back in the second half of the 19th century, John Kerr, in Glasgow, observed effects that were proportional to the square of the applied field. These effects, which can be associated with the birth of nonlinear optics, described how an isotropic transparent substance becomes birefringent when it is placed in an electric field: two papers were published in 1875, the first related to a solid dielectric [6] and the second to liquids [7]. Another pioneering work, predicting quantum two-photon phenomena, was performed at the end of the 1920s by Maria Goeppert-Mayer (a German physicist who moved to the United States in 1930) [8]. The field of nonlinear optics, however, became much more important and application-effective after the discovery of the laser and the wide availability of intense light beams. The center of the research moved from Europe to the United States, where around 1960, the seminal studies by Peter Franken [9] and Nicolaas Bloembergen (a Dutch physicist who had moved to the United States in 1945) [10,11] actually opened the route to the amazing development of science and technology related to nonlinear optical phenomena. The first nonlinear optics experiment in waveguides, showing second-harmonic generation (SHG) in GaAs waveguides, was reported in 1971 by Anderson and Boyd, from the North American Rockwell Science Center [12].

Since the efficiency of nonlinear interaction depends upon the interacting beam intensities (power/area) and is also proportional (either linearly or quadratically) to the interaction distance, it soon became clear that optical waveguides offer important advantages. First, the confinement of light intensity within an area comparable to the wavelength of light leads to an enhanced field strength and high power density; second, the diffractionless propagation in one or two dimensions results in interaction lengths over a distance (at least of the order of centimeters, if not longer) much longer than the one obtained within a bulk material. Definitely, waveguide geometries offer the best prospects for optimizing the efficiency of nonlinear devices. Therefore, the subject of nonlinear integrated optics (NLIO) and, more broadly, nonlinear integrated photonics (NLIP) has greatly expanded due to the development of novel light sources, advanced materials, and very effective guiding structures [13–22].

The materials of interest for realizing nonlinear optical devices are typically in bulk form with interaction lengths in the millimeter range to obtain efficient buildup of the nonlinear signal of interest. In nonlinear integrated photonics, nonlinear optical effects must be exploited within photonics structures with dimensions comparable to or much less than the incident light wavelength. In this scenario, nonlinear properties of optical materials have become of paramount importance, with a dual role: on one side as sources of detrimental effects (such as limiting effects in high-power fiber lasers and amplifiers) and on the other side as essential elements to achieve a number of functions (such as light generation and modulation). Ideally, the best material should have large nonlinear susceptibilities of second and third order, ultrafast response time, and low linear and nonlinear losses (e.g., due to two-photon absorption and free-carrier absorption) within the desired wavelength ranges and the availability of mature manufacturing processes. The latter are crucial for practical applications, requiring multiple active and passive components.

The paper is organized as follows. In the next section, the fundamental aspects of the most common parametric NLO effects, that is, second-order nonlinear processes (second-harmonic generation (SHG), sum/difference frequency generation (SFG/DFG), and optical parametric amplification (OPA)) and third-order nonlinear processes (third-harmonic generation (THG), four-wave mixing (FWM), Kerr effect, SRS, and SBS) are described. As optical materials with favorable properties are the foundation to promote integrated photonic devices with large bandwidth, high efficiency, and flexibility in high-volume chip-scale fabrication, Section 3 is dedicated to optical materials of interest for nonlinear photonics.

2. Fundamental Phenomena

The first described nonlinear phenomenon was the electro-optical Pockels effect in 1906 by the German physicist Friedrich Pockels. In this effect, occurring only in non-centrosymmetric materials, the refractive index of a medium is modified in proportion to the applied electric field strength, which can be applied to the medium either longitudinally or transversely to the light beam. Transverse voltage requirements can be reduced by lengthening the crystal. The electro-optical Pockels effect is used in many applications, for example, EO modulation, high-speed optical shutters, electro-optical detection, and electro-optical switching. The most important material for these applications is lithium niobate ($LiNbO_3$), which has been used for decades in long-haul telecommunication [23].

Unlike linear effects, nonlinear phenomena are subject to symmetry constraints. Second-order nonlinear optical interaction can occur only in noncentrosymmetric materials (they do not display inversion symmetry); this means that in centrosymmetric materials, the nonlinear optical susceptibility $\chi^{(2)}$ is zero, whereas third-order nonlinear interaction can occur in all materials. A fundamental feature of nonlinear processes is that they can give rise to the exchange of energy among electromagnetic fields with different frequencies. In this regards, nonlinear processes can be divided in two classes [24–26]:

In parametric processes, which include both second-order (second-harmonic generation (SHG) and sum or difference frequency generation (SFG and DFG)) and third-order (third-harmonic generation (THG) and four-wave mixing (FWM)) nonlinear phenomena, energy conservation and momentum conservation (i.e., the phase matching) must be satisfied. The fundamental point is that the material's energy does not take part in the processes, and the energy transfer can occur only among waves. Parametric processes are related to non- or near-resonant interactions, where the initial and final quantum states are the same; this means that the transitional electron populations on the energy levels return to the initial condition, that is, there is no real absorption of photons. In a parametric process, the population can be removed from the ground state only for those brief intervals of time when it resides in a virtual level. Therefore, their lifetimes are extremely short (less than a femtosecond). They are always described by a real susceptibility.

To enhance the efficiency of parametric process, the condition of phase matching must be satisfied. This means that the phases of the incident and resultant waves must be in synchronism. Generally, phase matching becomes a major challenge and a constraint that limits the practical applications of the parametric process. On the other hand, because the phase matching condition can be satisfied by only one of the frequency components of the nonlinear polarization, it can work also as an effective method to select a single nonlinear interaction among many possible interactions.

In non-parametric processes, the energy of the whole system, including the material, is conservative, and the energy transfers between waves and materials exist simultaneously. As a result, part of the wave energy transfers to other energy types in the material, such as the optical phonon in the case of stimulated Raman scattering (SRS), the acoustic phonon in the case of stimulated Brillouin scattering (SBS), and the electron level in two-photon absorption (TPA). Non-parametric processes are related to resonant interactions, involving real energy levels with different initial and final quantum states, whose population are changed. In these processes, there is energy transfer from the photons to the host medium with a relative longer lifetime. We note that non-parametric processes are described by a complex susceptibility.

2.1. Second-Order Phenomena

In these nonlinear effects, the nonlinear polarization ($P^{(2)}$), which is proportional to the product of two optical fields through the second-order nonlinear susceptibility, radiates electric fields at the nonlinear frequency, growing linearly with distance propagation. The second-order nonlinearities are of great importance for nonlinear photonic applications, providing valuable options for optical frequency conversion, to generate new wavelengths and to amplify weak optical signals [27,28].

In the history of nonlinear optics, the discovery of second-harmonic generation (SHG) marked the birth of the field. SHG can be regarded as a wave-mixing process, where an optical wave is mixed with itself, generating a new wave at twice the frequency. As a consequence, part of the energy of the optical wave at frequency ω propagating through the material is converted to that of a wave at 2ω.

The basic results of two waves mixing in a nonlinear medium is that either the sum (i.e., sum frequency generation (SFG)) or the difference (difference-frequency generation (DFG)) between the input frequencies can be generated. These generations are achieved when a strong pump and a signal are injected into a second-order nonlinear medium and a specific phase-matching condition, selecting only one process, is satisfied. SFG is useful for converting an infrared beam into a more easily detectable visible beam by mixing visible and infrared light or for generating ultraviolet light. In difference-frequency generation (DFG), a weak input optical signal at frequency ω_{signal} is amplified by a strong laser pump with higher frequency ($\omega_{pump} > \omega_{signal}$), while an idler wave is generated The signal amplification is the fundamental feature of DFG and the most important difference with SFG. Since the signal is amplified by the DFG process, which is a parametric one, DFG is also known as optical parametric amplification. It is worth noting that standard optical amplifiers work only at frequencies corresponding to transitions among specific energy levels, while parametric amplifiers, using transparent crystals, can amplify, in principle, any frequency within the transparency window of the crystal, provided that phase matching can be achieved. In the special case where $\omega_{pump} = \omega_{signal}$, we have the exact reverse of SHG. This process is called degenerate parametric amplification.

An amplifier becomes an oscillator when a positive feedback is provided, so the parametric amplifier can be transformed into a parametric oscillator by using an optical cavity. In parametric oscillation, when a strong pump ω_{pump} injects into an optical cavity, including a nonlinear crystal and providing resonance for the signal or the idler (or both), the parametric gain will cause simultaneous oscillations at both the signal and the idler frequencies. The parametric oscillation occurs through amplification of noise photons initiated by parametric fluorescence, and its tuning range can be notably wide. This is of great technical importance as it provides a means for generating intense coherent tunable radiation in the infrared.

Due to the material dispersion of nonlinear media that impacts seriously the phase-matching condition, the second-order nonlinear optical processes, usually, exhibit limited spectral bandwidths. For bulk samples, the phase matching is typically obtained exploiting birefringence in anisotropic nonlinear crystals or with periodically poled nonlinear crystals (e.g., lithium niobate). In such nonlinear crystals, there is a stringent compromise between phase-matching bandwidth and total conversion efficiency, which are inversely and directly proportional, respectively, to the material's thickness. In addition, the limited availability of high-quality non-centrosymmetric crystalline materials in integrated photonic platforms makes second-order nonlinearities less common as compared with third-order ones [27,28]. In Figure 1, the geometry of interactions and the energy level diagrams of the considered second-order nonlinear phenomena are reported.

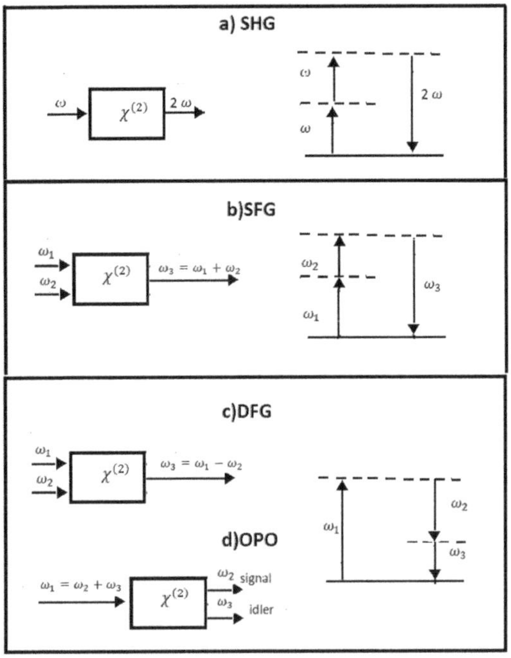

Figure 1. (**a**) Second-harmonic generation, (**b**) sum-frequency generation, (**c**) difference-frequency generation, and (**d**) optical parametric oscillations. In the left column, the geometry of interactions is shown, while in the right column, the energy level diagrams describing the interactions are represented.

2.2. Third-Order Phenomena

Third-order effects can be induced by monochromatic and polychromatic fields. In the simplest case, when the applied field is monochromatic, two phenomena are achieved: third-harmonic generation (THG) and intensity-dependent refractive index or Kerr effect.

THG can be considered as the third-order equivalent of SHG. According to the photon description, in this process, three photons of frequency ω are destroyed, and one photon of frequency 3ω is created. The main application of THG is the realization of ultraviolet (UV) sources, where there are few choices for lasers. However, the practical use of THG requires a material having a large χ_3, a transparency window including both χ_1 and χ_3, and the possibility of achieving phase matching, too. In Figure 2, the geometry of the interaction and the energy level diagram of THG are reported.

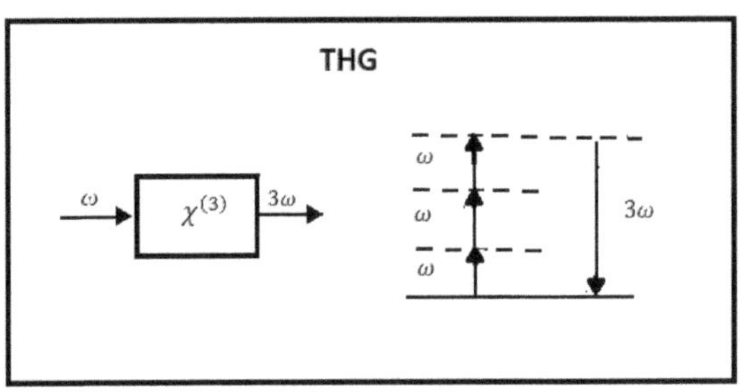

Figure 2. Third-harmonic generation. On the left is the geometry of the interaction, and on the right is the energy level diagram describing the interaction.

For a monochromatic incident field, third-order effects produce a material polarization: $P_3 = \varepsilon_0(\chi_3 E^2)E$ that leads to a term in which the factor $(\chi_3 E^2)$ oscillates at the fundamental frequency of the incident light. This causes the refractive index of materials to be described by $n(I) = n_0 + \gamma * I$, where n_0 stands for the linear refractive index, while the nonlinear refractive index γ is related to the real $\left(Re\, \chi^{(3)}\right)$ part of the third-order nonlinear susceptibility. The intensity-dependent index of refraction induces the phase effects affecting the beam propagation and leading to a number of related phenomena, i.e., self-focusing, self-phase modulation (SPM), and cross-phase modulation (CPM). All these effects are described by the real part of the susceptibility χ_3 and are fundamental for nonlinear photonic applications because they lead to field-dependent modifications of certain material properties, e.g., the refractive index of the medium, which make available a number of devices in which it is possible to control light by light.

Self-focusing is an induced lens effect. Due to the nonuniform transverse intensity distribution of propagating beam, the optical Kerr effect produces a transverse refractive index profile that follows the intensity profile of the beam. If n_2 is positive, a greater index of refraction is induced on axis than in the wings of the beam, which causes the rays to curve toward each other, creating a positive lens that tends to focus the beam.

In the general case, the applied field consists of three different frequency components, and the third-order nonlinear polarization contains a number of contributions with different frequencies. Among them are $\omega_4 = \omega_1 + \omega_2 + \omega_3$, $\omega_4 = \omega_1 + \omega_2 - \omega_3$, $\omega_4 = \omega_1 - \omega_2 + \omega_3$, and $\omega_4 = -\omega_1 + \omega_2 + \omega_3$. These represent four possible mixing processes, called non-degenerate four-wave mixing (NDFWM). In basic terms, NDFWM is the mixing of three different frequencies $\omega_1 \neq \omega_2 \neq \omega_3$, which interact simultaneously with the medium, to produce a fourth wavelength such that the respective frequencies obey one of the previous four equalities. In Figure 3, we represent two of the four possible mixing processes. In Figure 3, interactions and energy level diagrams of two of the possible mixing process are reported.

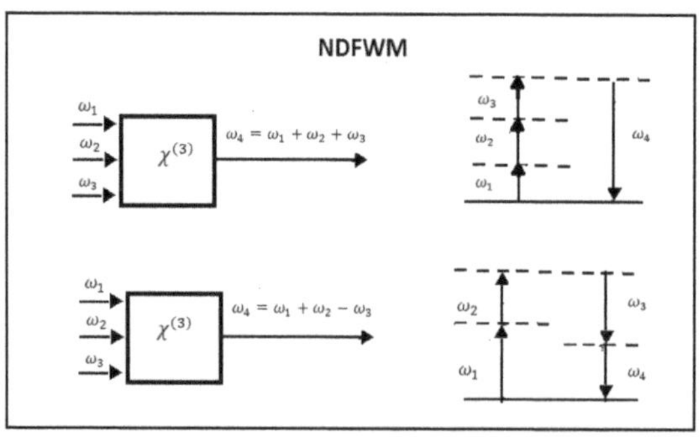

Figure 3. Non-degenerate four-wave mixing. Two of the possible mixing process that can occur when three input waves interact in a medium. On the left are the geometries of the interactions, and on the right are the energy level diagrams describing the interactions.

The conjugated phase of an incident wave can be created by the degenerate FWM. In this process, all the three mixed frequencies have the same frequencies $\omega_1 = \omega_2 = \omega_3$. They interact in a nonlinear medium, which is illuminated by two strong counterpropagating pump waves and by a signal wave. As a result, a new wave having the same frequency $(\omega = 2\omega - \omega)$ is created that is the phase conjugate of the signal.

Note that for the special case of the dual degenerate FWM (DDFWM) when $\omega_1 = \omega_2 = \omega_p$, two frequencies, related to a pump (ω_p) and a weaker signal (ω_s), can produce a third one, which is known as the idler frequency, such that $\omega_i = 2\omega_p - \omega_s$ holds. In other words, in the DDFWM, two equal pump photons (ω_p) are annihilated to create two output photons called the signal (ω_s) and idler (ω_i). In Figure 4, interactions and energy level diagrams for the degenerate and dual degenerate FWM are reported.

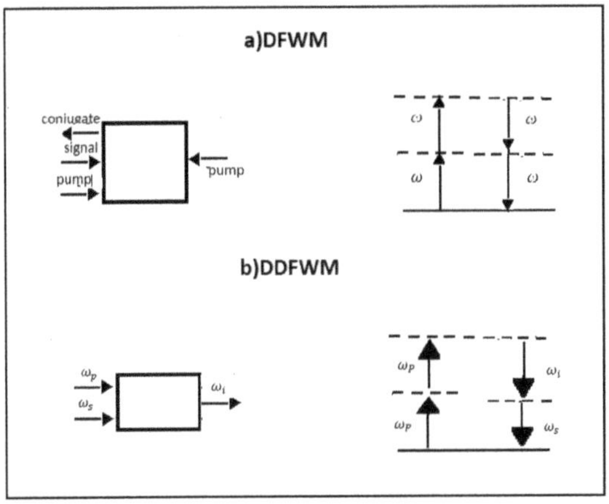

Figure 4. (**a**) Degenerate FWM and (**b**) dual degenerate FWM. On the left are the geometries of the interactions, and on the right are the energy level diagrams describing the interactions.

We note that in the DDFWM, conservation of energy requires the idler and signal being equally separated from the pump since $\omega_p + \omega_p = \omega_s + \omega_i \Rightarrow \omega_p - \omega_s = \omega_i - \omega_p$. In the case of the non-degenerate FWM, the process features two different input photons (ω_p and ω_s) and again an idler (ω_i) and signal ($\omega_{s'}$) photon as the output. In this case, conservation of energy requires $\omega_p + \omega_s = \omega_{s'} + \omega_i \Rightarrow \omega_p - \omega_i = \omega_{s'} - \omega_s$, which implies that the signal and idler photons have the same detuning from the two input photons.

An SRS phenomenon occurs in the presence of a high-energy transfer from a high-power pump beam to a probe beam (copropagating or counterpropagating) [29,30]. In particular, this energy exchange occurs when the frequency difference between the pump and the Stokes laser beams matches a given molecular vibrational frequency of the sample under test. The SRS effect occurs in the form of a gain of the Stokes beam power (stimulated Raman gain (SRG)) and a loss of the pump beam power (stimulated Raman loss (SRL)). SRS depends on the pump intensity and on a gain coefficient. The latter is proportional to the spontaneous Raman scattering cross section and inversely proportional to the linewidth of the corresponding Raman line. Because of its coherent nature, the molecular bonds oscillate in phase and interfere constructively inside the focus area of the laser beam. As a consequence, an SRS signal, which is orders of magnitude bigger than spontaneous Raman scattering, is generated (about 20–30% of the incident laser radiation can efficiently be converted into SRS). Due to its Raman-shifted output, SRS is a workable method for generating coherent radiation at new frequencies. SRS permits, in principle, the amplification in a wide interval of wavelengths, from the ultraviolet to the infrared. Since the Raman frequency of a medium is usually fixed, a tunable pump laser is required to achieve Raman source tunability [31,32]. In Figure 5, the SRG and SRL modalities and level diagrams for SRS phenomena are reported.

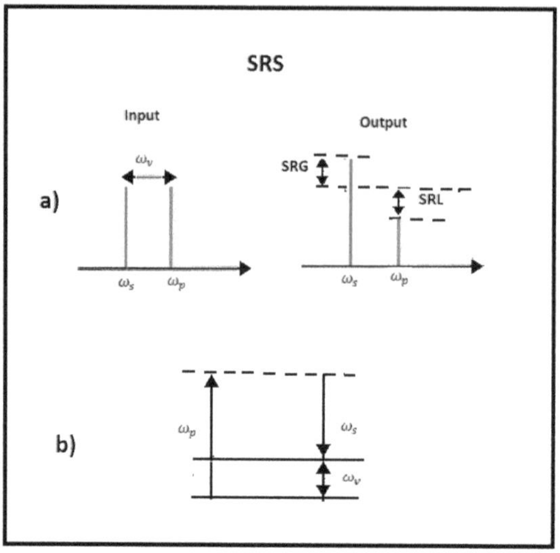

Figure 5. Stimulated Raman scattering. (**a**) SRS modalities: SRG, stimulated Raman gain; SRL, stimulated Raman loss. (**b**) The energy level diagram describing the SRS interaction.

In the last two decades, SRS in nanophotonics has received considerable attention [33–45]. Raman lasers in high-Q resonators have been long investigated because they allow attaining quite low pump threshold powers and high Stokes output powers, when the cavities are designed to have high Q factors at both pump and Stokes wavelengths. A high Q factor over a broad wavelength range is necessary if the Raman lasers are to have a wide tuning range at every resonant wavelength. Single-mode waveguide resonators have good mode confinements for enhanced optical nonlinearities, but they suffer from high propagation losses (typically about 2 dB/cm). On the other hand, conventional multimode racetrack resonators, consisting of one multimode bus waveguide and one multimode racetrack, offer smaller propagation losses for the fundamental modes, but their drawback is the reduction in Q factors of the fundamental mode-based resonances. In ref. [46], a new approach using multimode silicon concentric racetracks, allowing high Q factors to be maintained in every single FSR over a broad optical bandwidth, was presented. Based on this resonator, a widely tunable Raman lasing spanning from 1325 to 1841 nm was experimentally demonstrated [46].

Brillouin scattering describes the scattering of an optical wave from an acoustic wave, which can be a longitudinal pressure/density wave in a solid, gas, or liquid; an acoustic surface wave; or a transverse acoustic wave. When an optical wave is scattered from the acoustic wave, a frequency-shifted optical wave is generated, called the Stokes wave, when lower in frequency than the pump wave, and the anti-Stokes wave, when at a higher frequency. This process can be stimulated, which leads to an exponential gain of the optical Stokes wave. A small optical seed that counterpropagates the optical pump creates an optical beat pattern. The small seed can be a laser coupled from the opposite side into the medium than the pump or originate from the scattering of the pump from thermal phonons; see Figure 6. When the frequency separation of the pump and the seed matches the Brillouin frequency shift (BFS) in the medium, then the optical beat pattern reinforces the acoustic wave via electrostriction, which, in return, amplifies the seed/Stokes wave via scattering of the optical pump from the moving acoustic wave. This feedback can create strong amplification of the initially weak seed/Stokes wave [47]. In Figure 6, level diagrams for SBS phenomena are reported.

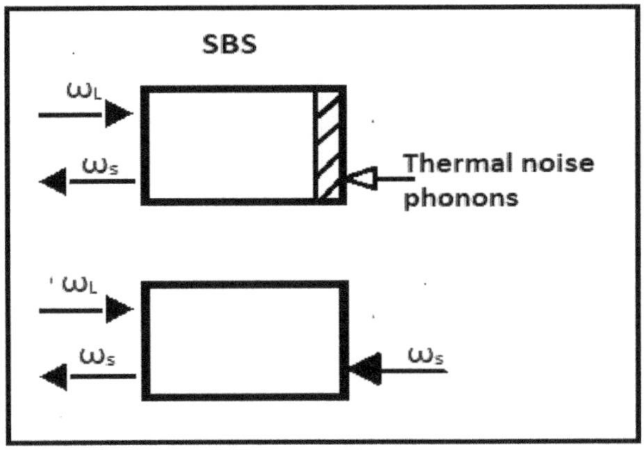

Figure 6. SBS generator on the top and SRS amplifier on the bottom.

A new trend of Brillouin research focuses on integrated photonic platforms. Microscale waveguides, engineered to guide acoustic and optical waves, enable new ways to control and manipulate optical signals and promise many novel applications in a compact and small footprint. Recently, two parallel trends for chip-scale platforms emerged: (1) SBS in high-Q resonators to enable SBS enhancement and (2) low-loss on-chip waveguides that guide both optical and acoustic modes. In the optical domain, SBS is used to achieve functionalities such as non-reciprocity and optical isolation, nonlinear mode conversion, delay and storage of optical signals, and ultra-narrowband lasing in resonators. A deeper understanding of the fundamental interactions between light and sound has drawn substantial interest in recent years toward optomechanical cavities that aim to address fundamental scientific questions, for example, the observation of quantum effects in macroscopic systems [48].

In nonlinear regimes, the optical absorption of materials is described by $\alpha(I) = \alpha_0 + \delta * I$, where α_0 stands for the linear absorption, while nonlinear absorption δ is related to the imaginary $\left(Im\ \chi^{(3)}\right)$ of the third-order nonlinear susceptibility. Nonlinear absorption phenomena refer to the change in transmittance of a material as a function of intensity. They involve two-photon absorption (TPA), saturable absorption (SA), or multiphoton absorption (MPA) [24,25].

Two-photon absorption (TPA) is a third-order nonlinear optical interaction, in which two photons are simultaneously absorbed through a virtual intermediate state to produce a real excitation. We can distinguish between two types of TPA: a degenerate process, in which two photons with the same wavelength are absorbed, and a nondegenerate process, in which one photon with wavelength λp from a pump source and one photon from a signal source with wavelength λs are both absorbed. The latter can be applied to modulate a probe signal through a process called cross-absorption modulation (XAM). Nonlinear absorption and related phenomena such as two-photon absorption and saturable or reverse saturable absorption have some important applications in light generation (Q-switching and mode-locking lasers), in laser modulation, taking advantage of cross-modulation (XAM), and in nonlinear detection of light by TPA. In Figure 7, level diagrams for DTPA and NDTPA are reported.

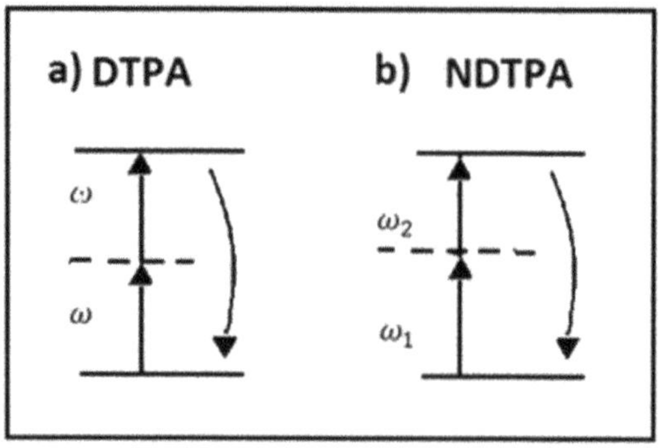

Figure 7. The energy level diagram describing the interactions: (**a**) degenerate TPA and (**b**) nondegenerate TPA.

In contrast to the TPA process where the absorption increases with light intensity reverse saturable), SA exhibits the opposite trend. Due to the saturation of excited electrons filling the conduction band and hence preventing further transitions due to Pauli blocking, in saturable absorption, when the frequency of incident light is near an absorption resonance of the material, as the intensity increases, an absorption reduction is obtained. The SA is useful for applications such as mode-locked fiber lasers and all-optical modulators.

Multiphoton absorption is a process in which an atom or molecule undergoes a transition from a ground state to an excited state by means of the simultaneous absorption of N photons. Multiphoton transitions take place via "virtual" energy levels. This phenomenon can be used for the three-dimensional microfabrication of photonics devices.

2.3. *Ultrafast Phenomena*

Nowadays, photonics is the technology of choice for the transmission and routing of vast amounts of very high-speed data through optical fibers [49,50]. The fabrication of low-loss single-mode optical fibers has made possible optical communication links with demonstrated bandwidths exceeding several terahertz, which are used to transmit data over hundreds of kilometers, without the need of any regeneration stage. As the demand for bandwidth increases, communication systems are forced to use higher bit rates and, hence, to require shorter pulses, for which waveguide dispersion tailoring becomes fundamental.

We note that even for the case of the medium with a linear response, the shape of a laser pulse can be modified by means of propagation effects, such as dispersion of the group velocity within the medium. Given a single input pulse at the carrier frequency ω_0, each spectral component of the input field propagates as a plane wave and acquires a slightly different phase shift because of the frequency dependence of the propagation constant β. It is useful to expand β in a Taylor series around the carrier frequency ω_0, and depending on the pulse bandwidth, the second-order dispersion term (β_2), called the group velocity dispersion (GVD), and the third- or higher-order dispersion terms must be considered. After transmission, even if the initial pulse is unchirped, the transmitted pulse becomes chirped, and the sign of linear chirp depends on the sign of GVD; if $\beta_2 > 0$, an up chirp is obtained and a broadening of pulse in time is achieved, while if $\beta_2 < 0$, a down chirp is obtained. The latter can be used for the compression of laser pulses.

When short optical pulses propagate through third-order nonlinear optical media and the dispersive term can be neglected, their spectral content can be modified by a nonlinear optical process named self-phase modulation The optical Kerr effect produces a time-dependent change in the refractive index, which induces a time-variable phase shift φ of the propagating wave. As a result of the of nonlinear varying phase, the spectrum of the transmitted pulse will be typically broader than the incident pulse, i.e., new frequency components are created, keeping the temporal shape unaltered [49–52]. As an example, in Figure 8, self-phase modulation achieved in high-index silica glass is reported.

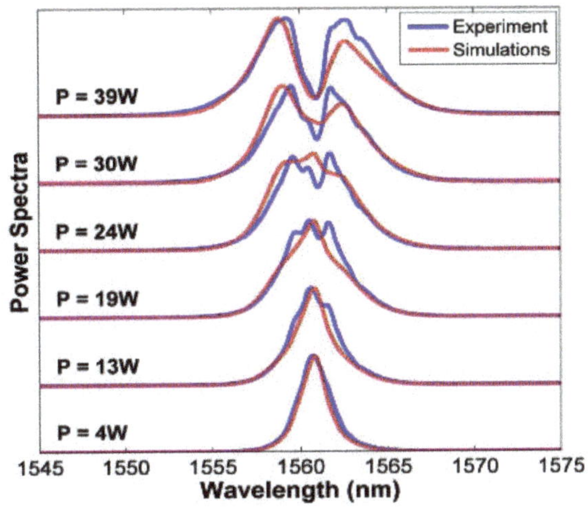

Figure 8. Efficient self-phase modulation, using ~1 ps pulses near 1560 nm, was achieved in low loss (<0.06 dB/cm) and 45 cm long spiral integrated waveguide of high-index silica glass [49]. In this figure, experimentally measured output power spectra (solid blue lines) and theory (dashed red lines) for different input (coupled) power levels are reported. Reprinted with permission from [52] © The Optical Society.

The intensity dependence of the refractive index leads to another nonlinear phenomenon known as cross-phase modulation (CPM). When two or more optical pulses propagate simultaneously, the cross-phase modulation is always accompanied by SPM and occurs because the nonlinear refractive index seen by an optical beam depends not only on the intensity of that beam but also on the intensity of the other copropagating beams. In fact, CPM converts power fluctuations in a particular wavelength channel to phase fluctuations in other copropagating channels. The result of CPM may be an asymmetric spectral broadening and a distortion of the pulse shape. The CPM phenomenon can be used for optical switching, using interferometric methods. In an interferometer, a weak signal pulse can be divided equally between two arms, it experiences identical phase shifts in each arm, and it is transmitted through constructive interference. When a pump pulse at different wavelength is injected into one of the arms, it will change the signal phase through CPM phenomenon in that arm. If the CPM-induced phase shift is large (close to π), this phase shift results in destructive interference and hence no transmission of signal pulse [49,50].

An interesting nonlinear phenomenon is supercontinuum (SC) generation [53,54]. An ultrabroadband emission is produced by a material upon irradiation with an intense laser source, due to a complicated interplay among dispersive term and NLO processes, such as self-phase modulation, stimulated Raman scattering, and soliton effects [50,51]. In Figure 9, as an example, measured SC in a tellurite microstructured optical fiber (MOF) is reported.

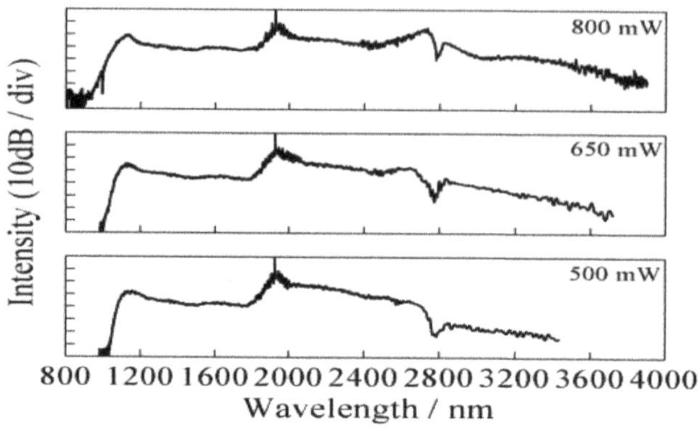

Figure 9. Measured SC in a tellurite microstructured optical fiber (MOF) at the pump wavelengths of ~1958 nm with average pump powers of ~500, 650, and 800 mW. With the average pump power increasing to ~800 mW, the broadband mid-infrared SC generation with the spectrum from ~900 to 3900 nm is observed. Reprinted with permission from [54] © The Optical Society.

In optics, an optical wave packet (a pulse or a beam) has a natural tendency to spread as it propagates in a medium, due either to chromatic dispersion or to spatial diffraction. When this natural broadening is compensated by a nonlinear process, a stable self-localized wave packet is formed. Such a self-trapped wave packet, whether in time or space or both, is known as an optical soliton. Solitons are localized wave entities that can propagate in nonlinear media, exhibiting invariant or recurrent propagation behavior. This means that the mutual compensation of dispersion and nonlinear effects can take place continuously with the distance as in the case of "classical" solitons or periodically as in the case of dispersion-managed solitons [55,56].

The temporal broadening of a pulse in a dispersive material due to chromatic dispersion, i.e., linear group velocity dispersion (GVD), can be compensated by the narrowing associated to self-phase modulation (SPM), so that a narrow pulse can propagate without temporal spreading. We note that when the normal dispersion is positive ($\beta_2 > 0$), both the pulse shape and spectrum change as the pulse propagates through the fiber/waveguide, and the combined effects of GVD and SPM cause a broadening of the pulse both in time and frequency. Whereas for anomalous dispersion ($\beta_2 < 0$), where β_2 and n_2 can have opposite signs, the group velocity dispersion induces a down chirp, while SPM nonlinearity induces an up chirp, so these two effects can compensate each other. This could be surprising since GVD affects the pulse in the time domain, while the SPM effect is in the frequency domain. However, a small time-dependent phase shift added to a Fourier transform-limited pulse does not change the spectrum to first order. If this phase shift is canceled by GVD in the same fiber, the pulse does not change its shape or its spectrum as it propagates [55]. Temporal solitons have been observed in laser resonators [57], microresonators [58,59], and microcombs [60]. In each of these cases, nonlinear compensation of GVD is provided by the Kerr effect.

Optical spatial solitons are shape-invariant self-guided beams of light, i.e., their beam diameter remains invariant during propagation in nonlinear media, thanks to a dynamical balance between diffraction and a self-focusing effect. In other words, a laser beam can produce its own dielectric waveguide and propagate without spreading. To obtain this result, it is necessary to compensate the spatial broadening of the beam due to diffraction during the propagation with the self-focusing. At low intensities, the nonlinear Kerr effect can be neglected, and the beam spreads under diffraction inducing a positive wavefront curvature. In contrast, an intense single beam with a well-shaped intensity profile at its waist generates a refractive index distribution looking like a graded index thick lens when the n_2 coefficient is positive, and as a consequence of balance with diffraction, a constant transverse dimension of the beam during propagation is maintained. An analysis of this phenomenon shows that it occurs when the power carried out by the beam is exactly equal to the critical power for self-trapping [61,62]. As an example, Kerr spatial solitons in slab chalcogenide waveguides at near-IR wavelengths have been observed [63].

Distortionless propagation, i.e., a mechanism capable of holding together the beam in the space and time domain, has a great technological relevance for optical telecommunications [55,58]. The use of fundamental solitons as an information bit can solve the problem of dispersion because in a fiber channel, nonlinear phase modulation can compensate for linear group dispersion leading to pulses that propagate without changing the temporal shape and spectrum. Chip-scale devices that support optical solitons harness high field confinement and flexibility in dispersion engineering for significantly smaller footprints and lower operating powers compared with fiber-based equivalents [64–67]. In Figure 10, a temporally bright spatiotemporal pulse-train soliton is reported as an example [67].

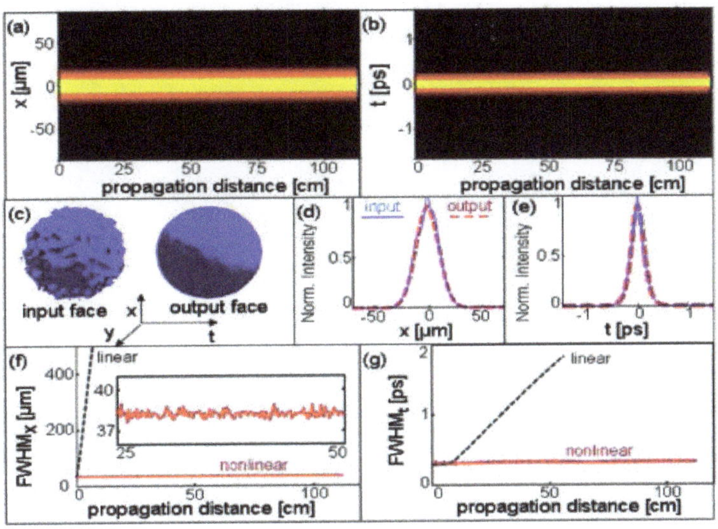

Figure 10. Temporally bright spatiotemporal pulse-train soliton. Intensity of the soliton in plane xz (**a**) and t-z (**b**) demonstrating the stationary propagation. (**c**) Intensity in x-y-t at the input and output faces of the sample. Intensity profiles in the x direction (**d**) and t direction (**e**) at the input and output faces of the sample. Width (FWHM) of the intensity in the x direction (**f**) and t direction (**g**) versus propagation distance for linear and nonlinear propagations. Reprinted with permission from [67] © The Optical Society.

3. Photonic Nonlinear Materials

In this section, we classify photonic nonlinear materials in three groups: semiconductors, glassy material and lithium niobate, and innovative materials.

3.1. Semiconductors

In the last decades, silicon photonics has received considerable attention. The main advantage of silicon photonics has been its compatibility with mature CMOS technology, offering structure sizes down to 10 nm at low cost. Silicon photonics has proven to be a promising technology (and in some cases a very effective solution) for a variety of applications; data center transceiver application has been the most successful, but consumer health and photonic computing are now emerging as rapidly growing fields. Other important applications, such as free-space communication, chemical and biomolecular sensing, and infrared spectroscopy, would push the development of integrated photonics toward the mid-infrared region, but another shortcoming of SOI platform is related to the loss of the buried oxide layer of the SOI waveguides at mid-infrared wavelengths. Silicon photonics also extends to highly integrated multifunctional devices that perform both optical and electrical operations on a single low-cost chip [68–70].

On the other hand, the major challenge in the development of active silicon elements is due to the indirect bandgap of silicon, which makes spontaneous emission unlikely and thus prevents lasing. This physical limitation with respect to light emission is a serious drawback when compared with integrated optical devices based on III-V direct bandgap materials, which are capable of efficiently emitting light. Overcoming this limitation in silicon is considered to be the holy grail of silicon photonics. Unfortunately, for a variety of reasons, room temperature electrically injected silicon-based lasers remain elusive to date. Another important shortcoming is that silicon does not absorb light at wavelengths above the material's bandgap ($\sim$1.1 μm), so light detection is prevented in the range of interest for telecommunications [68–72].

When going to nonlinear optical applications, several shortcomings are long known, such as the high optical absorption at telecommunication wavelengths and the inherent lack of second-order optical susceptibility ($\chi^{(2)}$) due to Si centrosymmetric lattice structure. The latter, of course, limits the use of Si photonics for standard applications, such as the development of electro-optic silicon modulators based on second-order nonlinear effects. In recent years, many types of third-order nonlinear optical phenomena have been investigated in silicon-based photonic devices. Substantial progress has been achieved, e.g., in the field of Raman amplification, while nonlinear effects such as two-photon absorption (TPA), self-phase modulation (SPM), cross-phase modulation (XPM), continuum generation, four-wave mixing (FWM), and the optical Kerr effect have also been successfully demonstrated and thoroughly investigated, on time scales ranging from the femtosecond to the nanosecond regime. However, even if silicon has large Si third-order optical susceptibility ($\chi^{(3)}$) compared with silica fibers or a silica integrated optical platform [73], the exploitation of silicon photonics devices is limited, at least at telecom wavelengths, by the presence of two-photon and free-carrier absorptions (TPA and FCA), at the required high optical intensities [68–74].

In order to address the shortcomings of silicon photonics, several materials have been investigated based on the heterogeneous integration of other material systems on silicon substrates, with the common requirement to remain compatible with the complementary metal-oxide-semiconductor (CMOS) technology [75]. The typical materials adopted for these heterogeneous integrations include GeSi, Ge-on-Si, silicon nitride (SiN), amorphous silicon (a-Si), silicon nanocrystal (Si-nc) [34–42], and silicon carbide (SiC). While germanium's cutoff wavelength of ~1.8 μm appeals as a perfect choice, in order to realize photodetectors, SiN, a-Si, Si-nc, and SiC are considered as appealing material for nonlinear photonic applications.

SiN has been routinely used in fabrication processes and final device structures for electronic integrated circuits. Due to its wide transparency window from visible to mid-infrared, progress on advanced SiN fabrication processes has enabled ultra-low propagation losses in waveguide structures. Due to the larger energy bandgap, SiN shows negligible TPA in the telecom bands, while the Kerr coefficients (n_2) of stoichiometric and non-stoichiometric SiN are about an order of magnitude and about three times smaller than silicon [76], respectively. However, the combination of linear and nonlinear optical properties makes photonic signal processing based on silicon nitride a promising research area. Silicon-rich nitride waveguides for ultra-broadband nonlinear signal processing (e.g., wavelength conversion and radio-frequency spectrum analyzer) have been demonstrated [77,78]. Recently, large second-harmonic generation enhancement in Si3N4 waveguides has been also demonstrated [79,80].

A drawback of silicon-rich and ultra-silicon-rich nitride films, realized by CVD, is the presence of N–H bonds and Si–H bonds, which possess absorption overtones close to 1.55 μm. In order to reduce Si–H related loss, precursor gases, which do not contain any Si–H bonds, can be considered. Deuterated silane (SiD_4) is chemically almost identical to SiH_4, but the Si–D bonds have an absorption overtone close to 2 μm. In ref. [81], the growth and optical characterization of silicon-rich nitride films grown using SiD_4 gas were reported. The absence of Si–H bonds and a reduction of propagation losses in waveguides compared with a conventional SRN film were demonstrated. A nonlinear refractive index of 980×10^{-20} m^2 W^{-1}, about two orders of magnitude larger than in stoichiometric silicon nitride, was demonstrated. The bandgap of 1.9 eV indicates that two-photon absorption is absent at a wavelength of 1.55 μm.

Amorphous silicon (a-Si) has been identified as a possible solution to overcome the main limitations showed by the SOI platform. Indeed, a-Si shows enhanced nonlinear performance with respect to crystalline silicon, exhibiting an enhanced Kerr response. The Kerr coefficient (n2) is about an order of magnitude greater than that of silicon [82–84].

Silicon carbide (SiC) exhibits unique optical properties that can be utilized for novel photonic devices. SiC is a wide bandgap compound semiconductor, so it is a transparent material from the UV to the infrared, and it is a meta-material in the mid-infrared range. Improving the quality factor (Q) of the microring and microdisk resonators in the SiC on insulator (SiCOI) platform, NLO effects based on high second-order and third-order nonlinear coefficients have been recently observed also by nanostructuring SiC [80,81]. The third-order nonlinearity n_2 of 60×10^{-20} m^2/W at 1550 nm for 4H-SiC [85,86] was reported, while in a-SiC, a value of the Kerr nonlinearity of $n_2 = 480 \times 10^{-20}$ m^2/W was estimated [87], which is of the same order of magnitude as silicon. The unique combination of excellent electronic, photonic, and spintronic properties of SiC has prompted research to develop novel devices and sensors in the quantum technology domain, too [87–89].

Tantalum pentoxide (Ta_2O_5) is a CMOS-compatible material that has been very recently considered as a possible alternative for nonlinear applications at telecom-wavelength applications. Thanks to its extremely large bandgap value (3.8 eV), no TPA effect is present at the 1550 nm wavelength region, while its n_2 is an order of magnitude lower than that of silicon [90–92].

Among the III–V integrated nonlinear photonic platforms considered to date, AlGaAs is the most studied waveguide platform [93]. AlGaAs has been termed "the silicon of nonlinear optics" for a number of reasons. The bandgap energy of AlGaAs can be modified by altering the aluminum concentration. It exhibits a wide transparency window from near- to mid-infrared (MIR, 0.9–17 μm), low linear and nonlinear propagation losses in the telecom spectral range (1400–1600 nm), a strong electro-optic effect, a large thermo-optic coefficient (two times larger than that of Si) enabling efficient thermal tuning, and a large second-order nonlinear susceptibility ($\chi^{(2)}$ over 200 pm/V) and third-order optical nonlinearity (Kerr coefficient $n_2 = 1500 \times 10^{-20}$ m^2/W) [94,95]. Depending on the structure and the composition, values of the nonlinear refractive index n_2 measured in (Al)GaAs waveguides range from 210×10^{-20} m^2/W at 1600 nm in an $Al_{0.23}Ga_{0.77}As$ ridge waveguide [96] up to 5500×10^{-20} m^2/W at 1505 nm in a GaAs/AlAs structure constituting six layers (a 0.6 μm thick GaAs/AlAs superlattice core layer with 75 periods of 14:14 monolayers each, two 0.3 μm thick $Al_{0.56}Ga_{0.44}As$ buffer layers, and two $Al_{0.60}Ga_{0.40}As$ cladding layers (the upper being 0.8 μm thick and the lower 4.0 μm, with a 0.1 μm GaAs cap) [97]. The n_2 value of that multilayer structure is impressive; for comparison, in a graphene/Si hybrid waveguide, the effective Kerr coefficient n_2 was calculated to be ~2000×10^{-20} m^2/W, five times higher than that of the Si waveguide alone [98].

The advances in the fabrication process made it possible to minimize the propagation loss in AlGaAs waveguides, e.g., in the AlGaAs-on-insulator (AlGaAs-OI) material platform, down to values less than 1 dB/cm, not so different from those achievable in SOI waveguides [93]. To date, in AlGaAs waveguide platforms, second- and third-order nonlinear optical phenomena, such as SHG, SFG, DFG, spontaneous parametric down-conversion (SPDC), FWM, 2PA, SPM, XPM, spontaneous four-wave mixing (SFWM), spatial soliton formation, SRS, and supercontinuum generation (SCG), have been experimentally investigated. AlGaAs waveguide platforms can be used for a wide range of integrated nonlinear photonic devices, including all-optical signal processing in optical communication networks and Kerr frequency microcomb and integrated quantum photonic circuits.

Recently, efforts have been made to realize photonic devices in gallium phosphide (GaP). The motivations were (1) GaP is nearly lattice-matched to silicon, in principle enabling wafer-scale production; (2) GaP has negligible two-photon absorption (TPA) for wavelengths above 1.1 μm; (3) among visibly transparent III–V materials, GaP has the largest refractive index ($n_0 > 3$), enabling strong optical confinement and implying a large $\chi^{(3)}$ nonlinearity [$n_2 = 1.100 \times 10^{-20}$ m^2 W^{-1}]; and (4) the non-centrosymmetric crystal structure of GaP yields a large $\chi^{(2)}$ nonlinearity. Thus, a variety of GaP nanophotonic devices have been fabricated and studied, including one-dimensional (1D) and 2D photonic crystals, microdisks, and strip waveguides. Among the different applications, particular attention has been paid to realizing frequency doublers from telecommunication to visible wavelengths and solid-state quantum emitters [99].

A crucial point is that GaP must be integrated onto a low index material—ideally by a method compatible with wafer-scale production—then patterned into devices with sufficiently low propagation loss to permit net optical gain. In ref. [99], a GaP-on-insulator platform for integrated nonlinear photonics was proposed, making use of direct wafer-bonding to integrate high-quality, epitaxially grown GaP onto SiO_2. The large index contrast between GaP and SiO_2, in conjunction with highly anisotropic, low-roughness dry etching, allowed the realization of single-mode strip waveguides with low propagation losses. As a demonstration of the capabilities of the platform, frequency comb generation in GaP microresonators with a threshold power as low as 3 mW and with the simultaneous formation of doubled combs at visible wavelengths was observed [99].

3.2. Glassy Materials and Lithium Niobate

Glasses are interesting non-linear optical materials, being isotropic and transparent in a wide spectral range, combining a low cost of fabrication with high optical quality, manufacturable not only as bulk shapes, or fibers, but also as thin films (e.g., nonlinear planar waveguides) [100,101]. The peculiar characteristics of glass materials are that their optical properties can be adjusted through doping and compositional changes to fit the specified requests of each application. Glasses are non-crystalline (or amorphous) materials with short-range order, so due to their isotropic structure, they have inversion symmetry and do not exhibit second-order nonlinearity, $\chi^{(2)}$, or Pockels effect. However, second-order nonlinearity can be achieved by an appropriate modification, obtained, for example, by the application of both heat and electric fields (thermal poling). In silicate glasses, $\chi^{(2)}$ also appeared by the introduction of optical non-linear nanocrystals within a glass matrix, obtained by precipitation of crystallites of non-centrosymmetric compounds [102]. This strategy gives rise to transparent crystallized glasses (glass-ceramics). The nonlinear optical properties of glasses have been used in several technological applications with a broad spectrum of phenomena, such as optical frequency conversion, optical solitons, phase conjugation, and Raman amplification [43–46].

The optical fiber based on fused silica glass is one of the most important platforms for guided-wave optics. For its excellent properties including extremely low losses between ~0.2 and 2 μm, low dispersion, and low nonlinearity, it has been the backbone of the contemporary telecom infrastructure. Due to the low Kerr coefficient [103], standard step-index fused silica fiber is not the most suitable for nonlinear optics. To overcome these limitations, new glasses for optical device applications and photonics have been investigated. These include fluoride glasses, tellurite glasses, aluminosilicates, phosphate glasses, borate glasses, and chalcogenide glasses. Glasses based on heavy metal oxides, such as Sb, Bi, Pb, W, Ga, Ge, and Te, allow applications such as optical switches due to their characteristics of low linear and nonlinear loss, large Kerr nonlinearity, and ultra-fast response. High-index glass (Hydex™) is a special type of doped fused silica glass with a refractive index in the range from 1.5 to 1.9. It has been used as an alternative material for CMOS-compatible low-loss optical waveguides, but it has a relatively low

Kerr nonlinearity [52,104]. Chalcogenide glasses are formed by the chalcogen elements S, Se, and Te and the addition of other elements such as Ge, As, Sb, Ga, etc. They have excellent transparency in the mid-IR region and are relatively easy to process, so they are the basis for the manufacture of devices operating in the mid-infrared region, where conventional silica glass shows strong absorption. Chalcogenide glass is one of the most interesting materials for nonlinear photonics since it possesses large optical nonlinearities in the infrared spectrum [105,106].

The field of third-order nonlinear phenomena of glasses has been mainly focused on two main groups: resonant and non-resonant [100,101]. Non-resonant phenomena occur when the light excitation falls in the transparent wavelengths range of the glass longer than its electronic absorption edge. As no electronic transitions take place, the process can be seen as lossless, and an ultrafast glass response due to third-order electronic polarization is assured. Examples are, in general, high-refractive-index and high-dispersion glasses such as heavy metal oxide glasses or chalcogenide glasses. Resonant phenomena occur when the frequencies of optical field are near the electronic absorption edge. Metallic nanoparticle doped glasses and semiconductors nanocrystals (quantum dots (QDs)) such as CdS, CdSe, CdTe, PbS, CuCl, etc. doped glasses are suitable materials for resonant NLO devices with response times on the ps domain. Great interest has driven the study of the third-order nonlinear susceptibility of metal particles embedded in dielectric matrices, such as glasses [107], which are influenced not only by the type and size of the metal particles but also by the metal-dielectric constant. The most significant effect of the confinement of metal particles on the optical properties of nanocomposite glasses is the appearance of the surface plasmon resonance, which deeply enhances the glass $\chi^{(3)}$ responses with picosecond temporal responses. QD-doped glasses can be prepared through the dispersion of a nanocrystalline phase in a glass matrix. This approach, through the reduction of bulk size to nanometric scale or quasi-zero-dimensional quantum dots, allows the change of the electronic properties of glasses accordingly with enhanced nonlinearity compared with the corresponding bulk semiconductors [100,101].

Lithium niobate (LN) is a transparent material in a wide wavelength range from 0.5 to 4 μm. Due to the intrinsic birefringence, its large EO coefficient (r_{33} = ≈31 pm/V and r_{13} = 8 pm/V), and the capacity of modulation bandwidths in excess of 100 GHz, LN has been recognized for a long time in the photonic industry as the best material for EO modulation. Having a non-centrosymmetric structure, LN shows both $\chi^{(2)}$ and $\chi^{(3)}$ optical nonlinear responses. Lithium niobate has a high second-order optical nonlinearity ($d_{33} = -33$ pm/V at $\lambda = 1.064$ μm) enabling parametric wavelength conversion and optical signal generation [105,106]. Due to its ferroelectricity, LN can be periodically poled (PPLN) with alternate domains of inverted electric dipole orientation to meet efficient quasi phase-matching conditions [108–116].

Table 1 presents the nonlinear refractive indices n_2, two-photon absorption coefficients β_{TPA} and figures of merit FOM of various materials. The wavelength at which the measurement had been made is also indicated.

Table 1. Nonlinear refractive indices n_2, two-photon absorption coefficients β_{TPA}, and figures of merit (FOMs) of various materials.

Material	$n_2\ \left(10^{-20}\,\frac{m^2}{W}\right)$	$\beta_{TPA}\ \left(\frac{cm}{GW}\right)$	Wavelength (nm)	FOM = $n_2/\beta_{TPA}\lambda$
SiO_2 [103]	3	Negligible	1053	-
Hydex [52]	13	Negligible	1560	-
Stoichiometric SiN [76]	28	Negligible	1550	-
Lithium niobate [108]	39	Negligible	1064	-
4H-SiC [86]	60	Negligible	1545	-
Ta_2O_5 [91]	72	Negligible	800	-

Table 1. *Cont.*

Material	n_2 $\left(10^{-20}\frac{m^2}{W}\right)$	β_{TPA} $\left(\frac{cm}{GW}\right)$	Wavelength (nm)	FOM = $n_2/\beta_{TPA}\lambda$
Non-stoichiometric SiN [76]	140	Negligible	1550	-
Chalcogenides [106]	370	Negligible	1550	-
Crystalline Si [73]	400	0.8	1540	0.32
a-SiC [87]	480	Negligible	1550	-
Deuterated silicon-rich nitride [81]	980	Negligible	1550	-
GaP [99]	1100	Negligible	1550	-
AlGaAs [94]	1500	0.05	1550	19
Graphene/Si hybrid waveguide [98]	2000	0.5	1548	2.6
a-Si:H [82]	4200	4.1	1550	0.7
GaAs/AlAs superlattice waveguides [97]	5500	4	1505	0.9

3.3. *Innovative Materials*

Most photonic applications would strongly benefit from tunable and reconfigurable properties of materials. In order to deal with this challenge, recently, phase-change materials (PCMs) have become a popular method of optical tunability without any moving parts. PCMs are a class of materials with unique physical properties: their structural arrangement can be controllably modified back and forth on a fast timescale using a thermal, electrical, or optical excitation [117,118]. For some of these materials, the crystallographic re-arrangement translates into a large refractive index modification ($\Delta n \geq 1$). Such a large and fast refractive index modulation is a long-sought effect for photonics: an enabling technology to control and tune in real time the optical properties of devices at the nanoscale. Among PCMs, vanadium dioxide (VO_2) is a prototypical example of functional materials showing large modifications in their physical properties upon specific external excitation [119].

Research in 2D layered materials (2DLMs) has started with the discovery of graphene [120]. The 2DLMs commonly refer to crystals composed of few layers of atoms, whose electrons move freely only on two dimensions (scale: 1–100 nm). It has been realized that the van der Waals layered materials with atomic thickness can not only exist stably but also exhibit unique and technically useful properties including small size effect, surface effect, macro quantum tunnel effect, and quantum effect. The 2D materials meet several requirements of an ideal nonlinear optical material: large and ultrafast nonlinear optical response, ultrafast photoexcitation dynamics, broadband and tunable optical absorption, saturable absorption characteristics, ultrafast recovery time, strong interlayer coupling, and large optical and thermal damage threshold [120–144]. The 2DLMs have been successfully employed in all-optical modulators, as broadband efficient and versatile saturable absorbers in passive mode locking and Q-switching, in wavelength converters, and in optical limiters [144–146].

The remarkable properties of 2D forms of graphite have attracted a rapidly growing research interest. In graphene, the conduction band and valence band converge at the Dirac point, which indicates that graphene owns a gapless semimetallic band structure. However, graphene is widely studied due to its broadband absorption at optical and terahertz (THz) frequencies [121–124]. Graphene exhibits saturable absorption behavior [125,126], which plays an important role in lasing applications and optical limiting devices.

Recently, graphene oxide (GO), with its large optical nonlinearity, high flexibility in altering its properties, facile solution-based synthesis process, and high compatibility with integrated device fabrication, offers competitive advantages for industrial manufacturing beyond the laboratory, which is a challenge for the majority of 2D materials.

The bandgap of GO can be engineered, and its value spans between 2.1 and 3.6 eV, yielding an absorption that is over two orders of magnitude lower than that of graphene in the telecom band. Similar to graphene, GO films exhibit strong anisotropy in their optical absorption in a broad band from the visible to the infrared regions. This property is useful for implementing polarization selective devices with wide operation bandwidths. Due to heterogenous atomic structure, GO exhibits noncentrosymmetry, yielding a large second-order optical nonlinearity, which is absent in pristine graphene that has a centrosymmetric structure. We note that at the moment, applications of second-order optical nonlinearity to chip-scale devices are still in their infancy.

The absolute value of n_2 for GO is about 10 times lower than that of graphene, but it is about four orders of magnitude higher than that of nonlinear bulk materials such as silicon and chalcogenide glasses. Enhanced FWM in GO hybrid integrated devices was demonstrated using Hydex waveguides [57] and Si_3N4 waveguides. Most of the state-of-the-art GO nonlinear integrated photonic devices incorporate GO films with little modification or optimization of their properties. However, GO's properties can be significantly changed, offering a high degree of flexibility in engineering its capabilities for different nonlinear optical processes [127,128].

Other classes of recently discovered 2DLMs include transition metal chalcogenides (TMCs) [129,130], transition metal-dichalcogenides (TMDs, e.g., MoS_2) [131–137], black phosphorus (BP) [138–140], and hexagonal boron nitride (h-BN) [141,142], which have broadened the set of specific properties and functionalities of 2D materials.

TMCs are semiconductor materials with the formula MX, where M represents the transition metal from group 4 to 7 (such as Mo, W, Re, and Ga) and X is the chalcogenide element, which is usually one of the three elements S, Se, and Te. TMCs exhibit the typical layer-dependent tunable bandgap, spanning from 1 eV for bulk or ML cases to 3 eV for single-layer (SL) configurations [131,132]. Transition metal dichalcogenides (TMDs) have the formula MX_2 (such as MoS_2, $MoSe_2$, $MoTe_2$, WS_2, WSe_2, and TiS_2). The single-layer TMD structure can be roughly represented by the three-layer covalent bonding in the form of X–M–X, in which metal atoms M are sandwiched between two layers of chalcogen atom X, and the chemical bond M–X provides the connection endowing the whole layer with strong stability, the thickness of which is usually 6–7 Å. Compared with graphene, the 2D TMDs are more suitable for optical applications due to their direct and tunable bandgaps and relatively high carrier mobility [131,132]. The inherent 2D and semiconductor character of the TMDs inevitably lead to the excitonic effects, typical of the optical response of the semiconductors. The optical properties of the 2D MoS_2 material are sensitive to the number of layers and interlayer distance, which makes it difficult to anticipate the optical properties of these materials. Several theoretical and experimental investigations have proved that TMDs (e.g., MoS_2 and WS_2) exhibit nonlinear optical response well beyond their lower energy bandgap (~1 eV) [133,137].

BP is a thickness-dependent direct bandgap semiconductor that could be widely tunable from 0.3 eV (bulk) to 2 eV (single layer (SL)). It bridges the gap between zero bandgap graphene and relatively wide bandgap TMCs, making it suitable for broadband optoelectronic applications, particularly in the IR and mid-IR regions [138–140].

Hexagonal boron nitride (h-BN) is a graphene analogue having a crystallographic appearance with boron and nitrogen atoms in place of carbon. h-BN has a large bandgap of ~6 eV. BP also exhibits saturable absorption followed by an optical limiting response upon varying the pump fluence. It finds nonlinear optics applications in UV to NIR regions [141,142].

According to their predominant electromagnetic response, materials or structures with near-zero parameters at a given frequency can be classified as epsilon-near-zero (ENZ), $\varepsilon \cong 0$; mu-near-zero (MNZ), $\mu \cong 0$; and epsilon-and-mu-near-zero (EMNZ), $\varepsilon \cong 0$ and $\mu \cong 0$ media. All aforementioned classes exhibit a near-zero index of refraction $n =$

$\sqrt{\varepsilon\mu} \cong 0$ at the frequency of interest and can be jointly addressed as zero-index media [147–150]. These materials or structures empower two of the main requirements to boost the nonlinear response of matter: phase matching [150] and high field intensities. Structures with near-zero parameters can provide large field intensity enhancements over large regions, while simultaneously providing phase matching [147–150]. ENZ materials are becoming a platform to obtain non-classical, non-reciprocal, and non-local responses of matter [147–150]. ENZ structures loaded with nonlinear media are particularly well suited to enable the coherent superposition of nonlinear effects over long distances, leading to enhanced nonlinear effects, in particular, Kerr nonlinearity, SHG, THG, and FWM [147–150]. Doped semiconductors such as transparent conducting oxides (TCOs), for example, aluminium-doped zinc oxide (AZO) [151], and ITO [152], exhibit a near-zero permittivity at near-infrared frequencies, with the additional advantages of being CMOS compatible and of providing tunable platforms, whose ENZ frequency can be adjusted by controlling the doping level.

4. Conclusions

In this review, first, the main aspects of the most common fundamental NLO effects are described, pointing out that various types of all-optical functionalities can be implemented by nonlinear integrated photonic devices, based mainly on second- and third-order phenomena. Then, since the field of nonlinear photonics has been the target of constant innovations based on the investigation of many optical materials, the nonlinear optical (NLO) properties of most important optical materials are described. Silicon and related materials, such as SiN, a-Si, and SiC; glasses, such as silica, high-index glass, and chalcogenide glasses; and III–V semiconductors, in particular, AlGaAs, lithium niobate (LN), and recently investigated materials such as tantalum pentoxide (Ta_2O_5) and vanadium dioxide (VO_2) are discussed, pointing out their pros and cons. Last, 2DLMs and zero-index media are described, too. We note that encouraging prospectives are provided by 2DLMs, but considering the production speed, yield, and quality, at the moment, the preparation of 2D materials is not sufficient to meet the requirements of industry or commercialization. In addition, the relatively low physical and/or chemical stability of some 2D materials makes them unsuitable for storage and the application of long-term stable optoelectronic devices.

A very interesting material platform, for instance, is scandium-doped aluminum nitride ($Al_{1-x}Sc_xN$), which maintains the CMOS compatibility of aluminum nitride (AlN) and exhibits a noticeable $\chi^{(2)}$ enhancement; for the $Al_{0.64}Sc_{0.36}N$ composition, the $\chi^{(2)}$ component d_{33} reaches a value of 62.3 $\pm$ 5.6 pm/V, which is 12 times stronger than that of the intrinsic AlN and twice as strong as that of lithium niobate [153]. Another class of recently investigated materials are the phase-change materials (PCMs). In principle, PCMs are very promising to enable dynamic modification to the physical properties of devices at the microscale and nanoscale [117–119].

For more information on photonic structures, their fabrication issues, and recent developments and new progress in nonlinear photonics devices and system integration, the interested readers are referred to a companion paper [154].

Author Contributions: Conceptualization, L.S. and G.C.R.; writing—original draft preparation, L.S. and G.C.R.; writing—review and editing, L.S. and G.C.R. All authors have read and agreed to the published version of the manuscript.

Funding: This research received no external funding.

Data Availability Statement: No new data were created.

Conflicts of Interest: The authors declare no conflict of interest.

References

1. "Photonics", Merriam-Webster Dictionary. Available online: https://www.merriam-webster.com/dictionary/photonics (accessed on 21 April 2020).
2. Krasnodebski, M. Throwing light on photonics: The genealogy of a technological paradigm. *Centaurus* **2018**, *60*, 3–24. [CrossRef]
3. Miller, S.E. Integrated Optics: An Introduction. *Bell Sys. Tech. J.* **1969**, *48*, 2059–2069. [CrossRef]

4. Koch, T.L.; Koren, U. Semiconductor photonic integrated circuits. *IEEE J. Quant. Electron.* **1991**, *27*, 641–653. [CrossRef]
5. Nagarajan, R.; Joyner, C.H.; Schneider, R.P.; Bostak, J.S.; Butrie, T.; Dentai, A.G.; Dominic, V. Large-scale photonic integrated circuits. *IEEE J. Sel. Topics Quant. Electron.* **2005**, *11*, 50–65. [CrossRef]
6. Kerr, J. XL. A new relation between electricity and light: Dielectrified media birefringent. *Lond. Edinb. Dublin Philos. Mag. J. Sci.* **1875**, *50*, 337–348. [CrossRef]
7. Kerr, J. LIV. A new relation between electricity and light: Dielectrified media birefringent. *Lond. Edinb. Dublin Philos. Mag. J. Sci.* **1875**, *50*, 446–458. [CrossRef]
8. Göppert, M. Über die Wahrscheinlichkeit des Zusammenwirkens zweier Lichtquanten in einem Elementarakt. *Naturwissenschaften* **1929**, *17*, 932. [CrossRef]
9. Franken, P.A.; Hill, A.E.; Peters, C.W.; Weinreich, G. Generation of Optical Harmonics. *Phys. Rev. Lett.* **1961**, *7*, 118. [CrossRef]
10. Armstrong, J.A.; Bloembergen, N.; Ducuing, J.; Pershan, P.S. Interactions between Light Waves in a Nonlinear Dielectric. *Phys. Rev.* **1962**, *127*, 1918. [CrossRef]
11. Bloembergen, N.; Pershan, P. Light Waves at the Boundary of Nonlinear Media. *Phys. Rev.* **1962**, *128*, 606–622. [CrossRef]
12. Anderson, D.B.; Boyd, J.T. Wideband CO_2 Laser Second Harmonic Generation Phase Matched in GaAs Thin-Film Waveguides. *Appl. Phys. Lett.* **1971**, *19*, 266. [CrossRef]
13. Stegeman, G.I.; Seaton, C.T. Nonlinear integrated optics. *J. Appl. Phys.* **1985**, *58*, R57–R78. [CrossRef]
14. Guo, Y.; Kao, C.K.; Li, H.E.; Chiang, K.S. *Nonlinear Photonics: Nonlinearities in Optics, Optoelectronics and Fiber Communications*; Springer: Berlin/Heidelberg, Germany, 2002.
15. Gibbs, H.M.; Khitrova, G.; Peyghambarian, N. *Nonlinear Photonics*; Springer: Berlin/Heidelberg, Germany, 1990.
16. Chen, Z.; Morandotti, R. *Nonlinear Photonics and Novel Optical Phenomena*; Springer: New York, USA, 2012.
17. Hendrickson, S.M.; Foster, A.C.; Camacho, R.M.; Clader, B.D. Integrated nonlinear photonics: Emerging applications and ongoing challenges. *J. Optical Soc. Am. B* **2014**, *31*, 3193–3203. [CrossRef]
18. Fathpour, S. Heterogeneous Nonlinear Integrated Photonics. *IEEE J. Quantum Electron.* **2018**, *54*, 1–16. [CrossRef]
19. Wabnitz, S.; Eggleton, B.J. *All-Optical Signal Processing*; Springer International Publishing: Cham, Switzerland, 2015.
20. Ji, X.; Liu, J.; He, J.; Wang, R.N.; Qiu, Z.; Riemensberger, J.; Kippenberget, T.J. Compact, spatial-mode-interaction-free, ultralow-loss, nonlinear photonic integrated circuits. *Commun. Phys.* **2022**, *5*, 84. [CrossRef]
21. Koshelev, K.; Kruk, S.; Melik-Gaykazyan, E.; Jae-Hyuck Choi, J.H.; Andrey Bogdanov, A.; Hong-Gyu Park, H.-G.; Kivshar, Y. Subwavelength dielectric resonators for nonlinear nanophotonics. *Science* **2020**, *367*, 288–292. [CrossRef]
22. Caspani, L.; Duchesne, D.; Dolgaleva, K.; Wagner, S.J.; Ferrera, M.; Razzari, L.; Pasquazi, A.; Peccianti, M.; Moss, D.J.; Aitchison, J.S.; et al. Optical frequency conversion in integrated devices. *J. Opt. Soc. Am. B* **2011**, *28*, A67–A82. [CrossRef]
23. Janner, D.; Tulli, D.; García-Granda, M.; Belmonte, M.; Pruneri, V. Micro-structured integrated electro-optic $LiNbO_3$ modulators. *Laser Photon. Rev.* **2009**, *3*, 301–313. [CrossRef]
24. Boyd, R.W. *Nonlinear Optics*; Academic Press: Cambridge, MA, USA, 2003.
25. Shen, Y.R. *The Principles of Nonlinear Optics*; Wiley: Hoboken, NJ, USA, 2003.
26. Agrawal, G.P.; Boyd, R.W. *Contemporary Nonlinear Optics*; Academic Press: Cambridge, MA, USA, 1992.
27. Langrock, C.; Kumar, S.; McGeehan, J.E.; Willner, A.E.; Fejer, M.M. All-Optical Signal Processing Using $\chi^{(2)}$ Nonlinearities in Guided-Wave Devices. *J. Light. Technol.* **2006**, *24*, 2579. [CrossRef]
28. Chang, L.; Boes, A.; Shu, H.; Xie, W.; Huang, H.; Qin, J.; Shen, B.; Wang, X.; Mitchell, A.; Bowers, J.E. Second Order Nonlinear Photonic Integrated Platforms for Optical Signal Processing. *IEEE J. Sel. Top. Quant. Electron.* **2021**, *27*, 1–11. [CrossRef]
29. Shen, Y.R.; Bloembergen, N. Theory of Stimulated Brillouin and Raman Scattering. *Phys. Rev.* **1965**, *137*, A1787–A1805. [CrossRef]
30. Eckhardt, G.; Bortfeld, D.P.; Geller, M. Stimulated emission of stokes and anti-stokes Raman lines from diamond, calcite, and sulfur single crystals. *Appl. Phys. Lett.* **1963**, *3*, 137–138. [CrossRef]
31. Sirleto, L.; Ferrara, M.A. Fiber Amplifiers and Fiber Lasers Based on Stimulated Raman Scattering: A Review. *Micromachines* **2020**, *11*, 247. [CrossRef]
32. Ferrara, M.A.; Sirleto, L. Integrated Raman Laser: A Review of the Last Two Decades. *Micromachines* **2020**, *11*, 330. [CrossRef]
33. Sirleto, L.; Vergara, A.; Ferrara, M.A. Advances in stimulated Raman scattering in nanostructures. *Adv. Opt. Photonics* **2017**, *9*, 169. [CrossRef]
34. Sirleto, L.; Ferrara, M.A.; Vergara, A. Toward an ideal nanomaterial for on-chip Raman laser. *J. Nonlinear Opt. Phys. Mater.* **2017**, *26*, 1750039. [CrossRef]
35. Sirleto, L.; Ferrara, M.A.; Nikitin, T.; Novikov, S.; Khriachtchev, L. Giant Raman gain in silicon nanocrystals. *Nat. Commun.* **2012**, *3*, 1220. [CrossRef]
36. Ferrara, M.A.; Sirleto, L.; Nicotra, G.; Spinella, C.; Rendina, I. Enhanced gain coefficient in Raman amplifier based on silicon nanocomposites. *Photon. Nanostructures Fundam Appl.* **2011**, *9*, 1–7. [CrossRef]
37. Sirleto, L.; Ferrara, M.A.; Nicotra, G.; Spinella, C.; Rendina, I. Observation of stimulated Raman scattering in silicon nanocomposites. *Appl. Phys. Lett.* **2009**, *94*, 221106. [CrossRef]
38. Ferrara, M.A.; Donato, M.G.; Sirleto, L.; Messina, G.; Santangelo, S.; Rendina, I. Study of strain and wetting phenomena in porous silicon by Raman scattering. *J. Raman Spectrosc.* **2008**, *39*, 199–204. [CrossRef]
39. Sirleto, L.; Ferrara, M.A.; Rendina, I.; Basu, S.N.; Warga, J.; Li, R.; Negro, L.D. Enhanced stimulated Raman scattering in silicon nanocrystals embedded in silicon-rich nitride/silicon superlattice structures. *Appl. Phys. Lett* **2008**, *93*, 251104. [CrossRef]

40. Sirleto, L.; Ferrara, M.A.; Rendina, I.; Jalali, B. Broadening and tuning of spontaneous Raman emission in porous silicon at 1.5 μm. *Appl. Phys. Lett.* **2006**, *88*, 211105. [CrossRef]
41. Sirleto, L.; Raghunathan, V.; Rossi, A.; Jalali, B. Raman emission in porous silicon at 1.54 μm. *Electron. Lett.* **2004**, *40*, 1221–1222. [CrossRef]
42. Sirleto, L.; Donato, M.G.; Messina, G.; Santangelo, S.; Lipovskii, A.; Tagantsev, D.; Pelli, S.; Righini, G.C. Raman gain in niobium-phosphate glasses. *Appl. Phys. Lett.* **2009**, *94*, 31105. [CrossRef]
43. Donato, M.G.; Gagliardi, M.; Sirleto, L.; Messina, G.; Lipovskii, A.; Tagantsev, D.; Righini, G.C. Raman optical amplification properties of sodium–niobium–phosphate glasses. *Appl. Phys. Lett.* **2010**, *97*, 231111. [CrossRef]
44. Sirleto, L.; Aronne, A.; Gioffré, M.; Fanelli, E.; Righini, G.C.; Pernice, P.; Vergara, A. Compositional and thermal treatment effects on Raman gain and bandwidth in nanostructured silica based glasses. *Opt. Mater.* **2013**, *36*, 408–413. [CrossRef]
45. Pernice, P.; Sirleto, L.; Vergara, A.; Aronne, A.; Gagliardi, M.; Fanelli, E.; Righini, G.C. Large Raman Gain in a Stable Nanocomposite Based on Niobiosilicate Glass. *J. Phys. Chem. C* **2011**, *115*, 17314–17319. [CrossRef]
46. Zhang, Y.; Zhong, K.; Zhou, X.; Tsang, H.K. Broadband high-Q multimode silicon concentric racetrack resonators for widely tunable Raman lasers. *Nat Commun* **2022**, *13*, 3534. [CrossRef]
47. Merklein, M.; Kabakova, I.V.; Zarifi, A.; Eggleton, B.J. 100 years of Brillouin scattering: Historical and future perspectives. *Appl. Phys. Rev.* **2022**, *9*, 041306. [CrossRef]
48. Eggleton, B.J.; Poulton, C.G.; Rakich, P.T.; Steel, M.J.; Bahl, G. Brillouin integrated photonics. *Nat. Photonics* **2019**, *13*, 664–677. [CrossRef]
49. Agrawal, G. *Nonlinear Fiber Optics*, 5th ed.; Academic Press: Cambridge, MA, USA, 2012.
50. Schneider, T. *Nonlinear Optics in Telecommunications*; Springer: Berlin/Heidelberg, Germany, 2004.
51. Alfano, R.R.; Shapiro, S.L. Observation of self-phase modulation and small-scale filaments in crystals and glasses. *Phys. Rev. Lett.* **1970**, *24*, 592–594. [CrossRef]
52. Duchesne, D.; Ferrera, M.; Razzari, L.; Morandotti, R.; Little, B.E.; Chu, S.T.; Moss, D.J. Efficient self-phase modulation in low loss, high index doped silica glass integrated waveguides. *Opt. Express* **2009**, *17*, 1865–1870. [CrossRef]
53. Dudley, J.M.; Genty, G.; Coen, S. Supercontinuum generation in photonic crystal fiber. *Rev. Mod. Phys.* **2006**, *78*, 1135–1184. [CrossRef]
54. Cheng, T.; Zhang, L.; Xue, X.; Deng, D.; Suzuki, T.; Ohishi, Y. Broadband cascaded four-wave mixing and supercontinuum generation in a tellurite microstructured optical fiber pumped at 2 μm. *Opt. Express* **2015**, *23*, 4125–4134. [CrossRef]
55. Kivshar, Y.; Agrawal, G.P. *Optical Solitons: From Fibres to Photonic Crystals*; Academy Press: San Diego, CA, USA, 2003.
56. Mitschke, F.; Mahnke, C.; Hause, A. Soliton Content of Fiber-Optic Light Pulses. *Appl. Sci.* **2017**, *7*, 635. [CrossRef]
57. Haus, H.A. Mode-locking of lasers. *IEEE J. Sel. Top. Quant. Electron.* **2000**, *6*, 1173–1185. [CrossRef]
58. Herr, T.; Brasch, V.; Jost, J.D.; Wang, C.Y.; Kondratiev, N.M.; Gorodetsky, M.L.; Kippenberg, T.J. Temporal solitons in optical microresonators. *Nat. Photon* **2014**, *8*, 145–152. [CrossRef]
59. Kippenberg, T.J.; Gaeta, A.L.; Lipson, M.; Gorodetsky, M.L. Dissipative Kerr solitons in optical microresonators. *Science* **2018**, *361*, aan8083. [CrossRef]
60. Xiang, C.; Liu, J.; Guo, J.; Chang, L.; Wang, R.N.; Weng, W.; Peters, J.; Xie, W.; Zhang, Z.; Riemensberger, J.; et al. Laser soliton microcombs heterogeneously integrated on silicon. *Science* **2021**, *373*, 99–103. [CrossRef]
61. Stegeman, G.I.; Segev, M. Optical Spatial Solitons and Their Interactions: Universality and Diversity. *Science* **1999**, *286*, 1518–1523. [CrossRef]
62. Chen, Z.; Segev, M.; Christodoulides, D.N. Optical spatial solitons: Historical overview and recent advances. *Rep. Prog. Phys.* **2012**, *75*, 086401. [CrossRef]
63. Chauvet, M.; Fanjoux, G.; Huy, K.P.; Nazabal, V.; Charpentier, F.; Billeton, T.; Boudebs, G.; Cathelinaud, M.; Gorza, S.P. Kerr spatial solitons in chalcogenide waveguides. *Opt. Lett.* **2009**, *34*, 1804–1806. [CrossRef]
64. Gurgov, H.C.; Cohen, O. Spatiotemporal pulse-train solitons. *Opt. Express* **2009**, *17*, 7052–7058. [CrossRef]
65. Wu, Y.-D. Nonlinear all-optical WDM based on spatial solitons in optical communication spectral regions. *J. Electromagn. Waves Appl.* **2014**, *28*, 2025–2033. [CrossRef]
66. Ma, M.; Chen, L.R. Harnessing mode-selective nonlinear optics for on-chip multi-channel all-optical signal processing. *APL Photonics* **2016**, *1*, 086104. [CrossRef]
67. Grubel, B.C.; Bosworth, B.T.; Kossey, M.R.; Cooper, A.B.; Foster, M.A.; Foster, A.C. Secure communications using nonlinear silicon photonic keys. *Opt. Express* **2018**, *26*, 4710–4722. [CrossRef]
68. Dekker, R.; Usechak, N.; Först, M.; Driessen, A. Ultrafast nonlinear all-optical processes in silicon-on-insulator waveguides. *J. Phys. D: Appl. Phys.* **2007**, *40*, R249. [CrossRef]
69. Osgood, R.M.; Panoiu, N.C.; Dadap, J.I.; Liu, X.; Chen, X.; Hsieh, I.-W.; Dulkeith, E.; Green, W.M.J.; Vlasov, Y.A. Engineering nonlinearities in nanoscale optical systems: Physics and applications in dispersion-engineered silicon nanophotonic wires. *Adv. Opt. Photonics* **2009**, *1*, 162–235. [CrossRef]
70. Leuthold, J.; Koos, C.; Freude, W. Nonlinear silicon photonics. *Nat. Photonics* **2010**, *4*, 535–544. [CrossRef]
71. Nikdast, M.; Pasricha, S.; Nicolescu, G.; Seyedi, A.; Liang, D. *Silicon Photonics for High-Performance Computing and Beyond*; CRC Press: Boca Raton, FL, USA, 2021.

72. He, L.; Zhang, F.; Zhang, H.; Kong, L.-J.; Zhang, W.; Xu, X.; Zhang, X. Topology-Optimized Ultracompact All-Optical Logic Devices on Silicon Photonic Platforms. *ACS Photonics* **2022**, *9*, 597–604. [CrossRef]
73. Dinu, M.; Quochi, F.; Garcia, H. Third-order nonlinearities in silicon at telecom wavelengths. *Appl. Phys. Lett.* **2003**, *82*, 2954–2956. [CrossRef]
74. Liang, T.K.; Tsang, H.K. Role of free carriers from two-photon absorption in Raman amplification in silicon-on-insulator waveguides. *Appl. Phys. Lett.* **2004**, *84*, 2745. [CrossRef]
75. Fathpour, S. Emerging heterogeneous integrated photonic platforms on silicon. *Nanophotonics* **2015**, *4*, 143–164. [CrossRef]
76. Krückel, C.J.; Fülöp, A.; Klintberg, T.; Bengtsson, J.; Andrekson, P.A.; Torres-Company, V. Linear and nonlinear characterization of low-stress high-confinement silicon-rich nitride waveguides. *Opt. Express* **2015**, *23*, 25827–25837. [CrossRef]
77. Dizaji, M.R.; Krückel, C.J.; Fülöp, A.; Andrekson, P.A.; Torres-Company, V.; Chen, L.R. Silicon-rich nitride waveguides for ultrabroadband nonlinear signal processing. *Opt. Express* **2017**, *25*, 12100–12108. [CrossRef]
78. Tana, D.T.H.; Ngb, D.K.T.; Choia, J.W.; Sahina, E.; Sohna, B.-U.; Chena, G.F.R.; Xinga, P.; Gaoa, H.; Caoa, Y. Nonlinear optics in ultra-silicon-rich nitride devices: Recent developments and future outlook. *Adv. Phys. X* **2021**, *6*, 1905544. [CrossRef]
79. Billat, A.; Grassani, D.; Pfeiffer, M.H.P.; Kharitonov, S.; Kippenberg, T.J.; Brès, C.-S. Large second harmonic generation enhancement in Si3N4 waveguides by all-optically induced quasi-phase-matching. *Nat. Commun.* **2017**, *8*, 1016. [CrossRef]
80. Lu, X.; Moille, G.; Rao, A.; Westly, D.A.; Srinivasanet, K. Efficient photoinduced second-harmonic generation in silicon nitride photonics. *Nat. Photonics* **2021**, *15*, 131–136. [CrossRef]
81. Chia, X.X.; Chen, G.F.R.; Cao, Y.; Xing, P.; Gao, H.; Ng, D.K.T.; Tan, D.T.H. Optical characterization of deuterated silicon-rich nitride waveguides. *Sci. Rep.* **2022**, *12*, 12697. [CrossRef]
82. Kuyken, B.; Ji, H.; Clemmen, S.; Selvaraja, S.K.; Hu, H.; Pu, M.; Galili, M.; Jeppesen, P.; Morthier, G.; Massar, S.; et al. Nonlinear properties of and nonlinear processing in hydrogenated amorphous silicon waveguides. *Opt. Express* **2011**, *19*, B146–B153. [CrossRef]
83. Narayanan, K.; Preble, S.F. Optical nonlinearities in hydrogenated-amorphous silicon waveguides. *Opt. Express* **2010**, *18*, 8998–9005. [CrossRef]
84. Grillet, C.; Carletti, L.; Monat, C. Amorphous silicon nanowires combining high nonlinearity, FOM and optical stability. *Opt. Express* **2012**, *20*, 22609–22615. [CrossRef]
85. Lu, X.; Lee, J.Y.; Feng, P.X.-L.; Lin, Q. High Q silicon carbide microdisk resonator. *Appl. Phys. Lett.* **2014**, *104*, 181103. [CrossRef]
86. Wang, C.; Fang, Z.; Yi, A.; Yang, B.; Wang, Z.; Zhou, L.; Shen, C.; Zhu, Y.; Zhou, Y.; Bao, R.; et al. High-Q microresonators on 4H-silicon-carbide-on-insulator platform for nonlinear photonics. *Light Sci. Appl.* **2021**, *10*, 139. [CrossRef]
87. Xing, P.; Ma, D.; Ooi, K.J.; Choi, J.W.; Agarwal, A.M.; Tan, D. CMOS-compatible PECVD silicon carbide platform for linear and nonlinear optics. *ACS Photonics* **2019**, *6*, 1162–1167. [CrossRef]
88. Castelletto, S.; Peruzzo, A.; Bonato, C.; Johnson, B.C.; Marina Radulaski, M.; Ou, H.; Florian Kaiser, F.; Wrachtrup, J. Silicon Carbide Photonics Bridging Quantum Technology. *ACS Photonics* **2022**, *9*, 1434–1457. [CrossRef]
89. Castelletto, S.; Rosa, L.; Johnson, B.C. Silicon Carbide for Novel Quantum Technologies Devices, Chapter 9. In *Advanced Silicon Carbide Devices and Processing*; Intech Open: London, UK, 2015.
90. Splitthoff, L.; Wolff, M.A.; Grottke, T.; Schuck, C. Tantalum pentoxide nanophotonic circuits for integrated quantum technology. *Opt. Express* **2020**, *28*, 11921–11932. [CrossRef]
91. Tai, C.-Y.; Wilkinson, J.S.; Perney, N.M.B.; Netti, M.C.; Cattaneo, F.; Finlayson, C.E.; Baumberg, J.J. Determination of nonlinear refractive index in a Ta_2O_5 rib waveguide using self-phase modulation. *Opt. Express* **2004**, *12*, 5110–5116. [CrossRef]
92. Wu, C.-L.; Chiu, Y.-J.; Chen, C.-L.; Lin, Y.-Y.; Chu, A.-K.; Lee, C.-K. Four-wave-mixing in the loss low submicrometer Ta_2O_5 channel waveguide. *Opt. Lett.* **2015**, *40*, 4528–4531. [CrossRef]
93. Mobini, E.; Espinosa, D.H.G.; Vyas, K.; Dolgaleva, K. AlGaAs Nonlinear Integrated Photonics. *Micromachines* **2022**, *13*, 991. [CrossRef]
94. Aitchison, J.S.; Hutchings, D.; Kang, J.; Stegeman, G.; Villeneuve, A. The nonlinear optical properties of AlGaAs at the half band gap. *IEEE J. Quantum Electron.* **1997**, *33*, 341–348. [CrossRef]
95. Ohashi, M.; Kondo, T.; Ito, R.; Fukatsu, S.; Shiraki, Y.; Kumata, K.; Kano, S. Determination of quadratic nonlinear optical coefficient of Al_xGa_{1-x} As system by the method of reflected second harmonics. *J. Appl. Phys.* **1993**, *74*, 596–601. [CrossRef]
96. Espindola, R.P.; Udo, M.K.; Ho, S.T. Nearly-degenerate frequency technique for simultaneous measurement of n(2) and α(2), and four-wave mixing gain coefficients in waveguides. *Opt. Commun.* **1995**, *119*, 682–692. [CrossRef]
97. Wagner, S.J.; Meier, J.; Helmy, A.S.; Aitchison, J.S.; Sorel, M.; Hutchings, D.C. Polarization-dependent nonlinear refraction and two-photon absorption in GaAs/AlAs superlattice waveguides below the half-bandgap. *J. Opt. Soc. Am.* **2007**, *24*, 1557–1563. [CrossRef]
98. Feng, Q.; Cong, H.; Zhang, B.; Wei, W.; Liang, Y.; Fang, S.; Wang, T.; Zhang, J. Enhanced optical Kerr nonlinearity of graphene/Si hybrid waveguide. *Appl. Phys. Lett.* **2019**, *114*, 071104. [CrossRef]
99. Wilson, D.J.; Schneider, K.; Hönl, S.; Anderson, M.; Baumgartner, Y.; Czornomaz, L.; Kippenberg, T.J.; Seidler, P. Integrated gallium phosphide nonlinear photonics. *Nat. Photonics* **2020**, *14*, 57–62. [CrossRef]
100. Vasconcelos, H.C. Optical Nonlinearities in Glasses. In *Nonlinear Nanophotonics and Novel Materials for Nonlinear Optics*; IntechOpen: London, UK, 2022.
101. Yamane, M.; Asahara, Y. Nonlinear optical glass. In *Glasses for Photonics*; Cambridge University Press: Cambridge, UK, 2000.

102. Guignard, M.; Nazabal, V.; Zhang, X.; Smektala, F.; Moréac, A.; Pechev, S.; Zeghlache, H.; Kudlinski, A.; Martinelli, G.; Quiquempois, Y. Crystalline phase responsible for the permanent second-harmonic generation in chalcogenide glass-ceramics. *Opt. Mater.* **2007**, *30*, 338–345. [CrossRef]
103. Milam, D. Review and assessment of measured values of the nonlinear refractive-index coefficient of fused silica. *Appl. Opt.* **1998**, *37*, 546–550. [CrossRef] [PubMed]
104. Ferrera, M.; Duchesne, D.; Razzari, L.; Peccianti, M.; Morandotti, R.; Cheben, P.; Janz, S.; Xu, D.-X.; Little, B.E.; Chu, S.; et al. Low power four wave mixing in an integrated, micro-ring resonator with Q = 1.2 million. *Opt. Express* **2009**, *17*, 14098–14103. [CrossRef]
105. Eggleton, B.J.; Luther-Davies, B.; Richardson, K. Chalcogenide photonics. *Nat. Photonics* **2011**, *5*, 141–148. [CrossRef]
106. Choi, J.; Han, Z.; Sohn, B.-U.; Chen, G.F.R.; Smith, C.; Kimerling, L.C.; Richardson, K.A.; Anuradha, M.; Agarwal, A.M.; Tan, D.T.H. Nonlinear characterization of GeSbS chalcogenide glass waveguides. *Sci. Rep.* **2016**, *6*, 39234. [CrossRef]
107. Kim, K.-H.; Husakou, A.; Herrmann, J. Linear and nonlinear optical characteristics of composites containing metal nanoparticles with different sizes and shapes. *Opt. Express* **2010**, *18*, 7488–7496. [CrossRef] [PubMed]
108. Bache, M.; Schiek, R. Review of measurements of Kerr nonlinearities in lithium niobate: The role of the delayed Raman response. *arXiv* **2012**, arXiv:1211.1721.
109. Beyer, O.; Maxein, D.; Buse, K.; Sturman, B.; Hsieh, H.T.; Psaltis, D. Femtosecond time-resolved absorption processes in lithium niobate crystals. *Opt. Lett.* **2005**, *30*, 1366–1368. [CrossRef]
110. Wang, J.; Nuccio, S.; Wu, X.; Yilmaz, O.F.; Zhang, L.; Fazal, I.; Yang, J.Y.; Yue, Y.; Willner, A.E. 40 Gbit/s optical data exchange between wavelength-division-multiplexed channels using a periodically poled lithium niobate waveguide. *Opt. Lett.* **2010**, *35*, 1067–1069. [CrossRef] [PubMed]
111. Wang, J.; Fu, H.; Geng, D.; Willner, A.E. Single-PPLN-assisted wavelength-/time-selective switching/dropping/swapping for 100-GHz-spaced WDM signals. *Opt. Express* **2013**, *21*, 3756–3774. [CrossRef] [PubMed]
112. Poberaj, G.; Hu, H.; Sohler, W.; Günter, P. Lithium niobate on insulator (LNOI) for micro-photonic devices. *Laser Photonics Rev.* **2012**, *6*, 488–503. [CrossRef]
113. Luo, R.; Jiang, H.; Rogers, S.; Liang, H.; He, Y.; Lin, Q. On-chip second-harmonic generation and broadband parametric down-conversion in a lithium niobate microresonator. *Opt. Express* **2017**, *25*, 24531–24539. [CrossRef]
114. Bazzan, M.; Sada, C. Optical waveguides in lithium niobate: Recent developments and applications. *Appl. Phys. Rev.* **2015**, *2*, 40603. [CrossRef]
115. Boes, A.; Corcoran, B.; Chang, L.; Bowers, J.; Mitchell, A. Status and potential of lithium niobate on insulator (LNOI) for photonic integrated circuits. *Laser Photon. Rev.* **2018**, *12*, 1700256. [CrossRef]
116. Vazimali, M.G.; Fathpour, S. Applications of thin-film lithium niobate in nonlinear integrated photonics. *Adv. Photonics* **2022**, *4*, 034001. [CrossRef]
117. Yang, Z.; Ko, C.; Ramanathan, S. Oxide electronics utilizing ultrafast metalinsulator transitions. *Annu. Rev. Mater. Res.* **2011**, *41*, 337–367. [CrossRef]
118. Raoux, S. Phase change materials. *Annu. Rev. Mater. Res.* **2009**, *39*, 25–48. [CrossRef]
119. Cueff, S.; John, J.; Zhang, Z.; Parra, J.; Sun, J.; Orobtchouk, R.; Ramanathan, S.; Sanchis, P. VO2 nanophotonics. *APL Photonics* **2020**, *5*, 110901. [CrossRef]
120. Novoselov, K.S.; Geim, A.K.; Morozov, S.V.; Jiang, D.; Zhang, Y.; Dubonos, S.V.; Grigorieva, I.V.; Firsov, A.A. Electric field effect in atomically thin carbon films. *Science* **2004**, *306*, 666–669. [CrossRef]
121. Cheng, J.L.; Vermeulen, N.; Sipe, J.E. Third order optical nonlinearity of graphene. *New J. Phys.* **2014**, *16*, 053014. [CrossRef]
122. An, Y.Q.; Nelson, F.; Lee, J.U.; Diebold, A.C. Enhanced optical second-harmonic generation from the current-biased graphene/SiO_2/Si (001) structure. *Nano Lett.* **2013**, *13*, 2104–2109. [CrossRef]
123. Feng, M.; Zhan, H.; Chen, Y. Nonlinear optical and optical limiting properties of graphene families. *Appl. Phys. Lett.* **2010**, *96*, 033107. [CrossRef]
124. Hendry, E.; Hale, P.J.; Moger, J.; Savchenko, A.K.; Mikhailov, S.A. Coherent nonlinear optical response of graphene. *Phys. Rev. Lett.* **2010**, *105*, 097401. [CrossRef]
125. Zhang, H.; Bao, Q.; Tang, D.; Zhao, L.; Loh, K. Large energy soliton erbium doped fiber laser with a graphene-polymer composite mode locker. *Appl. Phys. Lett.* **2009**, *95*, 141103. [CrossRef]
126. Zhang, H.; Tang, D.Y.; Zhao, L.M.; Bao, Q.L.; Loh, K.P. Large energy mode locking of an erbium-doped fiber laser with atomic layer graphene. *Opt. Express* **2009**, *17*, 17630–17635. [CrossRef] [PubMed]
127. Zhang, Y.; Wu, J.; Jia, L.; Qu, Y.; Yang, Y.; Jia, B.; Moss, D.J. Graphene Oxide for Nonlinear Integrated Photonics. *Laser Photonics Rev.* **2022**, 2200512. [CrossRef]
128. Wu, J.; Lin, H.; Moss, D.J.; Loh, K.P.; Jia, B. Graphene oxide for photonics, electronics and optoelectronics. *Nat. Rev. Chem.* **2023**. [CrossRef]
129. You, J.W.; Bongu, S.R.; Bao, Q.; Panoiu, N.C. Nonlinear optical properties and applications of 2D materials: Theoretical and experimental aspects. *Nanophotonics* **2019**, *8*, 63–97. [CrossRef]
130. Zhao, M.; Ye, Z.; Suzuki, R.; Ye, Y.; Zhu, H.; Xiao, J.; Wang, Y.; Iwasa, Y.; Zhang, X. Atomically phase-matched second-harmonic generation in a 2D crystal. *Light Sci. Appl.* **2016**, *5*, e16131. [CrossRef]

131. Lee, J.Y.; Shin, J.H.; Lee, G.H.; Lee, C.H. Two-dimensional semiconductor optoelectronics based on van der Waals heterostructures. *Nanomaterials* **2016**, *6*, 193. [CrossRef]
132. Xia, F.; Wang, H.; Xiao, D.; Dubey, M.; Ramasubramaniam, A. Two-dimensional material nanophotonics. *Nat. Photonics* **2014**, *8*, 899–907. [CrossRef]
133. Li, Y.; Rao, Y.; Mak, K.F.; You, Y.; Wang, S.; Dean, C.R.; Heinz, T.F. Probing Symmetry Properties of Few-Layer MoS2 and h-BN by Optical Second-Harmonic Generation. *Nano Lett.* **2013**, *13*, 3329–3333. [CrossRef] [PubMed]
134. Varma, S.J.; Kumar, J.; Liu, Y.; Layne, K.; Wu, J.; Liang, C.; Nakanishi, Y.; Aliyan, A.; Yang, W.; Ajayan, P.M.; et al. 2D TiS2 layers: A superior nonlinear optical limiting material. *Adv. Opt. Mater.* **2017**, *5*, 1700713. [CrossRef]
135. Kumar, N.; Najmaei, S.; Cui, Q.; Ceballos, F.; Ajayan, P.-M.; Lou, J.; Zhao, H. Second harmonic microscopy of monolayer MoS2. *Phys. Rev. B* **2013**, *87*, 161403. [CrossRef]
136. Wang, G.; Marie, X.; Gerber, I.C.; Amand, T.; Lagarde, D.; Bouet, L.; Vidal, M.; Balocchi, A.; Urbaszek, B. Giant enhancement of the optical second-harmonic emission of WSe2 monolayers by laser excitation at exciton resonances. *Phys. Rev. Lett.* **2015**, *114*, 097403. [CrossRef]
137. Zhou, X.; Cheng, J.; Zhou, Y.; Cao, T.; Hong, H.; Liao, Z.; Wu, S.; Peng, H.; Liu, K.; Yu, D. Strong second-harmonic generation in atomic layered GaSe. *J. Am. Chem. Soc.* **2015**, *137*, 7994–7997. [CrossRef] [PubMed]
138. Rodrigues, M.J.L.F.; de Matos, C.J.S.; Ho, Y.W.; Peixoto, H.; de Oliveira, R.E.P.; Wu, H.Y.; Castro Neto, A.H.; Viana-Gomeset, J. Resonantly Increased Optical Frequency Conversion in Atomically Thin Black Phosphorus. *Adv. Mater.* **2016**, *28*, 10693–10700. [CrossRef] [PubMed]
139. Lu, S.B.; Miao, L.L.; Guo, Z.N.; Qi, X.; Zhao, C.J.; Zhang, H.; Wen, S.C.; Tang, D.Y.; Fan, D.Y. Broadband nonlinear optical response in multi-layer black phosphorus: An emerging infrared and mid-infrared optical material. *Opt. Express* **2015**, *23*, 11183–11194. [CrossRef]
140. Xu, Y.; Jiang, X.F.; Ge, Y.; Guo, Z.; Zeng, Z.; Xu, Q.-H.; Zhang, H.; Yu, X.-F.; Fana, D. Size-dependent nonlinear optical properties of black phosphorus nanosheets and their applications in ultrafast photonics. *J. Mater. Chem. C* **2017**, *5*, 3007–3013. [CrossRef]
141. Ouyang, Q.; Zhang, K.; Chen, W.; Zhou, F.; Ji, W. Nonlinear absorption and nonlinear refraction in a chemical vapor deposition-grown, ultrathin hexagonal boron nitride film. *Opt. Lett.* **2016**, *41*, 1368–1371. [CrossRef]
142. Kumbhakar, P.; Kole, A.K.; Tiwary, C.S.; Biswas, S.; Vinod, S.; Taha-Tijerina, J.; Chatterjee, U.; Ajayan, P.M. Nonlinear optical properties and temperature-dependent UV-vis absorption and photoluminescence emission in 2D hexagonal boron nitride nanosheets. *Adv. Opt. Mater.* **2015**, *3*, 828–835. [CrossRef]
143. Xie, Y.; Zhang, B.; Wang, S.; Wang, D.; Wang, A.; Wang, Z.; Yu, H.; Zhang, H.; Chen, Y.; Zhao, M.; et al. Ultrabroadband MoS2 photodetector with spectral response from 445 to 2717 nm. *Adv. Mater.* **2017**, *29*, 1605972. [CrossRef]
144. Wei, R.; Zhang, H.; Hu, Z.; Qiao, T.; He, X.; Guo, Q.; Tian, X.; Chen, Z.; Qiu, J. Ultra-broadband nonlinear saturable absorption of high-yield MoS2 nanosheets. *Nanotechnology* **2016**, *27*, 305203. [CrossRef]
145. Wei, C.; Luo, H.; Zhang, H.; Li, C.; Xie, J.; Li, J.; Liu, Y. Passively Q-switched mid-infrared fluoride fiber laser around 3 μm using a tungsten disulphide (WS2) saturable absorber. *Laser Phys. Lett.* **2016**, *13*, 105108. [CrossRef]
146. Wang, Y.; Li, J.; Han, L.; Lu, R.; Hu, Y.; Li, Z.; Liu, Y. Q-switched Tm3+-doped fiber laser with a micro-fiber based black phosphorus saturable absorber. *Laser Phys.* **2016**, *26*, 065104. [CrossRef]
147. Liberal, I.; Engheta, N. Near-zero refractive index photonics. *Nat. Photonics* **2017**, *11*, 149–158. [CrossRef]
148. Argyropoulos, C.; Chen, P.Y.; D'Aguanno, G.; Engheta, N.; Alù, A. Boosting optical nonlinearities in ε-near-zero plasmonic channels. *Phys. Rev. B* **2012**, *85*, 45129. [CrossRef]
149. de Ceglia, D.; Campione, S.; Vincenti, M.A.; Capolino, F.; Scalora, M. Low-damping epsilon-near-zero slabs: Nonlinear and nonlocal optical properties. *Phys. Rev. B* **2013**, *87*, 155140. [CrossRef]
150. Suchowski, H.; O'Brien, K.; Wong, Z.J.; Salandrino, A.; Yin, X.; Zhang, X. Phase mismatch-free nonlinear propagation in optical zero-index materials. *Science* **2013**, *342*, 1223–1226. [CrossRef]
151. Caspani, L.; Kaipurath, R.P.M.; Clerici, M.; Ferrera, M.; Roger, T.; Kim, J.; Kinsey, N.; Pietrzyk, M.; Di Falco, A.; Shalaev, V. M.; et al. Enhanced Nonlinear Refractive Index in ε-Near-Zero Materials. *Phys. Rev. Lett.* **2016**, *116*, 233901. [CrossRef]
152. Alam, M.Z.; De Leon, I.; Boyd, R.W. Large optical nonlinearity of indium tin oxide in its epsilon-near-zero region. *Science* **2016**, *352*, 795–797. [CrossRef]
153. Yoshioka, V.; Lu, J.; Tang, Z.; Jin, J.; Olsson, R.H.; Zhen, B. Strongly enhanced second-order optical nonlinearity in CMOS-compatible $Al_{1-x}Sc_xN$ thin films. *APL Mater.* **2021**, *9*, 101104. [CrossRef]
154. Sirleto, L.; Righini, G.C. An Introduction to Nonlinear Integrated Photonics Devices: Structures and Devices. *Micromachines* **2023**. submitted.

micromachines

Review

An Introduction to Nonlinear Integrated Photonics: Structures and Devices

Luigi Sirleto [1,*] and Giancarlo C. Righini [2,*]

1 National Research Council (CNR), Institute of Applied Sciences and Intelligent Systems (ISASI), Via Pietro Castellino 111, 80131 Napoli, Italy
2 National Research Council (CNR), Institute of Applied Physics "Nello Carrara" (IFAC), Via Madonna del Piano 10, Sesto Fiorentino, 50019 Florence, Italy
* Correspondence: luigi.sirleto@cnr.it (L.S.); righini@ifac.cnr.it (G.C.R.)

Abstract: The combination of integrated optics technologies with nonlinear photonics, which has led to growth of nonlinear integrated photonics, has also opened the way to groundbreaking new devices and applications. In a companion paper also submitted for publication in this journal, we introduce the main physical processes involved in nonlinear photonics applications and discuss the fundaments of this research area. The applications, on the other hand, have been made possible by availability of suitable materials with high nonlinear coefficients and/or by design of guided-wave structures that can enhance a material's nonlinear properties. A summary of the traditional and innovative nonlinear materials is presented there. Here, we discuss the fabrication processes and integration platforms, referring to semiconductors, glasses, lithium niobate, and two-dimensional materials. Various waveguide structures are presented. In addition, we report several examples of nonlinear photonic integrated devices to be employed in optical communications, all-optical signal processing and computing, or in quantum optics. We aimed at offering a broad overview, even if, certainly, not exhaustive. However, we hope that the overall work will provide guidance for newcomers to this field and some hints to interested researchers for more detailed investigation of the present and future development of this hot and rapidly growing field.

Keywords: photonics devices; nonlinear photonics; integrated photonics; photonic structures; optical materials; all-optical signal processing; all-optical computing; all-optical communications; all-optical digital devices; signal amplification and frequency conversion; all-optical signal regeneration; super-continuum generation; microcomb generation

Citation: Sirleto, L.; Righini, G.C. An Introduction to Nonlinear Integrated Photonics: Structures and Devices. *Micromachines* **2023**, *14*, 614. https://doi.org/10.3390/mi14030614

Academic Editor: Aaron Hawkins

Received: 31 January 2023
Revised: 24 February 2023
Accepted: 3 March 2023
Published: 7 March 2023

1. Introduction

In recent decades, the overall trend of research and development (R&D) towards miniaturization of devices and systems begun with microelectronics has led to continuous growth in integrated photonics. Integrated optic and photonic chips have huge potential for low-cost, scalable production, and high-density components integration. A major challenge of integrated photonics has been constituted by the lack of a unique base material, playing a similar role as silicon in microelectronics. In contrast to electronic integrated circuits (ICs), photonic integrated circuits (PICs) are currently fabricated on different constituent materials. In order to integrate multiple photonic functions, different building blocks are fabricated using monolithic or hybrid integration technology over a single substrate and connected via waveguides [1–4]. Photonic building blocks can be passive (e.g., couplers, switches, modulators, and multiplexers) or active (e.g., amplifiers, detectors, and lasers); nonlinear functionalities may be easily added, exploiting the nonlinear properties of the materials and suitable light-confining structures. As efficiency of nonlinear optical interactions drastically scales up with the power density of an optical wave, it had to be expected that combining integrated photonics and nonlinear photonics would produce groundbreaking results [5–12].

In this review, we focus on three main areas of application: all-optical computing, all-optical processing, and nonlinear photonics sources.

In all-optical computing, the key strength provided by optical technologies is parallelism of information transfer and processing onto multiple wavelength channels. Parallel access to information points is permitted due to the capability to use light waves of distinct wavelengths within the same device. In addition, photon being the information carrier, there is no propagation delay in different parts of the optical system.

Implementation of logic operations using a photon signal is a very challenging frontier of research because of the fundamental requirement of very efficient light-control. Ultrafast all-optical switch is a fundamental component for all-optical computing. It can be defined as a structure with a pump light-controlling the ON/OFF transition of the signal light. Optical devices, performing digital functions, are expected to fulfill the following criteria: (1) ultra-compactness; (2) low power consumption (~fJ/bit); (3) high-speed operation (scalable beyond 100 Gb/s); (4) parallel operation on multiple wavelength channels, reducing the need for large fan-outs and redundant parallel processing structures; (5) preservation of information (e.g., phase) carried in the optical domain, usually lost in optical–electronic conversion; (6) the ability to transparently process a data channel regardless of its data rate or its modulation format; (7) scalability; and (8) cascading [13–18]. Unfortunately, as underlined by Grinblat et al. in 2020, at that date, no known structure behaved as an ideal switch [18].

The basic building block of an optical flip-flop device is the optical bistable switch. In this device, the output of the system takes on one of two possible states depending on the states of the inputs. Optical bistable operation permits optical read–write memory operation, opening the possibility of an integrated optical logic circuit on a single chip. A typical way of forming a bistable optical device is to place a saturable absorber inside a resonator. As the input intensity is increased, the field inside the cavity also increases, lowering the absorption of material and thus increasing the field intensity further. If the intensity of the incident field is subsequently lowered, the field inside the cavity tends to remain large because the absorption of the material system has again been reduced. However, even if single bistable switches have already been demonstrated on different platforms, it is worth noting that the next big challenge is realization of a complex system where several bistable switches are connected in tandem and in parallel [19–27].

Nowadays, almost all data flow, including internet, phone calls, etc., goes through fiber optic transmission lines, and the field of communications continues to expand to higher data rates and shorter delays to allow more capacity. The demands of the modern world are for high-speed optical communication and interconnection. The increase in the amount of data available presents both opportunities and problems, leading to "big data" (i.e., large, diverse sets of information that grow at ever-increasing rates) issues. Storing and handling big data may be difficult using traditional techniques; such a new challenge may be faced by using photonic devices for massive parallel processing and nonlinear photonic devices for ultrafast handling [28–30]. Today, in optical networks, data are encoded on photons for transmission, while information processing is often carried out through optical–electrical–optical (OEO) conversions. In the last 20 years, with the aim to assist/replace some of the electronic modules used in network routers, optical signal processing systems (OSPs) have been investigated. OSP refers to a broad range of techniques with the aim to process and manipulate the signal, i.e., amplitude, phase, wavelength, and polarization of optical waves, directly in the optical domain. Optical signal processing techniques employ a wide range of devices and various nonlinearities to achieve multiple network functionalities. Widely used functionalities demonstrated in nonlinear photonic circuits (PICs) include wavelength [31–35] and format conversions [36,37], routing [38,39], phase-sensitive amplification [40], optical multiplexing and demultiplexing [41–45], optical memory [46,47], all-optical tunable delay [48,49], and signal regeneration [50,51]. Thus far, most of the existing OSP research has relied on third-order nonlinearities, such as four-wave mixing (FWM), self-phase modulation (SPM), and cross-phase modulation (XPM) [52–59].

Current optical networks are mostly based on time-division-multiplexing (TDM) and wavelength-division-multiplexing (WDM). In the former, multiple relatively low-bit-rate streams of data with the same carrier frequency are interleaved to create a single high-bit-rate stream, while the latter involves simultaneous propagation of multiple data signals, each at a different wavelength in a single optical line. Wavelength-division-multiplexing (WDM) techniques offer very effective utilization of the fiber bandwidth directly in the wavelength domain. A first issue of current optical networks is wavelength-blocking. In order to overcome it, a fundamental piece is represented by wavelength conversion devices. They can be obtained using different nonlinear effects, such as FWM or XPM [31–35]. Another issue is to increase transmission bandwidth, and an option is to combine TDM and WDM. In this process, demultiplexing an ultra-high-data-rate time-multiplexed signal to speeds receivable through electronics is achieved by wavelength conversion. Specifically, information multiplexed in the time domain through optical time demultiplexing (OTDM) can be converted into parallel lower-data-rate wavelength or spatial channels. This process has been achieved utilizing FWM, whereby a relatively low repetition rate pump switches out temporally multiplexed channels by converting them to new wavelengths (the idler in the FWM process) [41–43]. There are, of course, other routes that can be followed to increase transmission bandwidth and/or to develop advanced processing chips. The goal of increasing bandwidth density of on-chip interconnects without increasing the number of waveguides, waveguide crossings, and chip footprint, for instance, may be reached by exploiting mode-division-multiplexing (MDM) in conjunction with WDM [44]. Polarization-division-multiplexing (PDM) and orthogonal frequency-division-multiplexing (OFDM) are two other effective methods to increase the spectral efficiency of a communication system [45].

When signal modulation rate increases, signal degradation in the optical channel caused by dispersion, nonlinearity, and noise becomes a critical issue. Conventionally, signal regeneration in an optical system is performed through optical–electrical–optical (OEO) conversion, in which a weak and distorted signal is detected, restored in electronics, and retransmitted onto an optical fiber. Regenerators are designed to increase system performance, reduce data degradation, and enhance signal-to-noise ratio (SNR) for higher link capacity. In general, regenerators perform three signal processing functions: (1) reamplifying, (2) reshaping, and (3) retiming the signal. When the data rate is becoming higher and higher (towards 100 Gb/s), optoelectronic regeneration schemes will be very hard to implement or even impossible. Thus, all-optical regeneration, either 2R (re-amplification, re-shaping) or 3R (re-amplification, re-shaping, re-timing), has become a key technology to improve signal quality [50,51].

Nonlinear optical sources provide an outstanding example of new possibilities offered by integrated nonlinear photonic chips, such as generation of new classes, named supercontinuum generation [60], and microcombs [61], which are capable of generating coherent, ultra-broadband light sources that cannot be produced in linear photonic systems. Supercontinuum generation is a device functionality that has important applications in many areas of photonic integrated circuits, particularly in WDM applications. As an example, it could be beneficial to use a single broadband laser source, select by filtering specific wavelength channels, and then modulate these channels instead of using a separate laser for each wavelength.

Optical frequency combs (OFCs) are often referred to as optical rulers: their spectra consist of a precise sequence of discrete and equally spaced spectral lines that represent precise marks in frequency. This discrete ensemble of equally spaced laser frequencies that distinguish OFCs from other light sources is the spectral counterpart of the regular train of short pulses emitted by mode-locked lasers. The OFCs solve the challenge of directly measuring optical frequencies and are now exploited as the most accurate time references available, ready to replace the current standard for time. Laser frequency combs can provide integrated sources of mutually coherent laser lines for terabit-per-second transceivers, parallel coherent light detection, or photonics-assisted signal processing.

Readers interested to delve into the main physical processes involved in nonlinear photonics applications and the most appealing materials are referred to a companion paper [62]. The focus here is to present some examples of nonlinear integrated photonic devices, with a brief overview of the types of optical integrated structures and the most common material fabrication platforms. In the next section, planar and three-dimensional geometry structures are introduced; in the third section, material platforms are discussed, and, in the last section, advances in nonlinear integrated photonic devices that provide an idea of the high potential for practical applications and future challenges are reported and discussed.

2. Integrated Photonic Structures

Influenced by the general trend of sciences towards the nano-world, development of integrated photonics, begun with cm-long circuits, went through microphotonics and finally came down to nanophotonics. Guided wave structures, resulting from the coupling between an electromagnetic field and some resonance, are the basic blocks that have accompanied this process.

2.1. Optical 2D and 3D Waveguides

Resonance can be geometric, as occurs in integrated optical waveguides where constructive interferences result in guiding light due to total internal reflection. Thus, optical waveguides can be classified depending on geometry. Planar (2D) dielectric waveguides are built of layers of high- and low-refractive-index materials, providing confinement only in vertical direction. Nonplanar (3D) waveguides, providing confinement in two directions, can have different cross-sections, such as ridge, rib, stripe-loaded, or buried; slot structures, moreover, can have different forms of guiding core surrounded by cladding material (see Figure 1 for some examples) [63]. Correspondingly, the most suitable fabrication methods must be selected (e.g., high- or low-energy diffusion, thin-film deposition, direct laser writing). The need of efficient coupling to other integrated components makes the channel waveguide the most commonly used structure. Numerical and analytical methods are widely available to optimize the design of waveguides and circuits [64].

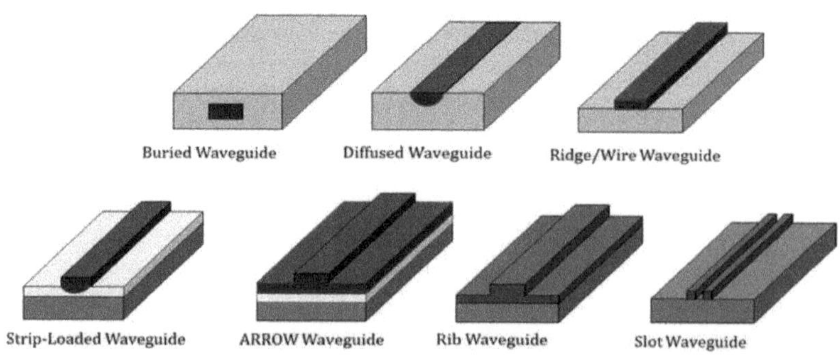

Figure 1. Schematic representation of various channel waveguides. Adapted from [63] under Creative Commons License 3.0.

The next step, from cm to micrometer scale, was motivated by the wish to investigate light behavior on the microscale and its interaction with micro-objects. The key challenges were a reduction in size of optical devices and improvement in their performances. The main goal was to go beyond the limit of integrated optics, offering a reliable platform for dense integration [65].

2.2. Microresonators and Photonic Crystals

During recent decades, the fast growth of micro-scale fabrication techniques has enabled successful demonstration of various types of microphotonic devices; special attention has been addressed to microcavities [66,67], such as whispering gallery modes resonators (WGMR) and photonic crystals (PhC) [68,69], which combine small modal volume with very high optical quality factor (Q). In these microphotonic devices, due to their reduced size, photons are trapped in small volumes close to the diffraction limit for sufficiently long times so that these photons strongly interact with the host material, creating enhanced nonlinear effect and significant reduction in their power threshold [66–69].

WGMRs have shown high mode stability and ultra-high-quality factors Q, up to 10^{11}. These resonators are excellent platforms for fundamental and applied studies of nonlinear processes due to their long photon lifetimes (temporal confinement) and their small mode volumes (spatial confinement) [70–74]. Temporal and spatial confinement have made possible optical frequency conversion with low-power continuous-wave (CW) lasers with powers ranging from micro-watts to milliwatts. However, the high circulating intensities inside a WGMR are not a sufficient condition for efficient harmonic generation, parametric, and hyper-parametric oscillations: these phenomena require fulfilling phase and mode matching and energy conservation conditions [71,75]. Figure 2 shows different types of optical microcavities. It must be underlined that, even if more difficult to integrate in a compact structure, microspherical resonators play an important role in nonlinear photonics [70,71,76].

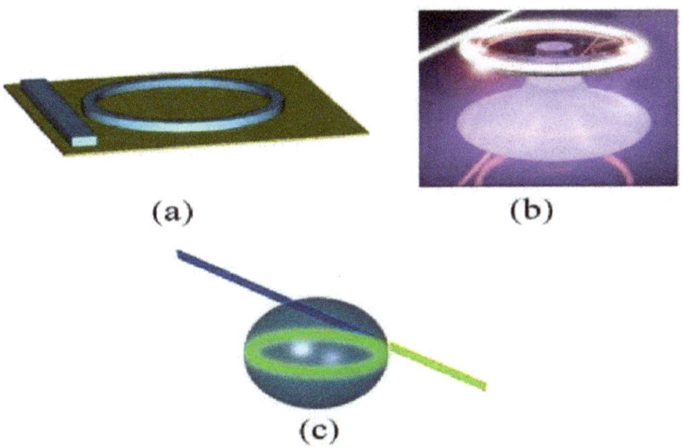

Figure 2. Different types of optical microcavities: (**a**) micro-ring resonator, (**b**) micro-toroid resonator, (**c**) micro-sphere resonator. Adapted from [76] under Creative Commons License 3.0.

Photonic crystals provide an excellent building block for photonic integrated circuits and for enhancing a variety of nonlinear optical processes as well [77–81]. A strong advantage of PhC cavities with ultra-small-mode volume V_m is that quality factors can be relaxed while still achieving excellent Q/V_m values; despite the fact that the quality factor is inversely proportional to the linewidth of the cavity, both high-bandwidth and low-threshold all-optical processes can be realized in a single system. Integration issues, of course, must be considered in the case of PhCs as well, especially for 3D structures [80].

Figure 3 shows the design of a dispersion-engineered slow-light photonic crystal waveguide (PhCW); the blue circles represent the air-holes in a standard single-line defect (W1) PhCW with a lattice constant a, while the red circles represent the engineered air-holes, with radius decreasing from R to R2 and horizontally shifted from the dashed circles by ΔX [81].

Guided waves can also result from coupling between an electromagnetic field and resonance related to material properties. For example, plasma resonance associated with the electron gas in a metal can be coupled to an electromagnetic field via the interaction between the field and the charges, leading to surface plasmon. These surface waves, propagating along the dielectric–conductor interface, are evanescently confined in the perpendicular direction due to very shallow penetration of the electromagnetic field into the metal. Plasmonic waveguiding allows breaking the diffraction limit of light, opening possibilities for subwavelength light confinement. To increase integration density and compactness of photonic structures, many different geometries of plasmonic waveguide have been investigated in recent years, including interferometers and ring resonators [82] and periodic metal structures [83]. However, in all cases, energy dissipation in closely spaced metal layers causes high losses, limiting the propagation length to a few micrometers.

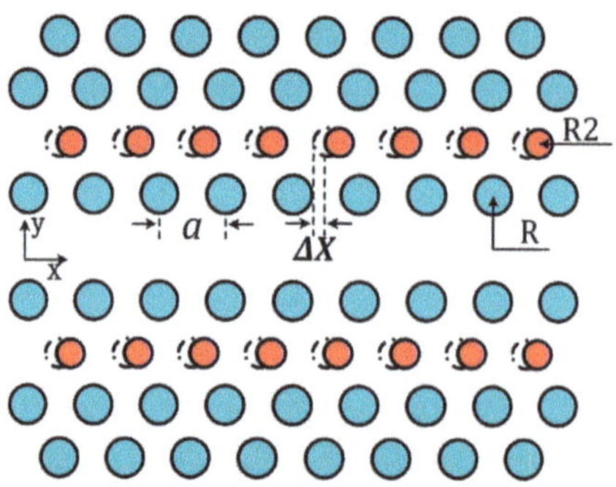

Figure 3. Structure of a photonic crystal waveguide, with air holes represented by blue and red circles. Simulations were made by considering standard silicon-on-insulator air-bridge PhCW of 220 nm thickness, lattice period 400 nm, and radius R of the first air-holes row 100 nm. By adjusting the radius R2 and lattice positions ΔX of the second air-holes rows, a very wide flat band larger than 50 nm could be obtained. Reproduced from [81] under Creative Commons License 3.0.

Nanophotonics is another fascinating field, investigating the light behavior on the nanometer scale and its interaction with nanometer-scale objects [84–86]. Major demand in the near future is expected for devices that should allow to control light with light in a very thin nanoscale layer or in a single nanoparticle of nonlinear material. During the past few decades, a significant number of nanomaterials have shown notable optical properties, motivating fabrication and design of nanoscale photonic devices [87,88].

Electromagnetic metamaterials are artificial media in which subwavelength electromagnetic constituents replace atoms as the basic structural elements to control light–matter interaction. Many novel phenomena related to metamaterials are due to optical magnetism, observed in specifically designed artificial subwavelength structures, even when such structures are made of non-magnetic materials. The most popular constitutive elements of metamaterials are made of metals, where free electrons oscillate back and forth, creating effective loops of current, thus inducing an efficient magnetic response. Plasmonic nanostructures support field localization due to localized surface-plasmon resonances with multipolar electric-type characteristics. The plasmonic resonances, generally, result in high field enhancement but at the price of higher absorption losses due to free-carrier absorption and reduced optical damage thresholds [89,90].

In order to control light–matter interactions at the nanoscale, the most disruptive strategy would be to replace metals by all-dielectric nanoparticles with high refractive index [91–93], providing fine control over the light features (amplitude, phase, and polarization). Dielectric nanoparticles support an alternative mechanism of light localization in subwavelength optical structures via low-order dipole and multipole Mie resonances [94]. They may generate a magnetic response via displacement current contribution, playing a crucial role in the realization of the unique functionalities of meta-atoms, also driving novel effects in the fields of metamaterials and nanophotonics.

It can be mentioned that important steps towards ultra-compactness of photonic circuits have been made very recently via either anisotropic metamaterials [95] or photonic inverse design [96–98]. As an example, a polarization beam splitter (PBS), which is an important device in PICs for multiplexing and demultiplexing polarizations, was designed and fabricated on a silicon-on-insulator (SOI) platform using a directional coupler with single-mode waveguides 500 nm wide; the coupling region was designed by tailoring the anisotropic metamaterial, which was composed of identical periodic subwavelength strips. The device, with a $2.5 \times 14\ \mu m^2$ footprint, exhibited a low insertion loss of 1 dB, high extinction ratio >20 dB, and wide operational bandwidth >80 nm [98].

There is strong interest in exploring interesting physical mechanisms to create the optical resonances in such structures, for example, using guided-mode resonances [99] and bound-states in continuum resonances [100], to enhance local electric fields and consequently amplify the nonlinear optical effects in metamaterials [12,101–105]. Bound states in the continuum (BIC), experimentally observed by Capasso et al. in semiconductor heterostructures grown by molecular-beam epitaxy [106] in 1992, are special wave solutions embedded in a radiative continuum, which, however, remain localized without coupling to the extended waves or radiation. BICs have been observed in photonics by Marinica et al. [107] by using simple structures, such as two arrays of identical dielectric gratings or two arrays of parallel dielectric cylinders [107]. BICs have then been studied in a wide range of material systems, including dielectric photonic crystals, optical fibers, and waveguides; an interesting review was published in 2016 by Hsu et al. [108]. Figure 4 shows an example of BIC structure based on two coupled polymer strip waveguides (WG1 and WG2) onto a lithium niobate (LN) thin film. Etching of LN is not necessary as the strips can be directly patterned through a single e-beam lithography process. When the parallel strip waveguides are placed closely (distance of the order of micrometer), the TE continuum modes exist, as shown in Figure 4b.

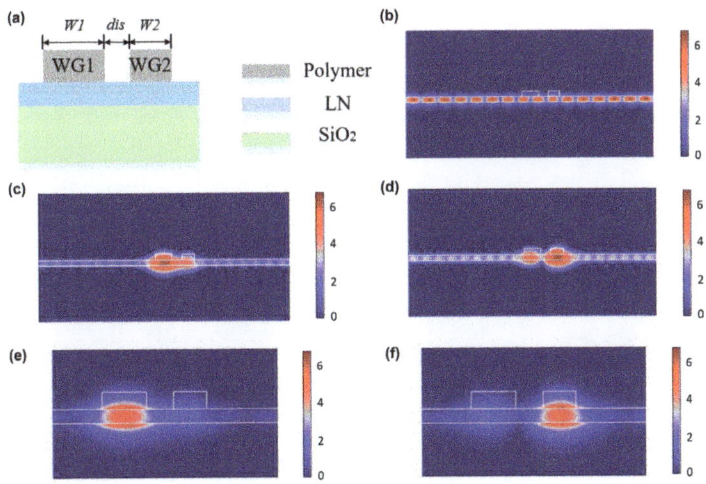

Figure 4. BIC waveguide structure based on polymer strip waveguides (WG1 and WG2) onto a lithium niobate film. Here, WG1 and WG2 are 500 nm thick and LN film is 300 nm thick. (**a**) Cross-section of the structure; (**b**) normalized electric field distribution of the TE continuous mode; (**c**) and (**d**) TM leaky modes. Under proper conditions, the coupling between the TE continuum modes and the TM bound modes can lead to well-confined BIC modes (see (**e**,**f**)). Reproduced from [109] under Creative Commons License.

Recently, there have been several works on highly efficient second harmonic generation using BIC structures in lithium niobate, which seem to have great potential for development of compact coherent light sources in a broad wavelength range [110–112].

Figure 5 presents a schematic illustration of some metastructures [102].

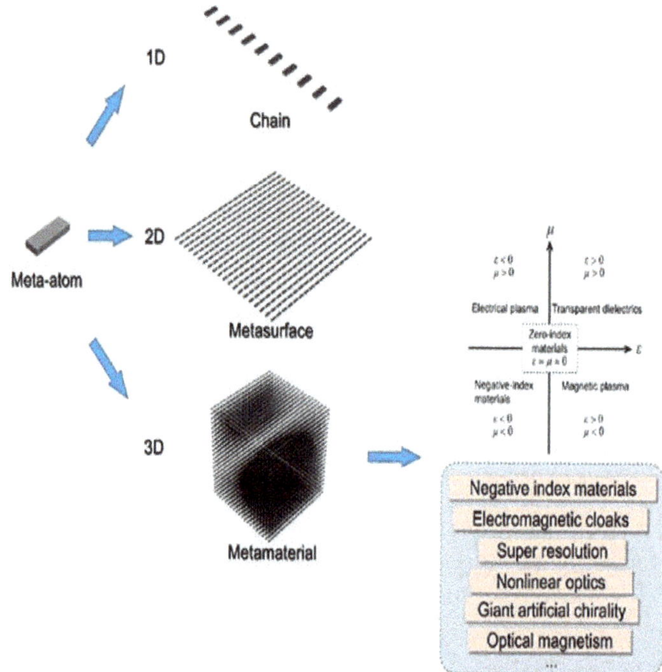

Figure 5. Schematic illustration of meta-atom, 1D chain, 2D metasurface, and 3D metamaterial. Inserts are the representation of the parameter space for permittivity ε and permeability μ and the typical examples of applications of metamaterials. Adapted from [102] under Creative Commons License 3.0.

3. Material Platforms for Nonlinear Integrated Photonics

As mentioned in the companion paper [62], numerous materials for integrated nonlinear optics have been investigated. Among them are Si and related materials, such as SiN, a-Si, and SiC; glasses, such as silica, high-index glass, and chalcogenide glasses; III–V semiconductors, in particular AlGaAs and lithium niobate (LN); and recently investigated materials, such as tantalum pentoxide (Ta_2O_5) and vanadium dioxide (VO_2) [62] and references therein. Correspondingly, several material platforms have been developed to achieve the goal of a dense components' integration. Each platform has its advantages and disadvantages and offers variable nonlinear efficiencies and integration densities depending on the values of the nonlinear coefficients and refractive index contrast. An additional limiting factor is related to losses that are determined by both the quality of the grown material and the maturity of the fabrication process, which affect, for instance, the roughness and absorption of the surface passivation layer. Although in the longer term the growth and fabrication quality for all these materials will eventually level up, there still remains one important limiting factor that clearly separates them regarding high power densities, namely two photon absorption (TPA). TPA is determined by the band gap of the materials and by the working wavelength, which is often in the telecom range due to the historically developed infrastructure of sources, detectors, etc. In this regard, small-band-gap semiconductors, such as Si [113–117] and GaAs [118,119], are fundamentally limited and there seems to be no way around this problem other than to change the working wavelength and all the surrounding infrastructure.

Silicon photonics research and development has much progressed, and both component performance and integration complexity have made significant steps forward in the past decade [113–117]. Nowadays, silicon-based platforms, in particular silicon-on-insulator (SOI), are among the most mature for PIC realization. The silicon-on-insulator (SOI) platform is a fabrication approach in which a thin silicon layer on top of an insulator layer resides on a silicon substrate. The functional optical elements are situated in the thin top-silicon layer, and the insulator is typically made from SiO_2. The SOI platform has become the foundation of silicon photonics for several reasons, including strong optical confinement of silicon due to the significant refractive index difference between silicon and SiO_2, which enables very compact optical devices, and demonstration of much more compact optical waveguides with lower optical propagation loss, better processing yield, low cost, etc.

Among the numerous nonlinear waveguide platforms that have been explored, the group of materials capable of combining both passive waveguides, for light steering and nonlinear manipulation, and active functionalities (laser sources, modulators, and detectors) monolithically on the same chip is III–V semiconductors [118,119]. This has been a strong driving force stimulating development of nonlinear optical devices based on III–V semiconductors and addressing challenges associated with loss mitigation and nonlinearity enhancement in these platforms. The main III–V integrated nonlinear photonic platforms considered to date are GaAs [76] and its AlGaAs [120] derivative, InP and InGaAsP, III-nitrides AlN [121,122], and GaN [123], as well as GaP [124] and its ternary derivative InGaP. We note that AlN [121,122] shows optical properties similar to SiN (refractive index, transparency, and n_2) but also possesses second-order nonlinearity.

Considering AlGaAs, the structures used in NLIP can be divided into three-layer, two-layer, and multi-layer platforms depending on the number of epitaxially grown layers with different material compositions. Figure 6 shows examples of these three AlGaAs platforms; the waveguides sketched in Figure 6h,i are often employed for phase matching of the $\chi^{(2)}$ processes.

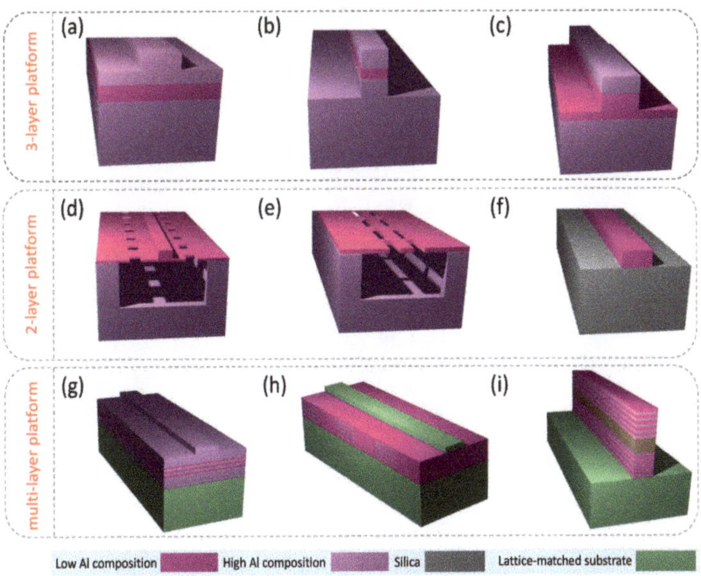

Figure 6. Schematics of AlGaAs platforms and waveguide geometries. (**a–c**) 3-layer platform with strip-loaded, nanowire, and half-core waveguides, respectively. (**d–f**) 2-layer platforms with suspended nanorib, suspended nanowire, and AlGaAs-OI waveguides, respectively. (**g–i**) Multi-layer platform with multi-quantum-well waveguide, modulated-$\chi^{(2)}$ waveguide, and Bragg-reflector waveguide, respectively. Reproduced from [120] under Creative Commons License.

In recent years, lithium niobate-on-insulator (LNOI) wafer fabrication process has been rapidly advancing [125–129]. This technology is revolutionizing the lithium niobate industry, enabling higher performance, lower costs, and entirely new devices and applications. Availability of LNOI wafers has sparked significant interest in the platform for integrated optical applications as LNOI offers the attractive material properties of lithium niobate while also offering stronger optical confinement and high optical element integration density that has driven the success of more mature silicon and silicon nitride (SiN) photonics platforms. Many of the key building blocks for highly integrated photonic integrated circuits (PICs) have been established on this platform, including low loss optical waveguides, electro-optical interfaces for ultra-fast modulation, and nonlinear optical elements and resonators. However, further work needs to be completed to make LNOI an attractive and competitive integrated optical platform: (i) optical interfacing to LNOI waveguides has to be improved, reducing fiber-to-chip coupling losses, for example by developing inverted tapers and waveguide tapers on LNOI; (ii) optical gain media need to be demonstrated in this platform, either via bonding already doped lithium niobate in the LNOI wafer fabrication process or by doping the LNOI wafer after fabrication using ion implantation techniques. A further approach could be heterogeneous integration of III–V lasers on LNOI waveguides, which needs to be investigated; (iii) development of photodetectors on LNOI waveguides requires further investigation [125–129].

Current trends in PIC material technologies, however, indicate that there is no waveguide material technology that can address the needs for all the potential applications of PICs. Furthermore, advanced PICs may require the best possible performance of many different photonic elements to achieve the desired functionality, which may not be possible with a single waveguide material technology. However, the fundamental size limit for photonic devices is significantly larger than that of electrical devices. Therefore, integrated photonics has, over the decades, developed at a considerably slower pace than integrated electronics, in integration density as well as total number of devices on a chip. The specific application requirements drive the choice of the most suitable substrate material as each material has its own specific advantageous features and limitations. A solution to overcome this limitation is to integrate different material technologies into a single PIC or package. This approach has the benefit that each material can be used to provide the photonic element functionality for which it is best suited without compromising the functionalities of the other elements in the system. Integration of the different material technologies can occur through two different routes: (i) hybrid integration and (ii) heterogeneous integration. The former is a process that connects two or more PICs or photonic device chips, usually from different material technologies into one single package. This process is, in general, performed at the packaging stage after fabrication of the PIC and photonic device chips. The latter, instead, is a process that combines two or more material technologies into a single PIC chip. This process is generally performed at the early- to mid stages of fabrication of the PIC chip, as in the case of unpatterned III–V thin-films integrated onto pre-processed silicon photonic wafers [130–132].

In order to address the shortcomings of the SOI platform, several novel waveguide platforms have been developing based on heterogeneous integration of other material systems on silicon substrates, with the common requirement of remaining compatible with the complementary metal-oxide-semiconductor (CMOS) technology [132]. As a more recent example, photonic components in SiC (specifically waveguides, 1D and 2D photonic crystal cavities, microdisk, and microring resonators), based on thin layers of SiC on insulator (SiCOI), have been implemented [133,134]. High nonlinearity and low loss were demonstrated in waveguides and ring resonators fabricated in amorphous SiC (a-SiC) grown directly on silica, using plasma-enhanced chemical vapor deposition, with loss as low as 3 dB/cm, providing a very scalable material growth [133]. The intrinsic quality factor of the microring resonator was around 160,000 [133]. Figure 7 shows the fabrication process flow of microring devices (a) and the SEM micrographs of a microring (b) and of the coupling area to a channel waveguide (c).

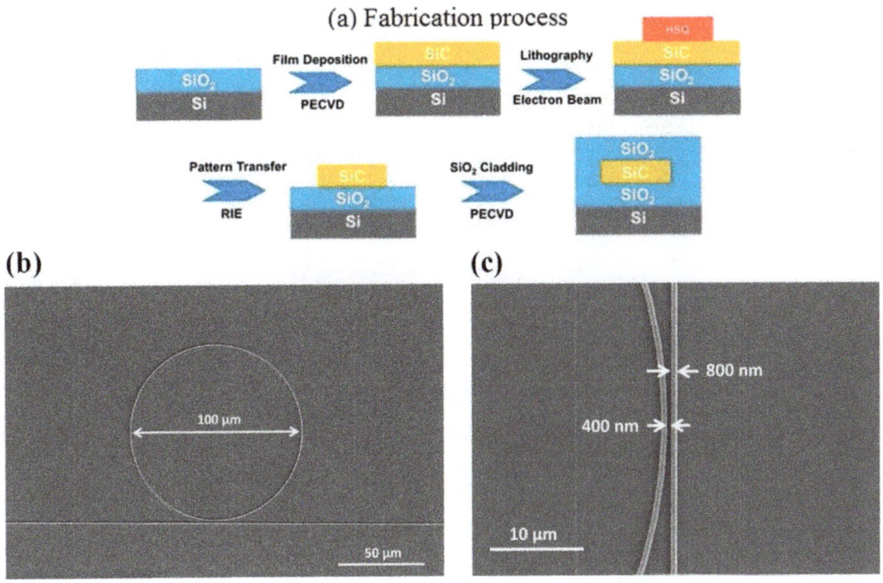

Figure 7. Fabrication of a microring in a 350 nm thick a-SiC film with silica cladding of 3 μm (bottom) and 2 μm (top). (**a**) Schematic process flow; (**b**) SEM micrograph of the microring having 100 μm diameter; (**c**) higher magnification SEM image of the coupling area between the ring and a ridge waveguide. Reprinted with permission from Xing et al., ACS Photonics [134]. Copyright 2019, American Chemical Society.

Much higher quality factors have been achieved by using the 4H polytype of silicon carbide (4H-SiC). For instance, a microring fabricated in high-purity semi-insulating 4H-SiC exhibited a $Q = 1.1 \times 10^6$, corresponding to a waveguide loss of 0.38 dB/cm [135]. The high field enhancement of this microring (55 μm diameter) enabled demonstration of optical parametric oscillation and optical microcombs. More recently, generation of an octave-spanning microcomb, covering a wavelength range from 1100 nm to 2400 nm, was achieved by using a 36 μm radius microring resonator fabricated in 4H-SiC on a silicon dioxide layer, with intrinsic quality factor above one million [136].

As a rule, photonic packaging of PICs is much more challenging and much more expensive (at least to date) than electronic packaging since it requires robust high-precision alignment of optical components and stringent real-time temperature control. This is one of the reasons why huge development of commercial integrated photonic devices is still lacking. Figure 8 may provide an idea of the technical complexity of the optical, electrical, thermal, and mechanical interconnections inside and outside a PIC [137].

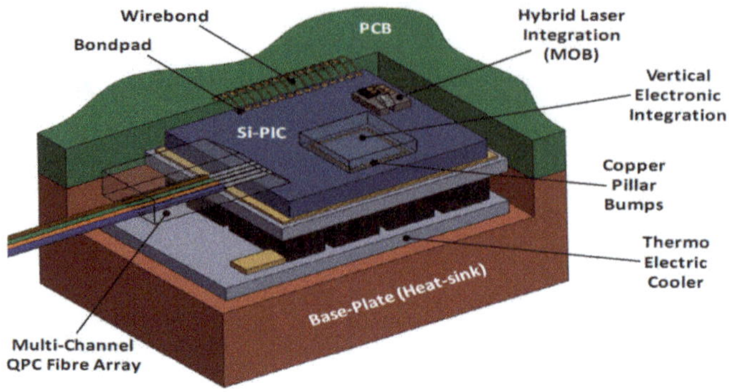

Figure 8. Schematic of a Si-PIC packaged with a multi-channel quasi-planar-coupled (QPC) fiber-array, a hybrid-integrated laser source based on a micro-optic bench (MOB), a vertically integrated electronic integrated circuit (EIC), and a thermo-electric cooler. Electrical connections between the PIC and the printed circuit board (PCB) are made by wire-bonds, while the connections between the PIC and EIC are made using copper pillar bumps (CPBs). Reproduced from [137] under Creative Commons (CC-BY) license.

4. Nonlinear Photonics Devices

Development of material platforms described in the previous section has enabled design and implementation of several effective integrated nonlinear devices. Here, we provide a brief, and certainly not exhaustive, overview of this field; for simplicity, we tried to classify the examples into groups characterized by a common goal, even if quite broad.

4.1. All-Optical Digital Devices: Switches, Gates, Flip-Flop Units, and Optical Transistors

Two methods are widely used in all-optical switches. The first one is based on linear interference between two light signals, whose relative optical phase difference determines the logic operation functions. Implementation of this method is simple, but, due to the lack of precise control of the optical phase difference, inherent instability, with low-intensity contrast of output logic states '1' and '0', is experienced. The second method is based on third-order nonlinear optical effects, i.e., on the effect of intensity-dependent refractive index. The advantage of this method is its strong universality, and, in principle, complex all-optical devices could be realized based on it. The main obstacle limiting its application is the intrinsic material bottleneck limitation due to the contradiction between the huge third-order nonlinear coefficient and the ultrafast response time (i.e., the larger the third-order nonlinear coefficient, the slower the response time).

An important enabling technology to implement all-optical switching operation is provided by high Q/V_m resonators, Q being the quality factor and V_m the mode volume of the cavity. The field intensity enhancement inside the cavity makes it possible to utilize optical nonlinearities at a low input power (a few hundred μW to mW level), achieving picosecond–femtosecond response time and lower energy consumption [13,138]. Microring resonators are among the most widely used configurations for realization of all-optical switch due to their simple design and easy experimental realization [14,16,22,139]. The switching principle, based on nonlinear effects, relies on a shift in resonant wavelength. It is the following: the refractive index variation in the nonlinear material changes the resonant frequency of the microring and/or coupling between the waveguide and the microring. As a consequence, switch control of the signal light output can be enabled. As an example, Figure 9 presents a schematic illustration of an all-optical switch based on a microring resonator.

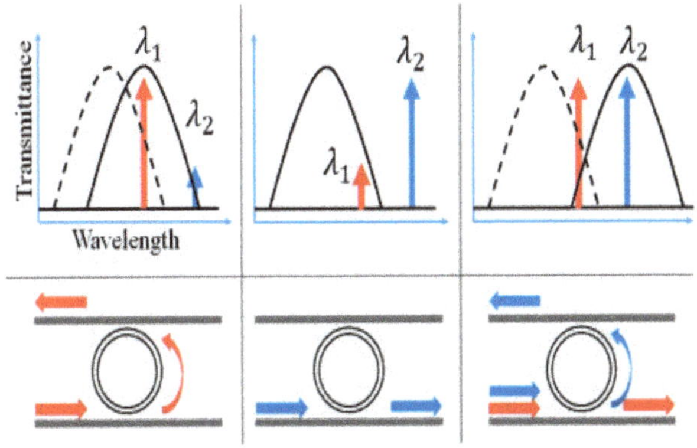

Figure 9. Schematic illustration of an all-optical switch made of an add–drop microring resonator. The dotted line represents the transmittance spectrum of a cold cavity. The solid line is the transmittance when inputs are applied. A resonant shift occurs due to the optical Kerr effect. Two different wavelengths, λ_1 and λ_2, are used for the operation. On the left column λ_1 will drop (high) when only λ_1 is inputted (high).) On the middle column λ_2 will not drop (low) when only λ_2 is inputted (high). On the right column λ_2 will drop (high) when both λ_1 and λ_2 are inputted (high). As a result, λ_2 can be switched off and on by turning λ_1 signal on and off. Reprinted with permission from [22] © The Optical Society.

Photonic crystal cavities are also an important tool in all-optical switching, mostly due to their ability to confine light to extremely small mode volumes [20,140–142]. Optical bistable switching by using a high-Q two-dimensional photonic crystal nanocavity was demonstrated experimentally using thermo-optic [20] and carrier [140] effects. In order to make possible an all-optical switch, PC transmission spectrum can change when the pump light is injected into the system and the propagation state can change from allowed to forbidden or from forbidden to allowed. An all-optical switch based on a photonic crystal nanocavity was demonstrated in 2015 [141]. The working principle of the device was based on the resonance wavelength change in the cavity by either the Kerr effect or free carriers. In 2020, Takiguchi et al. [142] demonstrated a picosecond all-optical switch using an InP/InAsP nanowire integrated in a silicon photonic crystal. The line-defect photonic crystal had a Q-factor of 25,000. The reaction time of the switch was 150 ps, while the switching energy was several hundred fJ.

Photonic crystal chips also lend themselves for large-scale and dense integration of optical memories; this result is obtained by introducing a wavelength-addressable serial integration scheme and exploiting the wavelength-division-multiplexing capability [140]. However, use of a high-Q cavity makes the operation slow because the light charging speed is slow for a cavity having a high Q. Although the carrier effect has enabled fast operation at a few ns [46], a tradeoff between low power and high speed due to the high Q of the cavity remains.

An approach based on nonlinear plasmonic slot waveguides was used to address the well-known tradeoffs between all-optical switching speeds and associated energy requirements [143]. Graphene's ultrafast saturable absorption (SA), leading to transmission of a signal pulse when a control pulse overlapped with it, was achieved, with an ultrafast response time of 260 fs.

All-optical logic gates and flip-flops are other fundamental elements for optical computing [22], even if still, at the beginning of the last decade, there were doubts about their feasibility and their possibility of competing with microelectronics [144]. Such skepticism was not fully justified, even if, still five years later, realization of an all-optical flip-flop required coupling two optical triode switches made by a nonlinear étalon and gradient-index lenses [19]; to date, however, several all-optical logic devices have been demonstrated [13,25,27,145–148].

Extremely-low-power all-optical bistability was demonstrated by utilizing silicon PhC nanocavities [46]. Bistability was obtained by the plasma effect of carriers generated by two-photon absorption, with a bistable threshold power of 0.4 mW, a set pulse energy of 74 fJ, and a switching speed of <100 ps.

Design and fabrication of a large-scale memory in InGaAsP/InP and Si devices were discussed in Ref. [140]. A large number (128) of serially integrated PhC Si-nanocavities were fabricated, and the experimental characterization showed that bistable memory operation was achieved for 105 bits in this chip, with bias power for this operation in the range 79–400 μW. The proposed device involves a parallel configuration, where side-coupled bit memory cavities are placed along a bus waveguide.

Electronic transistors, as the basic unit of logic circuits, have succeeded in supporting large-scale integrated circuits for computers. Any complex logic circuit can express a combination of three basic logic gates, AND, OR, and NOT, and these basic gates can be constructed with a universal transistor. Although electronic logic gates have enabled creation of integrated circuits with high density and functionality, optical logic gates cannot reach the far requirements of large-scale optical computing circuits even today [149]. On the other hand, optical transistors, as the core hardware of optical gates, so far have not been effectively exploited. Moreover, the universal optical transistor does not seem to exist or be practical for optical gates. Many different approaches to optical transistors have been proposed. Much early work was based on optical bistability [150] using nonlinear optical phenomena, mostly in resonators, whereas others worked on the optically controlled switching of light-exploiting single molecules [151] or quantum dots [152].

4.2. *All-Optical Processing*

4.2.1. Signal Amplification

Light amplification over a broad gain bandwidth, in principle, enables generating and processing an array of wavelength channels, leading to significant advancement for densely integrated photonic circuits. In the FWM conversion scheme, typically, a high-power pump laser at ω_p can be used to convert a signal frequency ω_s into a new frequency at $2\omega_p - \omega_s$. The main advantages of FWM converters are: (1) their sensitivity to both amplitude and phase information; (2) their ability to support ultra-high bit-rates beyond 160 Gbits/s; (3) simultaneous conversions of multiple input wavelengths to multiple output wavelengths; (4) their ability to support advanced modulation formats of input data signals [31,35]. A major limitation is that phase-matching must be maintained over the gain spectrum of interest, requiring careful dispersion engineering.

As of 2006, a silicon amplifier with net peak on/off gain of 1.9 dB and a broadband gain over a wavelength range of 28 nm through FWM was demonstrated in suitably designed SOI channel waveguides with a length 6.4 mm [153]. In addition, wavelength conversion in the range 1511–1591 nm with peak conversion efficiencies of +5.2 dB was reported as well. A few years later, on-Si-chip mid-infrared gain up to 25.4 dB around a wavelength of 2220 nm with a gain bandwidth exceeding 220 nm was demonstrated [154], providing zero-dispersion at 2260 nm and producing anomalous dispersion of 1000 ps nm^{-1} km^{-1} around the pump wavelength at 2170 nm. In 2017, an optical parametric gain of 42.5 dB, as well as cascaded four-wave mixing with gain down to the third idler, was experimentally demonstrated by FWM in an ultra-silicon-rich nitride (USRN) waveguide [155]. It was attributed to the high photon efficiency achieved through operating above the two-photon absorption edge, representing one of the largest optical parametric gains to date on a CMOS platform.

Another approach to achieve high gain is to exploit hybrid integration. As an example, an efficient semiconductor optical amplifier (SOA) may be introduced into a Si PIC by adding III–V materials as a chip over a selective area of the Si wafer [156]. An unsaturated gain of 12.25 dB/mm with 65 nm of 3 dB bandwidth was measured in an SOA with a 0.95 μm wide waveguide; the InP chip was directly bonded to individual dies on a processed SOI wafer.

4.2.2. Frequency Conversion

In FWM-based converters, the wavelength conversion from the pump to the idler is realized and the information optically encoded onto one wavelength channel can be simply copied onto another wavelength channel. Let us mention a few published results:

(i) C-band wavelength conversion in Si photonic wire waveguides with submicron cross-section was demonstrated by means of nondegenerate FWM (see Figure 10 of [157]). The nonresonant character of the FWM enabled demonstrating frequency tuning of the idler from ~20 GHz to >100 GHz, thus covering several C-band DWDM channels.
(ii) Conversion bandwidths greater than 150 nm and peak conversion efficiencies of −9.6 dB were also achieved via FWM and appropriate engineering in silicon nanowaveguides [158]. Furthermore, utilizing fourth-order dispersion, wavelength conversion across telecommunication bands from 1477 nm (S-band) to 1672 nm (U-band) was demonstrated with an efficiency of −12 dB.
(iii) A wavelength conversion bandwidth of 190 nm with an efficiency of 21 dB, obtained by FWM in polymer (PMMA)-cladded chalcogenide (As_2Se_3) hybrid microwires, was achieved [159]. Wavelength conversion combined with small footprint (10 cm length), low loss (<4 dB), ease of fabrication, and the transparency of As_2Se_3 from near-to-mid-infrared regions make the proposed device very promising.

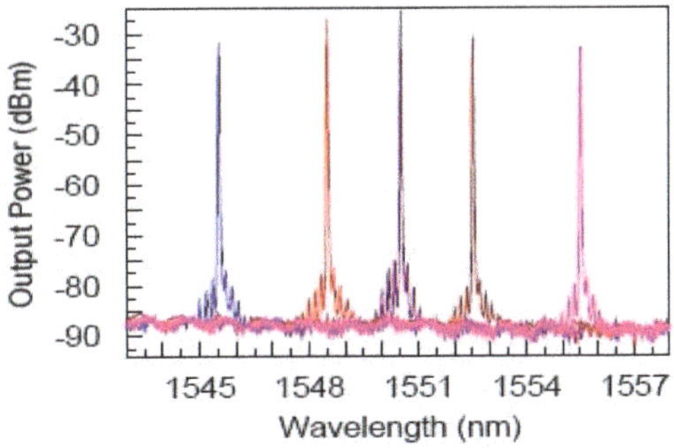

Figure 10. Output spectra from a single-mode silicon photonic wire waveguide with a cross-section of 220 nm × 445 nm and length of L = 4.2 mm fabricated on a SOI. pump (λp~1435 nm), and signal laser sources were multiplexed and launched into the waveguide using a tapered fiber in copropagating configuration; several signal wavelengths (λs = 1545.5, 1548.5, 1550.5, 1552.5, and 1555.5 nm) were used. On each side of the signal wavelength employed, newly generated satellite peaks are clearly seen. Reprinted with permission from [157] © The Optical Society.

The goal of achieving a broad wavelength conversion bandwidth may be achieved acting on group-velocity dispersion (GVD); flattened GVD is also important for effective supercontinuum generation. Silicon waveguides exhibit high nonlinearity, but dispersion flattening represents a difficult task due to the strong light confinement and high waveguide dispersion. One has mainly to work on the design of the cross-sectional dimension of the waveguide, and various solutions have been proposed. Staying with a rib waveguide allows only a few degrees of freedom so that sandwich and slot structures with greater flexibility in geometrical parameters have been designed and tested. Both sandwich and slot waveguides are realized by introducing slots between silicon strips and another material to fill the slot(s); the standard silicon-on-insulator (SOI) technology, however, makes easier fabricating the vertical-slot structure than the horizontal-sandwich one. An early work was based on a hybrid strip/slot silicon waveguide, capable of producing a flattened dispersion of 0 ± 16 ps/(nm·km) over a 553-nm wavelength range (1562 to 2115 nm) [160]. Shortly later, a horizontal double-slot Si waveguide was designed to achieve low and flattened dispersion: simulation indicated that two silica slots at the bottom of the waveguide could produce flattened dispersion from −26 to 21 ps/(nm·km) over a bandwidth of 802 nm, whereas a structure with two silica slots placed at both sides could achieve flattened dispersion from −17 to 23 ps/(nm·km) over a bandwidth of 878 nm (1498 to 2376 nm wavelength) [161]. In 2017, a new structure was proposed, where the field was mainly confined in the silicon core based on silica-filled vertical-dual slots; a spectrally flattened near-zero anomalous group-velocity dispersion covering the whole telecom wavelength range was possible, and, in a wavelength conversion experiment, a 3 dB bandwidth of 76 nm was measured, which was significantly broadened compared to the corresponding strip waveguide [162].

More recently, the problem of data transmission over the C band (1530–1565 nm) closely approaching the nonlinear Shannon capacity limit began to be considered. One of the novel ways to circumvent this limit is to extend WDM beyond the C band, e.g., to the O band (1260–1360 nm). It has been shown that a proper design of an SOI waveguide allows to shift wavelengths from the whole C band into the O band by FWM-based all-optical signal processing, without the need of changing pump laser wavelengths [163]. One year later, the same research group reported the experimental achievement of optical wavelength conversion over the C-to-O-band using multi-modal FWM in highly nonlinear SOI waveguides [164]. The authors claimed that it was an important step towards ultra-high-bandwidth optical communication networks, which utilize the entire low-attenuation infrared wavelength range of a standard single-mode fiber; as a further step towards a complete electronic–photonic-integrated circuit (EPIC), they envisaged the co-integration of the nonlinear signal processor with conventional transmitter- and receiver-units [164].

In order to face the difficulty of satisfying the energy and phase matching conditions in a microresonator, an interesting proposal was made by Marty et al. [165], who designed and fabricated a PhC microresonator exhibiting a constant free spectral range. An InGaP cavity, 200 μm long and 650 nm wide, was heterogeneously integrated on an SOI PIC. The structure is shown in Figure 11; the nanobeam cavity is fabricated on top of the Si waveguide using e-beam lithography and hole drilling by inductive coupling plasma (ICP) etching. A stimulated FWM with a −12 dB signal to idler conversion was measured.

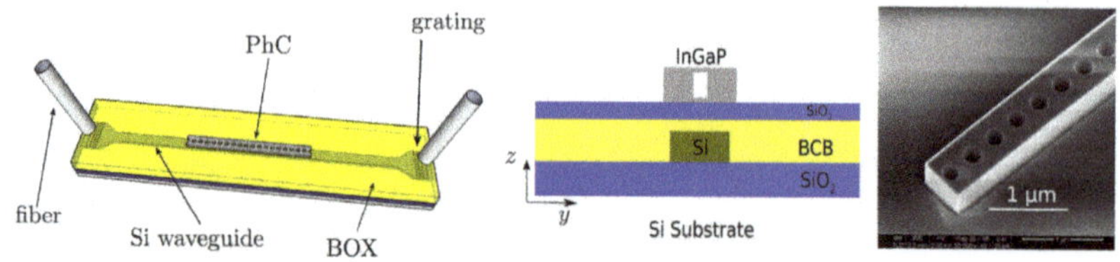

Figure 11. Photonic crystal (PhC) resonator integrated on an SOI platform on Si substrate: 3D sketch (**left**); YZ cross-section with the structure's layers (**center**); BCB is benzocyclobutene, the adhesive used for bonding; SEM image of the InGaP cavity, 650 nm wide, 290 nm thick (**right**). Reproduced with modifications from [165] under Creative Commons license.

Table 1 presents the characteristics of this PhC device in comparison to ring resonators in different materials (Q_{avg} is the geometrical average of the loaded Q for the different resonance modes; η_{NL} is the nonlinear FWM conversion efficiency). It clearly appears that the PhC on the SOI device, even if with lower conversion efficiency, has the advantage of requiring lower pump power and a remarkably smaller footprint [165].

Table 1. Comparison of continuous-wave FWM nonlinear efficiency conversion ηNL in some integrated devices.

Geometry	Material	On Chip Power (mW)	Footprint (μm^2)	Q_{avg}	η_{NL}(dB)
PhC on SOI [165]	InGaP	3	39	55,000	−12
Ring [166]	AlGaAs-O-I	7	929	44,000	−12
Ring-CROW(*) [167]	Si	100	4140	x	−21
Ring [168]	Graphene oxide on Hydex glass	158	x	50,000	−35
Ring [169]	Hydex	6	5730	10^6	−36
Ring [170]	Grahene	8	314	9000	−37

In order to increase the bandwidth density of on-chip interconnects without increasing the number of waveguides, waveguide crossings, and chip footprint, an option is to exploit mode-division-multiplexing (MDM) in conjunction with WDM [44]. A device based on multiple microring resonators in an SOI wafer, having 0.11 mm^2 footprint, operated with multiple co-propagating 10 Gb s^{-1} communication signals and reached up to 60 Gb s^{-1} of aggregate bandwidth. It was claimed that, properly designed with five spatial modes and 87 WDM channels, the device could be able to support an aggregate data rate up to 4.35 Tb s^{-1} [44].

Using polarization-division-multiplexing PDM, orthogonal-frequency-division-multiplexing OFDM transmission, and a polarization multiple-input multiple-output (MIMO) system at the receiver, a 16 × 52.5 Gb s^{-1} transmission over 4160 km of standard single-mode fiber was demonstrated with 50 GHz WDM channel spacing [45].

Advanced modulation formats, such as multiple (M)-order QAM (quadrature amplitude modulation), are fundamental for design of modern high-capacity optical transport networks [171]. In order to enable bandwidth efficient, high-speed transmission on a chip level, integrated optical components that can operate on these modulation formats providing fundamental optical functionalities, such as wavelength routing, wavelength conversion, format conversion, etc., are required. As an example, wavelength conversions of high-order OFDM m-QAM signal based on degenerate four-wave mixing (FWM) process has been demonstrated in a silicon waveguide [172].

Long et al. [173] also used a silicon waveguide: a continuous-wave (CW) pump and a four-channel WDM 16-QAM signal were simultaneously fed into the waveguide through a grating coupler. Owing to the FWM process, the idler takes the information carried by the input WDM 16-QAM signal at the output port. In this way, wavelength conversion of WDM 16-QAM signal using a silicon waveguide was achieved.

In [34], four-wave mixing (FWM)-based wavelength conversion of binary phase shift-keyed (BPSK) and quadrature phase shift-keyed (QPSK) signals at 20-Gb/s bit-rate in a 1-mm long amorphous silicon waveguide was demonstrated with a maximum FWM-efficiency of −26 dB.

The first demonstration of wavelength conversion of higher-order QAM signals at data rates above 100 Gb/s in silicon was reported by Adams et al. [174]. Error-free wavelength conversion of 28 GBaud 16-QAM single polarization (112 Gb/s) signals using FWM in a dispersion-engineered silicon nanowire (SNW) was demonstrated.

In XPM converters, a phase shift is induced on a signal by a pump when it propagates through a nonlinear device. The method is commonly applied along with an interferometric approach, which is able to translate at its output phase variations into amplitude variations. XPM-based wavelength conversion has the advantage of broadband operation, but the limitation is that it can applied only to amplitude modulation formats.

As an example, wavelength conversion in chalcogenide planar waveguides (As_2S_3) was achieved in the telecommunications C-band wavelength range via cross-phase modulation, with 5.4 ps optical pulses near 1550 nm [175]. In this device, a pulsed pump source, potentially containing digital data, is directed though the nonlinear medium along with a CW probe beam. The pump beam, inducing a transient chirp on the probe beam via XPM through Kerr nonlinearity, broadens the probe spectra generating sidebands. Using an optical filter, a single sideband can be selected so the output signal at the converted wavelength is modulated in time similarly to the pump pulse.

Astar et al. presented tunable wavelength conversion of a 10-Gb/s return-to-zero on–off-keyed (RZ-OOK) signal over a range of 20 nm, carried out in a silicon nanowire waveguide using cross-phase modulation (XPM), followed by a tunable filter [176].

All-optical four-channel format conversion from non-return-to-zero on–off-keyed (NRZ-OOK) to return-to-zero on–off-keyed (RZ-OOK) formats, based on semiconductor optical amplifiers (SOAs), was demonstrated in Ref. [177]. The corresponding monolithic InP-integrated chip is sketched in Figure 12: in its 2.0 × 4.6 mm^2 footprint, it includes a 2.0 mm long SOA1, used for nonlinear XPM and XGM effects, a tunable delay interferometer (DI), two short SOAs, and an arrayed waveguide grating (AWG). The DI, in turn, consists of two multi-mode interference (MMI) couplers and a 400 μm long SOA (SOA3) that compensates the loss introduced by the 500 μm long phase shifter (PS). An image of the chip obtained by a metallographic microscope is shown in Figure 13. The device was operating the conversion process at 40 Gb/s and with an average power penalty at BER (bit error rate) of 1 × 10^{-9} of the four channels less than 0.5 dB. The authors claimed that this chip has the potential to be used for other parallel all-optical signal processing, such as multichannel wavelength conversion, parallel signal regeneration, and so on [177].

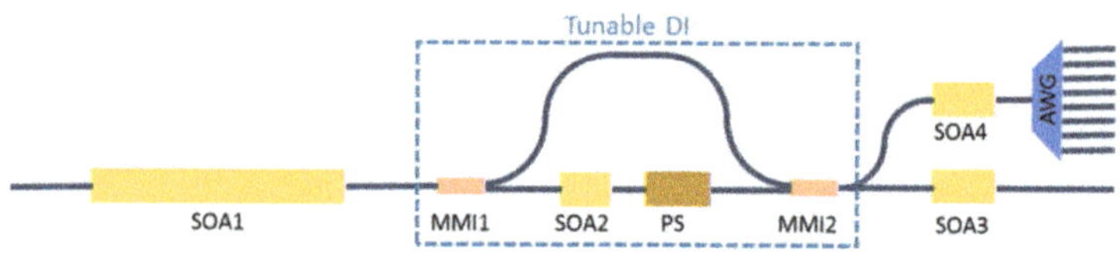

Figure 12. Schematic of an all-optical four-channel processor for NRZ to RZ format conversion to be integrated in an InP monolithic chip. MMI1 splits the signal and MMI2 recombines the signals from the two arms, which have a different phase shift. Operation wavelength is around 1570 nm, which is close to the gain peak of SOAs. Reproduced with permission from [177], 1943-0655 © 2023 IEEE.

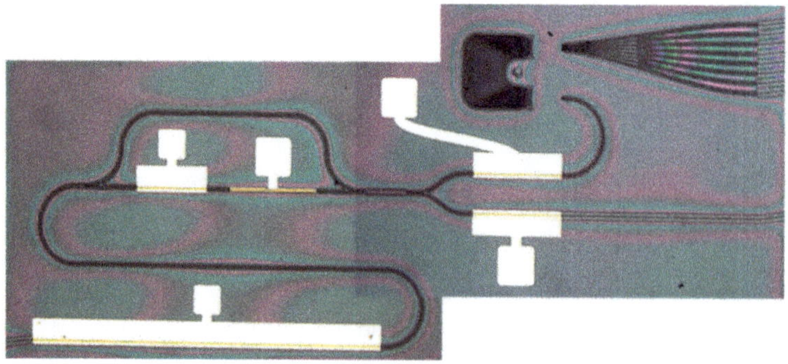

Figure 13. Image by a metallographic microscope of the fabricated InP monolithic chip according to the schematic in Figure 11. Reproduced with permission from [177], 1943-0655 © 2023 IEEE.

4.2.3. All-Optical Signal Regeneration

Most of the work completed so far on all-optical signal regeneration has exploited the nonlinear properties of optical fibers [178,179] or semiconductor optical amplifiers (OAS) [180–182]. Figure 14 shows the experimental setup for characterization of a polarization-independent differential phase shift keying (DPSK) regenerative wavelength converter [181]. A 10 Gb/s DPSK data stream at λ_{DATA} is generated by modulating the output of a tunable laser (TL) with a Mach–Zehnder modulator (MZM) driven by a bit pattern generator (BPG). The modulator output is then coupled with an amplified spontaneous emission (ASE) noise loading stage, constituted by an erbium-doped fiber amplifier (EDFA) and an optical filter (OF) with 1 nm bandwidth. Two TL pumps are coupled with the noisy data and injected into the SOA: the output converted data at λ_{FWM} are selected, and DPSK demodulation is made by using a standard delay-interferometer (DI). The dual co-polarized pumps allow FWM independence from input signal polarization and wavelength. In the system, which is suitable for photonic integration, wavelength conversion in a range up to 6 nm was measured [181].

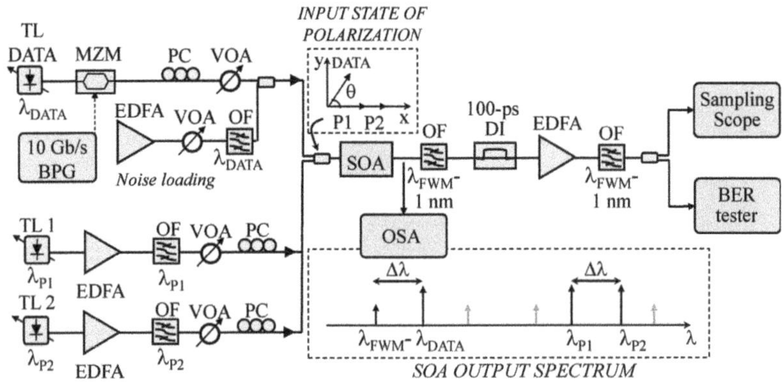

Figure 14. Sketch of the experimental setup for characterization of a SOA-based polarization-independent all-optical regenerator for DPSK data. The principles of operations are schematized within the two dashed boxes. (TL: tunable lasers; MZM: Mach–Zehnder modulator; BPG: bit pattern generator; EDFA: Erbium-doped fiber amplifier; PC: polarization controller; VOA: variable optical attenuator; OF: optical filter; DI: delay line.) Reproduced from [181] under Creative Commons license.

SOAs, however, are limited in speed due to carrier dynamics, and other attempts were made of realizing an integrated all-optical regenerator by means of other devices. One of the earliest results was obtained by B.J. Eggleton's group exploiting the strong Kerr nonlinearity of As_2S_3 chalcogenide glass waveguides (n_2~2×10^{-14} cm^2/W) [183]. The operation principle of the integrated 2R regenerator was based on a combination of nonlinear SPM-induced spectral broadening followed by spectral filtering; the device was constituted by a low loss As_2S_3 rib waveguide with integrated Bragg gratings. A nonlinear power transfer curve was demonstrated using 1.4 ps pulses, enough to suppress noise in an amplified transmission link [183]. Shortly thereafter, an SOI waveguide device was demonstrated to perform optical regeneration based on a Si nanowaveguide as the nonlinear medium for the SPM process integrated with a bandpass filter consisting of a waveguide ring resonator [184]. Compared with the chalcogenide device, which required ~50 W peak power, the SOI device operated at peak powers lower than 5 W. Since then, many other excellent results have been achieved; here, only a few of them are mentioned. On-chip four-level pulse–amplitude modulation (PAM-4) wavelength conversion and signal regeneration were demonstrated in a silicon waveguide, exploiting degenerate FWM [36]. Using a silicon nanowire waveguide, Geng et al. [185] experimentally demonstrated three-wavelength regeneration based on the clock-pump four-wave-mixing scheme.

A limitation of silicon devices lies in nonlinear absorption (two-photon absorption, TPA): it can, however, be circumvented by using reverse-biased p-i-n diode structures implemented across the waveguide [186]. With this approach, phase regeneration for DPSK signals using a Si waveguide with a reverse-biased p-i-n junction as nonlinear medium and continuous-wave pumping was demonstrated, with high wavelength conversion efficiency [187]. The Si nano-rib waveguides had width, height, and slab height of 500 nm, 210 nm, and 50 nm, respectively; doped regions were created at a distance of 350 nm from the waveguide through implantation of boron and arsenic with concentrations of

10^{18} cm^{-3} for the p- and n- regions, respectively. Keeping in mind development of 5G and 6G networks, data transmission rates as high as 100 Gb/s at the user must be considered; correspondingly, intensity modulation with the NRZ-OOK format will likely be a good solution for optical access networks. Wen et al. [180] have, therefore, investigated the feasibility of all-optical regeneration of these signals using the FWM process in a silicon waveguide with reverse-biased p-i-n junction. The Si rib waveguide is 4 cm long, 450 nm wide, and 70 nm thick, respectively. The p- and n- regions of the junction were 3.6 μm wide and separated by 1.35 μm; they had been obtained by implantation of boron and phosphorus ions, respectively. The achieved conversion efficiency was as high as −12 dB, and all-optical regeneration of NRZ-OOK signals with rates of 50 Gb/s and 100 Gb/s were experimentally demonstrated [188]. For completeness, it has also to be mentioned that signal regeneration was obtained by exploiting the nonlinear properties of lithium niobate, e.g., using nonlinear wave mixing in cascaded periodically poled lithium niobate (PPLN) waveguides [189–191] and in lithium niobate on insulator (LNOI) structures [35].

Before concluding our discussion on all-optical signal regeneration, we summarize some challenges for achieving efficient all-optical regenerators: (1) as optical networks are adopting flexible and modulation format variable transceivers, any dependence on modulation format will reduce the efficiency and upgradability of the regenerators; therefore, optical regeneration should be modulation format and baud-rate-transparent. (2) Optical regeneration should improve the optical SNR by amplifying only the data signal and reducing the signal's amplified spontaneous emission (ASE) noise arising from erbium-doped-fiber-amplifiers (EDFA). (3) The optical regenerator should mitigate any crosstalk and inter-symbol interference that arises in the different domains of the optical wave, such as wavelength, polarization, and spatial modes [50,51].

4.3. Nonlinear Sources

To generate SC from a chalcogenide glass fiber, a short pulse laser (i.e., fs laser) is often employed as the pump source operating at the NIR region where dispersion of chalcogenide glasses is positive. In SC generation under NIR pumping, the resultant spectra usually show limited bandwidth due to the limited pump power and materials dispersion. To this end, optical pumping at wavelengths longer than the zero-dispersion wavelength (ZDW) is often applied, where the dispersion of the chalcogenide glasses becomes negative. The SC generation is this case becomes much more efficient as the soliton effect becomes dominant, such as soliton splitting and self-frequency shift, which leads to much larger broadening of the SC spectra [192].

An experimental demonstration of a tunable continuum source in silicon photonic wires (SPWs) with a power-dependent broad spectrum was reported in [193]. As can be seen in Figure 15, a spectral broadening of more than 350 nm was observed upon propagation of ultrashort 1.3 μm wavelength optical pulses in a 4.7 mm long single-mode waveguide. Supercontinuum white light generation in plasmonic nanostructures, depending on the nonlocality of the electron response, has also been theoretically investigated [194].

Mode-locked lasers were initially used for comb generation. Thereafter, comb emission was demonstrated in continuous-wave (cw) laser-pumped resonators through cascaded third-order parametric processes [195]. In such Kerr resonators, a first pair of sidebands is generated around the pump frequency by cavity modulation instability or degenerate four-wave mixing (FWM); subsequently, cascaded four-wave mixing processes lead to formation, around the pump frequency, of a uniform frequency comb, where self- and cross-phase modulation act to compensate for the unequal cavity mode spacing induced by the group velocity dispersion (GVD) [196]. Because of the relatively low strength of third-order nonlinearity, generation of Kerr combs requires small interaction volumes and high-Q resonators. For these reasons, small resonators are particularly suited to reach broadband comb generation with quite moderate pump power [62]. Advancements in the fabrication technology of optical micro-cavities may enable realizing ultra-fast and stable optical clocks and pulsed sources with extremely high repetition-rates in the form of compact and integrated devices. In this framework, demonstration of planar high-Q resonators, compatible with silicon technology [197,198], has revealed a unique opportunity for these devices to provide entirely new capabilities for photonic-integrated technologies.

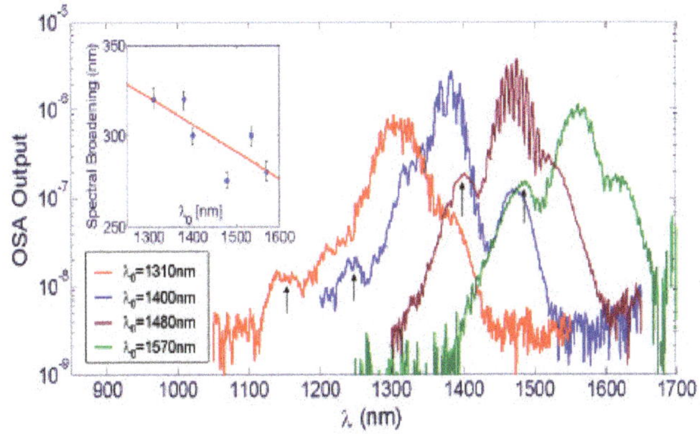

Figure 15. Supercontinuum generation in a 4.7 mm long silicon photonic wire waveguide for several input central wavelengths at P 0 $\approx$ 1W. The inset shows that spectral broadening increases as λ_0 approaches the ZGVD wavelength of 1290 nm. Reprinted with permission from [193] © The Optical Society.

It can be recalled that electro-optic (EO) modulation, in materials with second-order nonlinearity, provides an interesting alternative to Kerr (third-order nonlinearity) resonators for generation of OFCs; the electrical controllability of EO combs guarantees greater versatility and also excellent comb stability and phase coherence. A microring resonator, with loaded Q~1.5×10^6, may, therefore, be fabricated in a thin-film lithium niobate (TFLN) platform, which is also characterized by ultra-low-loss optical waveguides [199]. This solution is depicted in Figure 16; a microwave signal, with modulation frequency equal to the free spectral range (FSR) of the optical resonator, couples light between different resonator modes. The input cw laser light is, therefore, modulated, giving rise to sidebands at the modulation frequency, which are then recirculated to be modulated again. At the output, an EO comb can be measured that spans more frequencies than the entire telecommunications L-band (over 900 comb lines spaced about 10 gigahertz apart), with the prospect of enabling octave-spanning by proper dispersion engineering [199]. A disadvantage of this single-resonator device is the low (~0.3%) comb conversion efficiency due to the low coupling efficiency of the light from the input waveguide to the microring EO-driven resonance. This limit was overcome by the same research group by using two mutually coupled resonators, again realized in the TFLN platform: the fabricated device proven to be capable of generating an on-chip EO comb with a line spacing of 30.925 GHz, a pump-to-comb conversion efficiency of 30%, and a wide comb span of 132 nm [200].

Early attempts of micro-comb-based communications used simple on–off keying (OOK) modulation [201,202], where information is carried by the optical intensity. In this case, low comb intensity noise is essential for achieving high performance. Error-free transmission for each individual line of a low-noise micro-comb, transmitted over tens of kilometers of single-mode fiber, has been demonstrated with a power penalty of less than 0.5 dB [202].

Coherent data transmission, which typically poses stringent requirements on the spectral purity of the optical carrier, has also been demonstrated with phase-locked micro-combs. In the first work reported by Pfeifle et al. in 2014 [203], a data stream of 392 Gbit s^{-1} was encoded on six lines of a micro-comb using quadrature phase-shift-keying (QPSK) and 16-state quadrature amplitude modulation (16QAM). A second experiment demonstrated the feedback stabilization of the comb and transmission of a 1.44 Tbit s^{-1} data stream over up to 300 km with twenty comb lines. These results showed that micro-combs can indeed meet the highly demanding requirements of coherent communications and thus offer an attractive route towards chip-scale terabit/s transceivers [204].

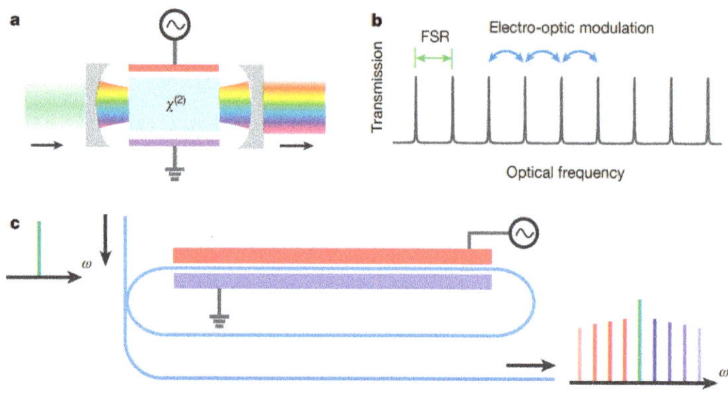

Figure 16. EO comb generator using a microring resonator. (**a**) Sketch of a bulk EO comb generator based on an EO ($\chi^{(2)}$) phase modulator inside a Fabry–Pérot resonator. A cw laser beam enters the resonator and an optical frequency comb is generated at the output. (**b**) EO comb generation principle; the microwave modulating signal has frequency equal to FSR of the Fabry–Pérot resonator. (**c**) Integrated EO comb generator, where a microring replaces the Fabry–Pérot resonator. The cw signal from input waveguide is coupled into the microring and EO-modulated at a frequency matching the FSR of the ring. The generated comb is then coupled back to the waveguide. Reproduced with permission from [199] © Springer Nature Ltd.

5. Conclusions

Despite the difficulties, impressive progress has been made in on-chip nonlinear optical processing with favorable performance, paving the way to full integration of complete optical communication and processing systems on a monolithic or hybrid chip. As an example, an excellent result concerns the high device integration density achieved by using optimized racetrack microresonators based on ultra-low-loss multimode Si_3N_4 photonic circuits. The optimized design uses adiabatic Euler bends in place of circular bends; it enables a device footprint as small as 0.21 mm^2, which can be very useful for several nonlinear photonic applications. Single-soliton generation of 19.8 GHz repetition rate was demonstrated; the soliton spectrum showed a 3-dB bandwidth of 16.3 nm, corresponding to a pulse duration of 156 fs [3]. Even better prospects exist for silicon carbide, which now appears able to fully compete with more mature nonlinear materials; as a proof, design and demonstration of octave-spanning microcombs have been reported in a 4H-SiC-on-insulator microring resonator with 36 μm radius [136].

All-optical computing, on the other hand, still presents many challenges due to the difficulty of implementing efficient all-optical logic devices. Photonic crystals seem to offer a suitable approach to realization of these devices, but, up to now, groundbreaking results have not been achieved. Other approaches make use of nonlinear optical fibers and semiconductor optical amplifiers, but, in both cases, size is an obstacle to chip integration. Plasmonic logic gates would have a desirable nanometer size, but their large optical loss makes them also unsuitable for use in integrated chips. As has been shown by various examples, PICs fabricated on a silicon-on-insulator platform may exhibit low power consumption and high transmission efficiency but occupy a relatively large footprint. A solution has, therefore, still to be searched. However, based on the linear interference approach, it is worth reporting here a very recent result obtained using a design based on topology optimization [205] and leading to implementation of an SOI chip containing seven major logic gates (AND, OR, NOT, NAND, NOR, XOR, and XNOR) and a half adder, with a footprint of only 1.3 × 4.5 μm^2 [205].

The evolving nano-devices would enable emerging advanced applications not only in optical processing and computing but also in metrology, single-molecule sensing, imaging, microscopy, mid-infrared photonics, terahertz generation, microwave photonics, and biomedicine. We expect that progress in nonlinear integrated photonics will occur along three paths: one involving new materials and new physics in nonlinear interactions; a second devoted to improving device performance using well-known and innovative guided-wave nonlinear optical schemes; and a third addressing new perspectives in quantum photonics. In this regard, being impossible to adequately cover the achievements in quantum optics in this short review, we just mention a few examples, all exploiting the exceptional characteristics of microring resonators, either on a silicon platform [206,207] or a 4H-SiC-on-insulator platform [208]. Quantum technologies may well exploit other materials as well, and we refer interested readers to an extended review on the 2023 roadmap for materials for quantum technologies, published in January 2023 [209].

Returning to use of Si or SiC, Guo et al. experimentally demonstrated nonclassical optical bistability in a photon-pair source using cw-laser-pumped spontaneous FWM in an all-pass microring resonator with 100 μm radius fabricated on an SOI platform [206]. Using similar microring (110 μm radius) with Q~5 $\times$ 10^4 and free spectral range (FSR)~100 GHz, matching the standard International Telecommunication Union (ITU) frequency grid to enable multichannel photon-pair generation compatible with commercial WDM devices, the same research group showed how frequency correlation can be applied to high-dimensional encoding for parallel quantum key distribution (QKD) [207].

More recently, various works have underlined the strong potential of 4H-SiC platform for on-chip quantum photonics; it was also demonstrated that implantation of silicon vacancy centers in SiC does not deteriorate their intrinsic spin-optical properties [210]. This result makes feasible development of large-scale quantum networks based on integrated quantum computational clusters with efficient spin–photon interfaces. Lukin et al. reported both strong enhancement of emission from a single color center in a nanophotonic cavity and efficient nonlinear frequency conversion in the same platform, also demonstrating full compatibility of thin-film SiC with industry-standard nanotechnologies and foundry production [208]. Still using a SiC microring structure, Guidry et al. studied the underlying quantum processes of soliton microcombs, investigating the quantum formation dynamics of dissipative Kerr soliton (DKS) states and their quantum correlations [211].

Obviously, theoretical advances are being made as well; as an example, Nozaki et al. have now (February 2023) published a paper where they propose to use a resonator with only passive optical components to scale up stimulated emission of polarization-entangled photon pairs [212]; they present only a proof-of-principle experimental demonstration with a double-pass polarization Sagnac interferometer and periodically poled KTP crystal, but use of waveguide structures should lead to realization of highly efficient and bright quantum-entangled photon sources needed for quantum information technologies.

Author Contributions: Conceptualization, L.S. and G.C.R.; writing—original draft preparation, L.S. and G.C.R.; writing—review and editing, L.S. and G.C.R. All authors have read and agreed to the published version of the manuscript.

Funding: This research received no external funding.

Conflicts of Interest: The authors declare no conflict of interest.

References

1. Nagarajan, R.; Joyner, C.H.; Schneider, R.P.; Bostak, J.S.; Butrie, T.; Dentai, A.G.; Dominic, V. Large-scale photonic integrated circuits. *IEEE J. Sel. Top. Quant. Electron.* **2005**, *11*, 50–65. [CrossRef]
2. Thylén, L.; Wosinski, L. Integrated photonics in the 21st century. *Photon. Res.* **2014**, *2*, 75–81. [CrossRef]
3. Ji, X.; Liu, J.; He, J.; Wang, R.N.; Qiu, Z.; Riemensberger, J.; Kippenberget, T.J. Compact, spatial-mode-interaction-free, ultralow-loss, nonlinear photonic integrated circuits. *Commun. Phys* **2022**, *5*, 84. [CrossRef]
4. Pelucchi, E.; Fagas, G.; Aharonovich, I.; Englund, D.; Figueroa, E.; Gong, Q.; Hannes, H.; Liu, J.; Lu, C.Y.; Matsuda, N.; et al. The potential and global outlook of integrated photonics for quantum technologies. *Nat. Rev. Phys.* **2022**, *4*, 194–208. [CrossRef]

5. Stegeman, G.I.; Seaton, C.T. Nonlinear integrated optics. *J. Appl. Phys.* **1985**, *58*, R57–R78. [CrossRef]
6. Gibbs, H.M.; Khitrova, G.; Peyghambarian, N. *Nonlinear Photonics*; Springer: Berlin/Heidelberg, Germany, 1990.
7. Chen, Z.; Morandotti, R. *Nonlinear Photonics and Novel Optical Phenomena*; Springer: New York, NY, USA, 2012.
8. Hendrickson, S.M.; Foster, A.C.; Camacho, R.M.; Clader, B.D. Integrated nonlinear photonics: Emerging applications and ongoing challenges. *J. Opt. Soc. Am. B* **2014**, *31*, 3193–3203. [CrossRef]
9. Smirnova, D.; Leykam, D.; Chong, Y.; Kivshar, Y. Nonlinear topological photonics. *Appl. Phys. Rev.* **2020**, *7*, 021306. [CrossRef]
10. Eggleton, B.J.; Vo, T.; Pant, R.; Schr, J.; Pelusi, M.; Choi, D.Y.; Madden, S.; Luther-Davies, B. Photonic chip based ultrafast optical processing based on high nonlinearity dispersion engineered chalcogenide waveguides. *Laser Photonics Rev.* **2012**, *6*, 97–114. [CrossRef]
11. Nonlinear integrated photonics: Current status and future trends, Special Issue. *Photonics Res.* **2018**, *6*, 346–484.
12. Koshelev, K.; Kruk, S.; Melik-Gaykazyan, E.; Jae-Hyuck Choi, J.H.; Andrey Bogdanov, A.; Hong-Gyu Park, H.-G.; Kivshar, Y. Subwavelength dielectric resonators for nonlinear nanophotonics. *Science* **2020**, *367*, 288–292. [CrossRef]
13. Vlachos, K.; Raffaelli, C.; Aleksic, S.; Andriolli, N.; Apostolopoulos, D.; Avramopoulos, H.; Erasme, D.; Klonidis, D.; Petersen, M.N.; Scaffardi, M.; et al. Photonics in switching: Enabling technologies and subsystem design. *J. Opt. Netw.* **2009**, *8*, 404–428. [CrossRef]
14. Lu, H.; Liu, X.; Wang, L.; Gong, Y.; Mao, D. Ultrafast all-optical switching in nanoplasmonic waveguide with Kerr nonlinear resonator. *Opt. Express* **2011**, *19*, 2910–2915. [CrossRef] [PubMed]
15. Sederberg, S.; Driedger, D.; Nielsen, M.; Elezzabi, A.Y. Ultrafast all-optical switching in a silicon-based plasmonic nanoring resonator. *Opt. Express* **2011**, *19*, 23494–23503. [CrossRef] [PubMed]
16. Pelc, J.S.; Rivoire, K.; Vo, S.; Santori, C.; Fattal, D.A.; Beausoleil, R.G. Picosecond all-optical switching in hydrogenated amorphous silicon microring resonators. *Opt. Express* **2014**, *22*, 3797–3810. [CrossRef]
17. Bohn, J.; Luk, T.S.; Tollerton, C.; Hutchings, S.W.; Brener, I.; Horsley, S.; Barnes, W.L.; Hendry, E. All-optical switching of an epsilon-near-zero plasmon resonance in indium tin oxide. *Nat. Commun.* **2021**, *12*, 1017. [CrossRef]
18. Grinblat, G.; Zhang, H.; Nielsen, M.P.; Krivitsky, L.; Berté, R.; Li, Y.; Tilmann, B.; Cortés, E.; Oulton, R.F.; Kuznetsov, A.I.; et al. Efficient Ultrafast All-Optical Modulation in a Nonlinear Crystalline Gallium Phosphide Nanodisk at the Anapole Excitation. *Sci. Adv.* **2020**, *6*, eabb3123. [CrossRef]
19. Tsuda, H.; Kurokawa, T. Construction of an all-optical flip-flop by combination of two optical triodes. *Appl. Phys. Lett.* **1990**, *57*, 1724. [CrossRef]
20. Notomi, M.; Shinya, A.; Mitsugi, S.; Kira, G.; Kuramochi, E.; Tanabe, T. Optical bistable switching action of Si high-Q photonic-crystal nanocavities. *Opt. Express* **2005**, *13*, 2678–2687. [CrossRef]
21. Wu, Y.-D.; Shih, T.T.; Chen, M.-H. New all-optical logic gates based on the local nonlinear Mach-Zehnder interferometer. *Opt. Express* **2008**, *16*, 248–257. [CrossRef]
22. Fushimi, A.; Tanabe, T. All-optical logic gate operating with single wavelength. *Opt. Express* **2014**, *22*, 4466–4479. [CrossRef]
23. Xiaoyu, Y.; Xiaoyong, H.; Hong, Y.; Qihuang, G. Ultracompact all-optical logic gates based on nonlinear plasmonic nanocavities. *Nanophotonics* **2017**, *6*, 365–376.
24. Jandieri, V.; Khomeriki, R.; Onoprishvili, T.; Werner, D.H.; Berakdar, J.; Erni, D. Functional all-optical logic gates for true time-domain signal processing in nonlinear photonic crystal waveguides. *Opt. Express* **2020**, *28*, 18317–18331. [CrossRef]
25. Anagha, E.G.; Jeyachitra, R.K. Review on all-optical logic gates: Design techniques and classifications—Heading toward high-speed optical integrated circuits. *Opt. Eng.* **2022**, *61*, 060902. [CrossRef]
26. Thankaraj, B.S.; Ramasamy, A. Revolution of optical computing logic gates based on its applications: An extensive survey. *Opt. Eng.* **2022**, *61*, 110901. [CrossRef]
27. Jiao, S.M.; Liu, J.W.; Zhang, L.W.; Yu, F.H.; Zuo, G.M.; Zhang, J.; Zhao, F.; Lin, W.; Shao, L. All-optical logic gate computing for high-speed parallel information processing. *Opto-Electron. Sci.* **2022**, *1*, 220010. [CrossRef]
28. Guo, Y.; Kao, C.K.; Li, H.E.; Chiang, K.S. *Nonlinear Photonics: Nonlinearities in Optics, Optoelectronics and Fiber Communications*; Springer: Berlin/Heidelberg, Germany, 2002.
29. Radic, S.; Moss, D.J.; Eggleton, B.J. Nonlinear optics in communications: From crippling impairment to ultrafast tools. In *Optical Fiber Telecommunications. Volume A: Components and Subsystems*; Kaminow, I.P., Li, T., Willner, A.E., Eds.; Elsevier: Amsterdam, The Netherlands, 2008.
30. Khulbe., M.; Kumar, S. Role of Nonlinear Optics in Big data transmission and Next Generation Computing Technologies. In Proceedings of the 9th International Conference on Cloud Computing, Data Science & Engineering (Confluence), Noida, India, 10–11 January 2019; pp. 234–238.
31. Yoo, S.J.B. Wavelength conversion technologies for WDM network applications. *J. Light. Technol.* **1996**, *14*, 955–966. [CrossRef]
32. Caspani, L.; Duchesne, D.; Dolgaleva, K.; Wagner, S.J.; Ferrera, M.; Razzari, L.; Pasquazi, A.; Peccianti, M.; Moss, D.J.; Aitchison, J.S.; et al. Optical frequency conversion in integrated devices. *J. Opt. Soc. Am. B* **2011**, *28*, A67–A82. [CrossRef]
33. Petrillo, K.G.; Foster, M.A. Full 160-Gb/s OTDM to 16x10-Gb/s WDM conversion with a single nonlinear interaction. *Opt. Express* **2013**, *21*, 508–518. [CrossRef] [PubMed]
34. Lacava, C.; Ettabib, M.; Cristiani, I.; Fedeli, J.; Richardson, D.; Petropoulos, P. Ultra-Compact Amorphous Silicon Waveguide for Wavelength Conversion. *IEEE Photonics Technol. Lett.* **2016**, *28*, 410–414. [CrossRef]
35. Zheng, Y.; Chen, X. Nonlinear wave mixing in lithium niobate thin film. *Adv. Phys. X* **2021**, *6*, 1889402. [CrossRef]

36. Long, Y.; Wang, A.; Zhou, L.; Wang, J. All-optical wavelength conversion and signal regeneration of PAM-4 signal using a silicon waveguide. *Opt. Express* **2016**, *24*, 7158–7167. [CrossRef]
37. Guo, B.; Wu, B.; Wang, Y.; Wen, F.; Geng, Y.; Zhou, H.; Qiu, K. On-chip Mach Zehnder interferometer-based all-optical amplitude regenerator for optical 16-QAM signals. *Opt. Express* **2021**, *29*, 27683–27695. [CrossRef] [PubMed]
38. Singh, L.; Iadicicco, A.; Agrawal, N.; Saha, C.; Chauhan, R. A compact formulation of all optical signal router by using plasmonic waveguides. *Opt. Quant. Electron.* **2022**, *54*, 478. [CrossRef]
39. Mateo, E.F.; Liñares, J. A phase insensitive all-optical router based on nonlinear lenslike planar waveguides. *Opt. Express* **2005**, *13*, 3355–3370. [CrossRef]
40. Yan, Z.; He, H.; Liu, H.; Iu, M.; Ahmed, O.; Chen, E.; Blakey, P.; Akasaka, Y.; Ikeuchi, T.; Helmy, A.S. $\chi^{(2)}$-based AlGaAs phase sensitive amplifier with record gain, noise, and sensitivity. *Optica* **2022**, *9*, 56–60. [CrossRef]
41. Pelusi, M.; Ta'Eed, V.; Lamont, M.; Madden, S.; Choi, D.-Y.; Luther-Davies, B.; Eggleton, B. Ultra-High Nonlinear As_2S_3 Planar Waveguide for 160-Gb/s Optical Time-Division Demultiplexing by Four-Wave Mixing. *IEEE Photonics Technol. Lett.* **2007**, *19*, 1496–1498. [CrossRef]
42. Palushani, E.; Hansen Mulvad, H.; Galili, M.; Hu, H.; Oxenlowe, L.K.; Clausen, A.T.; Jeppesen, P. OTDM-to-WDM conversion based on time-to frequency mapping by time-domain optical Fourier transformation. *IEEE J. Sel. Top. Quant. Electron.* **2012**, *18*, 681–688. [CrossRef]
43. Hansen Mulvad, H.C.; Palushani, E.; Hu, H.; Ji, H.; Lillieholm, M.; Galili, M.; Clausen, A.T.; Pu, M.; Yvind, K.; Hvam, J.M.; et al. Ultra-high-speed optical serial-to-parallel data conversion by time domain optical Fourier transformation in a silicon nanowire. *Opt. Express* **2011**, *19*, B825–B835.
44. Luo, L.W.; Ophir, N.; Chen, C.P.; Gabrielli, L.H.; Poitras, C.B.; Bergmen, K.; Lipson, M. WDM-compatible mode-division multiplexing on a silicon chip. *Nat. Commun.* **2014**, *5*, 3069. [CrossRef]
45. Jansen, S.L.; Morita, I.; Schenk, T.C.; Tanaka, H. Long-haul transmission of16×52.5 Gbits/s polarization-division- multiplexed OFDM enabled by MIMO processing (Invited). *J. Opt. Netw.* **2008**, *7*, 173–182. [CrossRef]
46. Tanabe, T.; Notomi, M.; Shinya, A.; Mitsugi, S.; Kuramochi, E. Fast bistable all-optical switch and memory on silicon photonic crystal on-chip. *Opt. Lett.* **2005**, *30*, 2575–2577. [CrossRef] [PubMed]
47. Hill, M.T.; Dorren, H.J.S.; Vries, T.; Leijtens, X.J.M.; Besten, J.H.; Smalbrugge, B.; Oei, Y.-S.; Binsma, H.; Khoe, G.-D.; Smit, M.K. A fast low-power optical memory based on coupled micro-ring lasers. *Nature* **2004**, *432*, 206–209. [CrossRef]
48. Kowsari, A.; Ahmadi, V.; Darvish, G.; Moravvej-Farshi, M.K. All-optical tunable delay line based on nonlinearities in a chalcogenide microfiber coil resonator. *J. Opt. Soc. Am. B* **2017**, *34*, 1199–1205. [CrossRef]
49. Liu, F.; Li, Q.; Zhang, Z.; Qiu, M.; Su, Y. Optically Tunable Delay Line in Silicon Microring Resonator Based on Thermal Nonlinear Effect. *IEEE J. Sel. Top. Quantum Electron.* **2008**, *14*, 706–712. [CrossRef]
50. Leclerc, O.; Lavigne, B.; Balmefrezol, E.; Brindel, P.; Pierre, L.; Rouvillain, D.; Seguineau, F. Optical regeneration at 40Gb/s and beyond. *J. Light. Technol.* **2003**, *21*, 2779–2790. [CrossRef]
51. Ciaramella, E. Wavelength Conversion and All-Optical Regeneration: Achievements and Open Issues. *J. Light. Technol.* **2012**, *30*, 572–582. [CrossRef]
52. Koos, C.; Jacome, L.; Poulton, C.; Leuthold, J.; Freude, W. Nonlinear silicon-on-insulator waveguides for all-optical signal processing. *Opt. Express* **2007**, *15*, 5976–5990. [CrossRef]
53. Willner, A.E.; Yilmaz, O.F.; Wang, J.; Wu, X.; Bogoni, A.; Zhang, L.; Nuccio, S.R. Optically Efficient Nonlinear Signal Processing. *IEEE J. Sel. Top. Quant. Electron.* **2011**, *17*, 320–332. [CrossRef]
54. Wabnitz, S.; Eggleton, B.J. *All-Optical Signal Processing*; Springer International Publishing: Cham, Switzerland, 2015.
55. Lacava, C.; Ettabib, M.A.; Petropoulos, P. Nonlinear Silicon Photonic Signal Processing Devices for Future Optical Networks. *Appl. Sci.* **2017**, *7*, 103. [CrossRef]
56. Wang, J.; Long, Y. On-chip silicon photonic signaling and processing: A review. *Sci. Bull.* **2018**, *63*, 1267–1310. [CrossRef]
57. Minzioni, P.; Lacava, C.; Tanabe, T.; Dong, J.; Hu, X.; Csaba, G.; Porod, W.; Singh, G.; Willner, A.E.; Almaiman, A.; et al. Roadmap on all-optical processing. *J. Opt.* **2019**, *21*, 063001. [CrossRef]
58. Chang, L.; Boes, A.; Shu, H.; Xie, W.; Huang, H.; Qin, J.; Shen, B.; Wang, X.; Mitchell, A.; Bowers, J.E.L. Second Order Nonlinear Photonic Integrated Platforms for Optical Signal Processing. *IEEE J. Sel. Top. Quant. Electron.* **2021**, *27*, 1–11. [CrossRef]
59. Huang, C.; Jha, A.; Ferreira de Lima, T.; Tait, A.N.; Shastri, B.J.; Prucnal, P.R. On-Chip Programmable Nonlinear Optical Signal Processor and Its Applications. *IEEE J. Sel. Top. Quantum Electron.* **2021**, *27*, 6100211. [CrossRef]
60. Dudley, J.M.; Genty, G.; Coen, S. Supercontinuum generation in photonic crystal fiber. *Rev. Mod. Phys.* **2006**, *78*, 1135–1184. [CrossRef]
61. Gaeta, A.L.; Lipson, M.; Kippenberg, T.J. Photonic-chip-based frequency combs. *Nat. Photonics* **2019**, *13*, 158–169. [CrossRef]
62. Sirleto, L.; Righini, G.C. An Introduction to Nonlinear Integrated Photonics Devices: Nonlinear effects and materials. *Micromachines* **2023**, *14*, 604. [CrossRef]
63. Selvaraja, S.K.; Sethi, P. Review on Optical Waveguides. In *Emerging Waveguide Technology*; You, K.Y., Ed.; IntechOpen: Rijeka, Croatia, 2018. [CrossRef]
64. Okamoto, K. *Fundamentals of Optical Waveguides*, 3rd ed.; Academic Press: Cambridge, MA, USA, 2022.
65. Tsuchizawa, T.; Yamada, K.; Fukuda, H.; Watanabe, T.; Takahashi, J.; Takahashi, M.; Shoji, T.; Tamechika, E.; Itabashi, S.; Morita, H. Microphotonics devices based on silicon microfabrication technology. *IEEE J. Sel. Top. Quant. Electr.* **2005**, *11*, 232–240. [CrossRef]

66. Gerard, J.-M.; Sermage, B.; Gayral, B.; Legrand, B.; Costard, E.; Thierry-Mieg, V. Enhanced Spontaneous Emission by Quantum Boxes in a Monolithic Optical Microcavity. *Phys. Rev. Lett.* **1998**, *81*, 1110–1113. [CrossRef]
67. Reithmaier, J.P.; Sęk, G.; Loer, A.; Hofmann, C.; Kühn, S.; Reitzenstein, S.; Keldysh, L.V.; Kulakovskii, V.D.; Reinecke, T.L.; Forchel, A. Strong coupling in a single quantum dot–semiconductor microcavity system. *Nature* **2004**, *432*, 197–200. [CrossRef]
68. Butt, M.A.; Khonina, S.N.; Kazanskiy, N.L. Recent advances in photonic crystal optical devices: A review. *Opt. Laser Technol.* **2021**, *142*, 107265. [CrossRef]
69. Bravo-Abad, J.; Rodriguez, A.; Bermel, P.; Johnson, S.G.; Joannopoulos, J.D.; Soljačić, M. Enhanced nonlinear optics in photonic-crystal microcavities. *Opt. Express* **2007**, *15*, 16161–16176. [CrossRef] [PubMed]
70. Righini, G.C.; Dumeige, Y.; Féron, P.; Ferrari, M.; Nunzi Conti, G.; Ristic, D.; Soria, S. Whispering gallery mode microresonators: Fundamentals and applications. *Riv. Nuovo Cim.* **2011**, *34*, 435–488.
71. Frigenti, G.; Farnesi, D.; Nunzi Conti, G.; Soria, S. Nonlinear Optics in Microspherical Resonators. *Micromachines* **2020**, *11*, 303. [CrossRef] [PubMed]
72. Lin, G.; Coillet, A.; Chembo, Y.K. Nonlinear photonics with high-Q whispering-gallery-mode resonators. *Adv. Opt. Photon.* **2017**, *9*, 828–890. [CrossRef]
73. Suhailin, F.H.; Healy, N.; Franz, Y.; Sumetsky, M.; Ballato, J.; Dibbs, A.N.; Gibson, U.J.; Peacock, A.C. Kerr nonlinear switching in a hybrid silica-silicon microspherical resonator. *Opt. Express* **2015**, *23*, 17263–17268. [CrossRef]
74. Strekalov1, D.V.; Marquardt, C.; Matsko, A.B.; Harald, G.L.; Schwefel, H.G.L.; Leuchs, G. Nonlinear and quantum optics with whispering gallery resonators. *J. Opt.* **2016**, *18*, 123002. [CrossRef]
75. Kuo, P.; Bravo-Abad, J.; Solomon, G. Second-harmonic generation using -quasi-phasematching in a GaAs whispering-gallery-mode microcavity. *Nat. Com.* **2014**, *5*, 3109. [CrossRef]
76. Rostami, A.; Ahmadi, H.; Heidarzadeh, H.; Taghipour, A. Microsphere and Fiber Optics based Optical Sensors. In *Optical Sensors—New Developments and Practical Applications*; Yasin, M., Harun, S.W., Arof, H., Eds.; IntechOpen: Rijeka, Croatia, 2014.
77. Slusher, R.E.; Eggleton, B.J. *Nonlinear Photonic Crystals*; Springer: Berlin/Heidelberg, Germany, 2003.
78. Vlasov, Y.A.; Boyle, M.O.; Hamann, H.F.; McNab, S.J. Active control of slow light on a chip with photonic crystal waveguides. *Nature* **2005**, *438*, 65–69. [CrossRef]
79. Soljačić, M.; Joannopoulous, J.D. Enhancement of nonlinear effect using photonic crystals. *Nat. Mater.* **2004**, *3*, 212–219. [CrossRef]
80. Zhang, Y.; Sheng, Y.; Zhu, S.; Xiao, M.; Krolikowski, W. Nonlinear photonic crystals: From 2D to 3D. *Optica* **2021**, *8*, 372–381. [CrossRef]
81. Pan, J.; Fu, M.; Yi, W.; Wang, X.; Liu, J.; Zhu, M.; Qi, J.; Yin, S.; Huang, G.; Zhu, S.; et al. Improving Low-Dispersion Bandwidth of the Silicon Photonic Crystal Waveguides for Ultrafast Integrated Photonics. *Photonics* **2021**, *8*, 105. [CrossRef]
82. Bozhevolnyi, S.I.; Volkov, V.S.; Devaux, E.; Laluet, J.-Y.; Ebbesen, T.W. Channel plasmon subwavelength waveguide components including interferometers and ring resonators. *Nature* **2006**, *440*, 508–511. [CrossRef] [PubMed]
83. Holmström, P.; Thylén, L.; Bratkovsky, A. Composite metal/ quantum-dot nanoparticle-array waveguides with compensated loss. *Appl. Phys. Lett.* **2010**, *97*, 073110. [CrossRef]
84. Prasad, P.N. *Nanophotonics*; Wiley: Hoboken, NJ, USA, 2004.
85. Haus, J.W. *Fundamentals and Applications of Nanophotonics*; Woodhead: Sawston, UK, 2016.
86. Zheludev, N.I. Nonlinear optics on the nano scale. *Contemp. Phys.* **2010**, *43*, 365–377. [CrossRef]
87. Suresh, S.; Arivuoli, D. Nanomaterials for nonlinear optical applications: A review. *Rev. Adv. Mater. Sci.* **2012**, *30*, 243–253.
88. Ganeev, R. *Nanostructured Nonlinear Optical Materials*, 1st ed.; Elsevier: Amsterdam, The Netherlands, 2018.
89. Panoiu, N.C.; Sha, W.E.I.; Lei, D.Y.; Li, G.-C. Nonlinear optics in plasmonic nanostructures. *J. Opt.* **2018**, *20*, 083001. [CrossRef]
90. Tuniz, A. Nanoscale nonlinear plasmonics in photonic waveguides and circuits. *Riv. Nuovo Cim.* **2021**, *44*, 193–249. [CrossRef]
91. Kuznetsov, A.I.; Miroshnichenko, A.E.; Brongersma, M.L.; Kivshar, Y.S.; Luk'yanchuk, B. Optically resonant dielectric nanostructures. *Science* **2016**, *354*, aag2472. [CrossRef]
92. Staude, I.; Schilling, J. Metamaterial-inspired silicon nanophotonics. *Nat. Photon.* **2017**, *11*, 274–284. [CrossRef]
93. Yang, Z.J.; Jiang, R.; Zhuo, X.; Xie, Y.-M.; Wang, J.; Lin, H.-Q. Dielectric nanoresonators for light manipulation. *Phys. Rep.* **2017**, *701*, 1–50. [CrossRef]
94. Kruk, S.; Kivshar, Y.S. Functional meta-optics and nanophotonics governed by Mie resonances. *ACS Photonics* **2017**, *4*, 2638–2649. [CrossRef]
95. Zhang, J.; Shi, X.; Zhang, Z.; Guo, K.; Yang, J. Ultra-compact, efficient and high-polarization-extinction-ratio polarization beam splitters based on photonic anisotropic metamaterials. *Opt. Express* **2022**, *30*, 538–549. [CrossRef] [PubMed]
96. Molesky, S.; Lin, Z.; Piggott, A.Y.; Jin, W.; Vukcović, J.; Rodriguez, A.W. Inverse design in nanophotonics. *Nat. Photonics* **2018**, *122*, 659–670. [CrossRef]
97. Vercruysse, D.; Sapra, N.V.; Yang, K.Y.; Vučković, J. Inverse-Designed Photonic Crystal Circuits for Optical Beam Steering. *ACS Photonics* **2021**, *8*, 3085–3093. [CrossRef]
98. Yang, K.Y.; Shirpurkar, C.; White, A.D.; Zang, J.; Chang, L.; Ashtiani, F.; Guidry, M.A.; Lukin, D.M.; Pericherla, S.V.; Yang, J. Multi-dimensional data transmission using inverse-designed silicon photonics and microcombs. *Nat. Commun.* **2022**, *13*, 7862. [CrossRef]
99. Quaranta, G.; Basset, G.; Martin, O.J.; Gallinet, B. Recent advances in resonant waveguide grating. *Laser Photonics Rev.* **2018**, *2*, 1800017. [CrossRef]

100. Yu, Z.; Xi, X.; Ma, J.; Tsang, H.K.; Zou, C.L.; Sun, X. Photonic integrated circuits with bound states in the continuum. *Optica* **2019**, *6*, 1342–1348. [CrossRef]
101. Kivshar, Y. All-dielectric meta-optics and non-linear nanophotonics. *Natl. Sci. Rev.* **2018**, *5*, 144–158. [CrossRef]
102. Wang, J.; Du, J. Metasurfaces for Spatial Light Manipulation. In *Metamaterials-Devices and Applications*; Borja, A.L., Ed.; Intech Open: Rijeka, Croatia, 2017.
103. Bonacina, L.; Brevet, P.F.; Finazzi, M.; Celebrano, M. Harmonic generation at the nanoscale. *J. Appl. Phys.* **2020**, *127*, 230901. [CrossRef]
104. Sain, B.; Meier, C.; Zentgraf, T. Nonlinear optics in all-dielectric nanoantennas and metasurfaces: A review. *Adv. Photonics* **2019**, *1*, 024002. [CrossRef]
105. Raghunathan, V.; Deka, J.; Menon, S.; Biswas, R.; A.S, L.K. Nonlinear Optics in Dielectric Guided-Mode Resonant Structures and Resonant Metasurfaces. *Micromachines* **2020**, *11*, 449. [CrossRef] [PubMed]
106. Capasso, F.; Sirtori, C.; Faist, J.; Sivco, D.L.; Sung-Nee, G.; Chu, S.-N.-G.; Cho, A.Y. Observation of an electronic bound state above a potential well. *Nature* **1992**, *358*, 565–567. [CrossRef]
107. Marinica, D.; Borisov, A.; Shabanov, S. Bound states in the continuum in photonics. *Phys. Rev. Lett.* **2008**, *100*, 183902. [CrossRef]
108. Hsu, C.W.; Zhen, B.; Stone, A.D.; Joannopoulos, J.D.; Soljačić, M. Bound states in the continuum. *Nat. Rev. Mater.* **2016**, *1*, 16048. [CrossRef]
109. Haoye, Q.; Xiaodong, S.; Haiyan, O. Exceptional points at bound states in the continuum in photonic integrated circuits. *Nanophotonics* **2022**, *11*, 4909–4917.
110. Ye, F.; Yu, Y.; Xi, X.; Sun, X. Second-harmonic generation in etchless lithium niobate nanophotonic waveguides with bound states in the continuum. *Laser Photon. Rev.* **2021**, *11*, 2100429.
111. Han, Z.; Ding, F.; Cai, Y.; Levy, U. Significantly enhanced second-harmonic generations with all-dielectric antenna array working in the quasi-bound states in the continuum and excited by linearly polarized plane waves. *Nanophotonics* **2021**, *10*, 1189–1196. [CrossRef]
112. Zhang, X.; He, L.; Gan, X.; Huang, X.; Du, Y.; Zhai, Z.; Li, Z.; Zheng, Y.; Chen, X.; Cai, Y.; et al. Quasi-Bound States in the Continuum Enhanced Second-Harmonic Generation in Thin-Film Lithium Niobate. *Laser Photonics Rev.* **2022**, *16*, 2200031. [CrossRef]
113. He, L.; Zhang, F.; Zhang, H.; Kong, L.-J.; Zhang, W.; Xu, X.; Zhang, X. Topology-Optimized Ultracompact All-Optical Logic Devices on Silicon Photonic Platforms. *ACS Photonics* **2022**, *9*, 597–604. [CrossRef]
114. Margalit, N.; Xiang, C.; Bowers, S.M.; Bjorlin, A.; Blum, R.; Bowers, J.E. Perspective on the future of silicon photonics and electronics. *Appl. Phys. Lett.* **2021**, *118*, 220501. [CrossRef]
115. Osgood, R.M.; Panoiu, N.C.; Dadap, J.I.; Liu, X.; Chen, X.; Hsieh, I.-W.; Dulkeith, E.; Green, W.M.J.; Vlasov, Y.A. Engineering nonlinearities in nanoscale optical systems: Physics and applications in dispersion-engineered silicon nanophotonic wires. *Adv. Opt. Photon.* **2009**, *1*, 162–235. [CrossRef]
116. Leuthold, J.; Koos, C.; Freude, W. Nonlinear silicon photonics. *Nat. Photonics* **2010**, *4*, 535–544. [CrossRef]
117. Nikdast, M.; Pasricha, S.; Nicolescu, G.; Seyedi, A.; Liang, D. *Silicon Photonics for High-Performance Computing and Beyond*; CRC Press: Boca Raton, FL, USA, 2021.
118. Saeidi, S.; Awan, K.M.; Sirbu, L.; Dolgaleva, K. Nonlinear photonics on-a-chip in III-V semiconductors: Quest for promising material candidates. *Appl. Opt.* **2017**, *56*, 5532–5541. [CrossRef] [PubMed]
119. Vyas, K.; Espinosa, D.H.G.; Hutama, D.; Jain, S.K.; Mahjoub, R.; Mobini, E.; Awan, K.M.; Lundeen, J.; Dolgaleva, K. Group III-V semiconductors as promising nonlinear integrated photonic platforms. *Adv. Phys. X* **2022**, *7*, 2097020. [CrossRef]
120. Mobini, E.; Espinosa, D.H.G.; Vyas, K.; Dolgaleva, K. AlGaAs Nonlinear Integrated Photonics. *Micromachines* **2022**, *13*, 991. [CrossRef]
121. Liu, X.; Sun, C.; Xiong, B.; Wang, L.; Wang, J.; Wang, Y.; Hao, Z.; Li, H.; Luo, Y.; Yan, J.; et al. Aluminum nitride-on-sapphire platform for integrated high-Q microresonators. *Opt. Express* **2017**, *25*, 587. [CrossRef]
122. Liu, J.; Weng, H.; Afridi, A.A.; Li, J.; Dai, J.; Ma, X.; Long, H.; Zhang, Y.; Lu, Q.; Donegan, J.F.; et al. Photolithography allows high-Q AlN microresonators for near octave-spanning frequency comb and harmonic generation. *Opt. Express* **2020**, *28*, 19270. [CrossRef]
123. Xiong, C.; Pernice, W.; Ryu, K.; Schuck, C.; Fong, K.Y.; Palacios, T.; Tang, H.X. Integrated GaN photonic circuits on silicon (100) for second harmonic generation. *Opt. Express* **2011**, *19*, 10462–10470. [CrossRef]
124. Lake, D.P.; Mitchell, M.; Jayakumar, H.; dos Santos, L.F.; Curic, D.; Barclay, P.E. Efficient telecom to visible wavelength conversion in doubly resonant gallium phosphide microdisks. *Appl. Phys. Lett.* **2016**, *108*, 031109. [CrossRef]
125. Poberaj, G.; Hu, H.; Sohler, W.; Günter, P. Lithium niobate on insulator (LNOI) for micro-photonic devices. *Laser Photonics Rev.* **2012**, *6*, 488–503. [CrossRef]
126. Luo, R.; Jiang, H.; Rogers, S.; Liang, H.; He, Y.; Lin, Q. On-chip second-harmonic generation and broadband parametric down-conversion in a lithium niobate microresonator. *Opt. Express* **2017**, *25*, 24531–24539. [CrossRef] [PubMed]
127. Bazzan, M.; Sada, C. Optical waveguides in lithium niobate: Recent developments and applications. *Appl. Phys. Rev.* **2015**, *2*, 40603. [CrossRef]
128. Boes, A.; Corcoran, B.; Chang, L.; Bowers, J.; Mitchell, A. Status and potential of lithium niobate on insulator (LNOI) for photonic integrated circuits. *Laser Photon. Rev.* **2018**, *12*, 1700256. [CrossRef]
129. Vazimali, M.G.; Fathpour, S. Applications of thin-film lithium niobate in nonlinear integrated photonics. *Adv. Photonics* **2022**, *4*, 034001. [CrossRef]

130. Kaur, P.; Boes, A.; Ren, G.; Nguyen, T.G.; Roelkens, G.; Mitchell, A. Hybrid and heterogeneous photonic integration. *APL Photonics* **2021**, *6*, 061102. [CrossRef]
131. Fathpour, S. Heterogeneous Nonlinear Integrated Photonics. *IEEE J. Quantum Electron.* **2018**, *54*, 6300716. [CrossRef]
132. Fathpour, S. Emerging heterogeneous integrated photonic platforms on silicon. *Nanophotonics* **2015**, *4*, 143–164. [CrossRef]
133. Xing, P.; Ma, D.; Ooi, K.J.; Choi, J.W.; Agarwal, A.M.; Tan, D. CMOS-compatible PECVD silicon carbide platform for linear and nonlinear optics. *ACS Photonics* **2019**, *6*, 1162–1167. [CrossRef]
134. Castelletto, S.; Rosa, L.; Johnson, B.C. Silicon carbide for novel quantum technologies devices. In *Advanced Silicon Carbide Devices and Processing*; Chapter 9; Saddow, S., La Via, F., Eds.; Intech Open: London, UK, 2015.
135. Guidry, M.A.; Yang, K.Y.; Lukin, D.M.; Markosyan, A.; Yang, J.; Fejer, M.M.; Vučković, J. Optical parametric oscillation in silicon carbide nanophotonics. *Optica* **2020**, *7*, 1139–1142. [CrossRef]
136. Cai, L.; Li, J.; Wang, R.; Li, Q. Octave-spanning microcomb generation in 4H-silicon-carbide-on-insulator photonics platform. *Photon. Res.* **2022**, *10*, 870–876. [CrossRef]
137. Carroll, L.; Lee, J.-S.; Scarcella, C.; Gradkowski, K.; Duperron, M.; Lu, H.; Zhao, Y.; Eason, C.; Morrissey, P.; Rensing, M. Photonic Packaging: Transforming Silicon Photonic Integrated Circuits into Photonic Devices. *Appl. Sci.* **2016**, *6*, 426. [CrossRef]
138. Sasikala, V.; Chitra, K. All optical switching and associated technologies: A review. *J. Opt.* **2018**, *47*, 307–317. [CrossRef]
139. Chhipa, M.K.; Madhav, B.T.P.; Suthar, B. An all-optical ultracompact microring-resonator-based optical switch. *J. Comput. Electron.* **2021**, *20*, 419–425. [CrossRef]
140. Kuramochi, E.; Nozaki, K.; Shinya, A.; Takeda, K.; Sato, T.; Matsuo, S.; Taniyama, H.; Sumikura, H.; Notomj, M. Large-scale integration of wavelength-addressable all-optical memories on a photonic crystal chip. *Nat. Photonics* **2014**, *8*, 474–481. [CrossRef]
141. Colman, P.; Lunnemann, P.; Yu, Y.; Mørk, J. Ultrafast coherent dynamics of a photonic crystal all-optical switch. *Phys. Rev. Lett.* **2016**, *117*, 233901. [CrossRef] [PubMed]
142. Takiguchi, M.; Takemura, N.; Tateno, K.; Nozaki, K.; Sasaki, S.; Sergent, S.; Kuramochi, E.; Wasawo, T.; Yokoo, A.; Shinya, A.; et al. All-optical InAsP/InP nanowire switches integrated in a Si photonic crystal. *ACS Photonics* **2020**, *7*, 1016–1021. [CrossRef]
143. Ono, M.; Hata, M.; Tsunekawa, M.; Nozaki, K.; Sumikura, H.; Chiba, H.; Notomi, M. Ultrafast and energy efficient all-optical switching with graphene-loaded deep-subwavelength plasmonic waveguides. *Nat. Photonics* **2020**, *14*, 37–43. [CrossRef]
144. Keyes, R.W. Optical logic-in the light of computer technology. *Opt. Acta Int. J. Opt.* **1985**, *32*, 525–535. [CrossRef]
145. Wang, C.; Li, Z.Y. Ultracompact linear on-chip silicon optical logic gates with phase insensitivity. *EPL* **2013**, *103*, 64001. [CrossRef]
146. Xiong, M.; Lei, L.; Ding, Y.; Huang, B.; Ou, H.; Peucheret, C.; Zhang, X. All-optical 10 Gb/s AND logic gate in a silicon microring resonator. *Opt. Express* **2013**, *21*, 25772–25779. [CrossRef]
147. Yuan, R.H.; Wang, C.; Li, Z. Design of on-chip optical logic gates in 2D silicon photonic crystal slab. *Opt. Rev.* **2020**, *27*, 277–282. [CrossRef]
148. Zarei, S.; Khavasi, A. Realization of optical logic gates using on-chip diffractive optical neural networks. *Sci. Rep.* **2022**, *12*, 15747. [CrossRef] [PubMed]
149. Miller, D.A. Are optical transistors the logical next step? *Nat. Photon.* **2010**, *4*, 3–5. [CrossRef]
150. Gibbs, H.M. *Optical Bistability: Controlling Light with Light*; Academic Press: Cambridge, MA, USA, 1985.
151. Hwang, J.; Pototschnig, M.; Lettow, R.; Zumofen, G.; Renn, A.; Götzinger, S.; Sandoghdar, V. A single-molecule optical transistor. *Nature* **2009**, *460*, 76–80. [CrossRef]
152. Fushman, I.; Englund, D.; Faraon, A.; Stoltz, N.; Petroff, P.; Vuckovic, J. Controlled Phase Shifts with a Single Quantum Dot. *Science* **2008**, *320*, 769–772. [CrossRef] [PubMed]
153. Foster, M.A.; Turner, A.C.; Sharping, J.E.; Schmidt, B.S.; Lipson, M.; Gaeta, A.L. Broad-band optical parametric gain on a silicon photonic chip. *Nature* **2006**, *441*, 960–963. [CrossRef] [PubMed]
154. Liu, X.; Osgood, R.M.; Vlasov, J.A.; Green, W.M.J. Mid-infrared optical parametric amplifier using silicon nanophotonic waveguides. *Nat. Photon.* **2010**, *4*, 557–560. [CrossRef]
155. Ooi, K.J.A.; Ng, D.K.T.; Wang, T.; Chee, A.K.L.; Ng, S.K.; Wang, Q.; Ang, L.K.; Agarwal, A.M.; Kimerling, L.C.; Tan, D.T.H. Pushing the limits of CMOS optical parametric amplifiers with USRN:Si_7N_3 above the two-photon absorption edge. *Nat. Commun.* **2017**, *8*, 13878. [CrossRef]
156. Davenport, M.L.; Skendzic, S.; Volet, N.; Bowers, J.E. Heterogeneous silicon/InP semiconductor optical amplifiers with high gain and high saturation power. In Proceedings of the Conference on Lasers and Electro-Optics (CLEO), San Jose, CA, USA, 5–10 June 2016; pp. 1–2. [CrossRef]
157. Espinola, R.L.; Dadap, J.I.; Osgood, R.M.; McNab, S.J.; Vlasov, Y. C-band wavelength conversion in silicon photonic wire waveguides. *Opt. Express* **2005**, *13*, 4341–4349. [CrossRef]
158. Foster, M.A.; Turner, A.C.; Salem, R.; Lipson, M.; Gaeta, A.L. Broad-band continuous-wave parametric wavelength conversion in silicon nanowaveguides. *Opt. Express* **2007**, *15*, 12949–12958. [CrossRef]
159. Ahmad, R.; Rochette, M. High efficiency and ultra-broadband optical parametric four-wave mixing in chalcogenide-PMMA hybrid microwires. *Opt. Express* **2012**, *20*, 9572–9580. [CrossRef] [PubMed]
160. Zhang, L.; Yue, Y.; Beausoleil, R.G.; Willner, A.E. Flattened dispersion in silicon slot waveguides. *Opt. Express* **2010**, *18*, 20529–20534. [CrossRef]
161. Bao, C.; Yan, Y.; Zhang, L.; Yue, Y.; Ahmed, N.; Agarwal, A.M.; Kimerling, L.C.; Michel, J.; Willner, A.E. Increased bandwidth with flattened and low dispersion in a horizontal double-slot silicon waveguide. *J. Opt. Soc. Am. B* **2015**, *32*, 26–30. [CrossRef]

162. Guo, K.; Lin, L.; Christensen, J.B.; Christensen, E.N.; Shi, X.; Ding, Y.; Rottwitt, K.; Ou, H. Broadband wavelength conversion in a silicon vertical-dual-slot waveguide. *Opt. Express* **2017**, *25*, 32964–32971. [CrossRef]
163. Kernetzky, T.; Ronniger, G.; Höfler, U.; Zimmermann, L.; Hanik, N. Numerical Optimization and CW Measurements of SOI Waveguides for Ultra-Broadband C-to-O-Band Conversion. In Proceedings of the European Conference on Optical Communication (ECOC), Bordeaux, France, 13–16 September 2021; pp. 1–4.
164. Hanik, N.; Kernetzky, T.; Jia, Y.; Höfler, U.; Freund, R.; Schubert, C.; Isaac Sackey, I.; Ronniger, G.; Zimmerman, L. Ultra-Broadband Optical Wavelength-Conversion using Nonlinear Multi-Modal Optical Waveguides. In Proceedings of the 13th International Symposium on Communication Systems, Networks and Digital Signal Processing (CSNDSP), Porto, Portugal, 20–22 July 2022; pp. 832–835.
165. Marty, G.; Combrié, S.; De Rossi, A.; Raineri, F. Hybrid InGaP nanobeam on silicon photonics for efficient four wave mixing. *APL Photonics* **2019**, *4*, 120801. [CrossRef]
166. Pu, M.; Hu, H.; Ottaviano, L.; Semenova, E.; Vukovic, D.; Oxenløwe, L.K.; Yvind, K. Ultra-efficient and broadband nonlinear AlGaAs-on-insulator chip for low-power optical signal processing. *Laser Photonics Rev.* **2018**, *12*, 1800111. [CrossRef]
167. Ong, J.R.; Kumar, R.; Mookherjea, S. Triply resonant four-wave mixing in silicon-coupled resonator microring waveguides. *Opt. Lett.* **2014**, *39*, 5653–5656. [CrossRef]
168. Wu, J.; Yang, Y.; Qu, Y.; Jia, L.; Zhang, Y.; Xu, X.; Chu, S.Y.; Little, B.E.; Morandotti, R.; Jia, B.; et al. 2D layered graphene oxide films integrated with micro-ring resonators for enhanced nonlinear optics. *Small* **2020**, *16*, 1906563. [CrossRef]
169. Ferrera, M.; Duchesne, D.; Razzari, L.; Peccianti, M.; Morandotti, R.; Cheben, P.; Janz, S.; Xu, D.-X.; Little, B.E.; Chu, S.; et al. Low power four wave mixing in an integrated, micro-ring resonator with Q = 1.2 million. *Opt. Express* **2009**, *17*, 14098–14103. [CrossRef]
170. Ji, M.; Cai, H.; Deng, L.; Huang, Y.; Huang, Q.; Xia, J.; Li, Z.; Yu, J.; Wang, Y. Enhanced parametric frequency conversion in a compact silicon-graphene microring resonator. *Opt. Express* **2015**, *23*, 18679–18685. [CrossRef]
171. Winzer, P.; Essiambre, R. Advanced Modulation Formats for High-Capacity Optical Transport Network. *J. Light. Technol.* **2010**, *24*, 4711–4728. [CrossRef]
172. Li, C.; Gui, C.; Xiao, X.; Yang, Q.; Yu, S.; Wang, J. On-chip all-optical wavelength conversion of multicarrier, multilevel modulation (OFDM m-QAM) signals using a silicon waveguide. *Opt. Lett.* **2014**, *39*, 4583–4586. [CrossRef] [PubMed]
173. Long, Y.; Liu, J.; Hu, X.; Wang, A.; Zhou, L.; Zou, K.; Zhu, Y.; Zhang, F.; Wang, J. All-optical multi-channel wavelength conversion of Nyquist 16 QAM signal using a silicon waveguide. *Opt. Lett.* **2015**, *40*, 5475–5478. [CrossRef]
174. Adams, R.; Spasojevic, M.; Chagnon, M.; Malekiha, M.; Li, J.; Plant, D.V.; Chen, L.R. Wavelength conversion of 28 GBaud 16-QAM signals based on four-wave mixing in a silicon nanowire. *Opt. Express* **2014**, *22*, 4083–4090. [CrossRef] [PubMed]
175. Ta'eed, V.G.; Lamont, M.R.E.; Moss, D.J.; Eggleton, B.J.; Choi, D.Y.; Madden, S.; Luther-Davies, B. All optical wavelength conversion via cross phase modulation in chalcogenide glass rib waveguides. *Opt. Express* **2006**, *14*, 11242–11247. [CrossRef]
176. Astar, W.; Driscoll, J.B.; Liu, X.; Dadap, J.I.; Green, W.M.J.; Vlasov, Y.A.; Carter, G.M.; Osgood, R.M. Tunable wavelength conversion by XPM in a silicon nanowire, and the potential for XPM-multicasting. *J. Light. Technol.* **2010**, *28*, 2499–2511. [CrossRef]
177. Huang, Z.; Cao, T.; Chen, L.; Yu, Y.; Zhang, X. Monolithic integrated chip with SOA and tunable DI for multichannel all-optical signal processing. *IEEE Photonics J.* **2018**, *10*, 6600709. [CrossRef]
178. Chow, C.; Liu, Y. Nonlinear Photonic Signal Processing Subsystems and Applications. In *Advances in Lasers and Electro Optics*; IntechOpen: Rijeka, Croatia, 2010. [CrossRef]
179. Matsumoto, M. Fiber-Based All-Optical Signal Regeneration. *IEEE J. Sel. Top. Quantum Electron.* **2012**, *18*, 738–752. [CrossRef]
180. Ribeiro, N.S.; Gallep, C.M.; Conforti, E. Wavelength Conversion and 2R-Regeneration in Simple Schemes with Semiconductor Optical Amplifiers. In *Advances in Lasers and Electro Optics*; IntechOpen: Rijeka, Croatia, 2010. [CrossRef]
181. Vercesi, V.; Porzi, C.; Contestabile, G.; Bogoni, A. Polarization-Independent All-Optical Regenerator for DPSK Data. *Photonics* **2014**, *1*, 154–161. [CrossRef]
182. Sobhanan, A.; Anthur, A.; O'Duill, S.; Pelusi, M.; Namiki, S.; Barry, L.; Venkitesh, D.; Agrawal, G.P. Semiconductor optical amplifiers: Recent advances and applications. *Adv. Opt. Photon.* **2022**, *14*, 571–651. [CrossRef]
183. Ta'eed, V.G.; Shokooh-Saremi, M.; Fu, L.; Moss, D.J.; Rochette, M.; Eggleton, B.J.; Ruan, Y.; Luther-Davies, B. Self-phase modulation-based integrated optical regeneration in chalcogenide waveguides. *IEEE J. Sel. Top. Quantum Electron.* **2006**, *12*, 360–370. [CrossRef]
184. Salem, R.; Foster, M.A.; Geraghty, D.F.; Gaeta, A.L.; Turner, A.C.; Lipson, M. Integrated Optical Regenerator on a Silicon Chip. In Proceedings of the Conference on Lasers and Electro-Optics (CLEO), Baltimore, MD, USA, 6–11 May 2007; pp. 1–2. [CrossRef]
185. Geng, Y.; Wu, B.; Wen, F.; Zhou, H.; Zhou, X.; Qiu, K. Clock-pump four-wave-mixing-based multichannel all-optical regeneration in silicon waveguide. *Opt. Eng.* **2017**, *56*, 117102. [CrossRef]
186. Gajda, A.; Zimmermann, L.; Jazayerifar, M.; Winzer, G.; Tian, H.; Elschner, R.; Richter, T.; Schubert, C.; Tillack, B.; Petermann, K. Highly efficient CW parametric conversion at 1550 nm in SOI waveguides by reverse biased p-i-n junction. *Opt. Express* **2012**, *20*, 13100–13107. [CrossRef] [PubMed]
187. Da Ros, F.; Vukovic, D.; Gajda, A.; Dalgaard, K.; Zimmermann, L.; Tillack, B.; Galili, M.; Petermann, K.; Peucheret, C. Phase regeneration of DPSK signals in a silicon waveguide with reverse-biased p-i-n junction. *Opt. Express* **2014**, *22*, 5029–5036. [CrossRef] [PubMed]

188. Wen, H.S.; Cui, J.B.; Zhou, H.; Chen, Y.F.; Jin, Y.; Xu, B.R.; Zhai, K.P.; Sun, J.Z.; Guo, Y.Y.; Wu, R.Y.; et al. 100 Gb/s NRZ OOK signal regeneration using four-wave mixing in a silicon waveguide with reverse-biased p-i-n junction. *Opt. Express* **2022**, *30*, 38077–38094. [CrossRef] [PubMed]
189. Liu, S.; Lee, K.J.; Parmigiani, F.; Kakande, J.; Gallo, K.; Petropoulos, P.; Richardson, D.J. Phase-regenerative wavelength conversion in periodically poled lithium niobate waveguides. *Opt. Express* **2011**, *19*, 11705–11715. [CrossRef]
190. Bogoni, A.; Wu, X.; Nuccio, S.R.; Willner, A.E. 640 Gb/s All-Optical Regenerator Based on a Periodically Poled Lithium Niobate Waveguide. *J. Light. Technol.* **2012**, *30*, 1829–1834. [CrossRef]
191. Cao, Y.; Ziyadi, M.; Mohajerin-Ariaei, A.; Almaiman, A.; Liao, P.; Bao, C.; Alishahi, F.; Falahpour, A.; Shamee, B.; Yang, J.-Y.; et al. Reconfigurable optical inter-channel interference mitigation for spectrally overlapped QPSK signals using nonlinear wave mixing in cascaded PPLN waveguides. *Opt. Lett.* **2016**, *41*, 3233–3236. [CrossRef]
192. Saini, T.S.; Sinha, R.K. Mid-infrared supercontinuum generation in soft-glass specialty optical fibers: A review. *Prog. Quant. Electron.* **2021**, *78*, 100342. [CrossRef]
193. Hsieh, I.-W.; Chen, X.; Liu, X.; Dadap, J.I.; Panoiu, N.C.; Chou, C.-Y.; Xia, F.; Green, W.M.; Vlasov, Y.A.; Osgood, R.M. Supercontinuum generation in silicon photonic wires. *Opt. Express* **2007**, *15*, 15242–15249. [CrossRef]
194. Krasavin, A.; Ginzburg, P.; Wurtz, G.; Zayats, A.V. Nonlocality-driven supercontinuum white light generation in plasmonic nanostructures. *Nat. Commun.* **2016**, *7*, 11497. [CrossRef] [PubMed]
195. Del'Haye, P.; Schliesser, A.; Arcizet, O.; Wilken, T.; Holzwarth, R.; Kippenberg, T.J. Optical frequency comb generation from a monolithic microresonator. *Nature* **2007**, *450*, 1214–1217. [CrossRef] [PubMed]
196. Kippenberg, T.J.; Holzwarth, R.; Diddams, S.A. Microresonator-based optical frequency combs. *Science* **2011**, *332*, 555–559. [CrossRef] [PubMed]
197. Ferdous, F.; Miao, H.; Leaird, D.E.; Srinivasan, K.; Wang, J.; Chen, L.; Varghese, L.T.; Weiner, A.M. Spectral line-by-line pulse shaping of on-chip microresonator frequency combs. *Nat. Photon.* **2011**, *5*, 770–776. [CrossRef]
198. Razzari, L.; Duchesne, D.; Ferrera, M.; Morandotti, R.; Chu, S.; Little, B.; Moss, D. CMOS-compatible integrated optical hyper-parametric oscillator. *Nat. Photon.* **2010**, *4*, 41–45. [CrossRef]
199. Zhang, M.; Buscaino, B.; Wang, C.; Shams-Ansari, A.; Reimer, C.; Zhu, R.; Kahn, J.M.; Lončar, M. Broadband electro-optic frequency comb generation in a lithium niobate microring resonator. *Nature* **2019**, *568*, 373–377. [CrossRef]
200. Hu, Y.; Yu, M.; Buscaino, B.; Sinclair, N.; Zhu, D.; Cheng, R.; Shams-Ansari, A.; Shao, L.; Zhang, M.; Kahn, J.M.; et al. High-efficiency and broadband on-chip electro-optic frequency comb generators. *Nat. Photon.* **2022**, *16*, 679–685. [CrossRef]
201. Levy, J.S.; Saha, K.; Okawachi, Y.; Foster, M.; Gaeta, A.; Lipson, M. High-performance silicon-nitride-based multiple-wavelength source. *IEEE Photon. Technol. Lett.* **2012**, *24*, 1375–1377. [CrossRef]
202. Wang, P.-H.; Ferdous, F.; Miao, H.; Wang, J.; Leaird, D.E.; Srinivasan, K.; Chen, L.; Aksyuk, V.; Weiner, A.M. Observation of correlation between route to formation, coherence, noise, and communication performance of Kerr combs. *Opt. Express* **2012**, *20*, 29284–29295. [CrossRef]
203. Pfeifle, J.; Brasch, V.; Lauermann, M.; Yu, Y.; Wegner, D.; Herr, T.; Hartinger, K.; Schindler, P.; Li, J.; Hillerkuss, D.; et al. Coherent terabit communications with microresonator Kerr frequency combs. *Nat. Photon.* **2014**, *8*, 375–380. [CrossRef]
204. Coillet, A.; Chembo, Y.K. On the robustness of phase-locking in Kerr optical frequency combs. *Opt. Lett.* **2014**, *39*, 1529–1533. [CrossRef] [PubMed]
205. Jensen, J.S.; Sigmund, O. Topology optimization for nano- photonics. *Laser Photonics Rev.* **2011**, *5*, 308–321. [CrossRef]
206. Guo, K.; Yang, L.; Shi, X.; Liu, X.; Cao, Y.; Zhang, J.; Wang, X.; Yang, J.; Ou, H.; Zhao, Y. Nonclassical Optical Bistability and Resonance-Locked Regime of Photon-Pair Sources Using Silicon Microring Resonator. *Phys. Rev. Appl.* **2019**, *11*, 034007. [CrossRef]
207. Shi, X.; Guo, K.; Christensen, J.B.; Castaneda, M.A.U.; Liu, X.; Ou, H.; Karsten, R. Multichannel Photon-Pair Generation with Strong and Uniform Spectral Correlation in a Silicon Microring Resonator. *Phys. Rev. Appl.* **2019**, *12*, 034053. [CrossRef]
208. Lukin, D.M.; Dory, C.; Guidry, M.A.; Yang, K.Y.; Mishra, S.D.; Trivedi, R.; Radulaski, M.; Sun, S.; Vercruysse, D.; Ahn, G.H. 4H-silicon-carbide-on-insulator for integrated quantum and nonlinear photonics. *Nat. Photonics* **2020**, *14*, 330–334. [CrossRef]
209. Becher, C.; Gao, W.; Kar, S.; Marciniak, C.D.; Monz, T.; Bartholomew, J.G.; Goldner, P.; Loh, H.; Marcellina, E.; Eng, J.G.K. 2023 roadmap for materials for quantum technologies. *Mater. Quantum. Technol.* **2023**, *3*, 012501. [CrossRef]
210. Babin, C.; Stöhr, R.; Morioka, N.; Linkewitz, T.; Steidl, T.; Wörnle, R.; Liu, D.; Hesselmeier, E.; Vorobyov, V.; Denisenko, A. Fabrication and nanophotonic waveguide integration of silicon carbide colour centres with preserved spin-optical coherence. *Nat. Mater.* **2022**, *21*, 67–73. [CrossRef]
211. Guidry, M.A.; Lukin, D.M.; Yang, K.Y.; Trivedi, R.; Vučković, J. Quantum optics of soliton microcombs. *Nat. Photon.* **2022**, *16*, 52–58. [CrossRef]
212. Nozaki, R.; Sato, Y.; Shimada, Y.; Suzuki, T.; Yasuno, K.; Ikai, Y.; Ueda, W.; Shimizu, K.; Yukawa, E.; Sanakaet, K. Enhancing the stimulated emission of polarization-entangled photons using passive optical components. *Phys. Rev. A* **2023**, *107*, 023707. [CrossRef]

micromachines

MDPI

Article

Meta-Atoms with Toroidal Topology for Strongly Resonant Responses

Odysseas Tsilipakos [1,*], Zacharias Viskadourakis [2], Anna C. Tasolamprou [2,3], Dimitrios C. Zografopoulos [4], Maria Kafesaki [2,5], George Kenanakis [2,*] and Eleftherios N. Economou [2,6]

1 Theoretical and Physical Chemistry Institute, National Hellenic Research Foundation, GR-11635 Athens, Greece
2 Institute of Electronic Structure and Laser, Foundation for Research and Technology-Hellas, GR-70013 Heraklion, Crete, Greece
3 Section of Electronic Physics and Systems, Department of Physics, National and Kapodistrian University of Athens, GR-15784 Athens, Greece
4 Consiglio Nazionale delle Ricerche, Istituto per la Microelettronica e Microsistemi (CNR-IMM), 00133 Rome, Italy
5 Department of Materials Science Technology, University of Crete, GR-70013 Heraklion, Crete, Greece
6 Department of Physics, University of Crete, GR-70013 Heraklion, Crete, Greece
* Correspondence: otsilipakos@eie.gr (O.T.); gkenanak@iesl.forth.gr (G.K.)

Abstract: A conductive meta-atom of toroidal topology is studied both theoretically and experimentally, demonstrating a sharp and highly controllable resonant response. Simulations are performed both for a free-space periodic metasurface and a pair of meta-atoms inserted within a rectangular metallic waveguide. A quasi-dark state with controllable radiative coupling is supported, allowing to tune the linewidth (quality factor) and lineshape of the supported resonance via the appropriate geometric parameters. By conducting a rigorous multipole analysis, we find that despite the strong toroidal dipole moment, it is the residual electric dipole moment that dictates the electromagnetic response. Subsequently, the structure is fabricated with 3D printing and coated with silver paste. Importantly, the structure is planar, consists of a single metallization layer and does not require a substrate when neighboring meta-atoms are touching, resulting in a practical, thin and potentially low-loss system. Measurements are performed in the 5 GHz regime with a vector network analyzer and a good agreement with simulations is demonstrated.

Keywords: toroidal dipole; metasurfaces; multipole expansion; broken symmetry; 3D printing; microwaves

Citation: Tsilipakos, O.; Viskadourakis, Z.; Tasolamprou, A.C.; Zografopoulos, D.C.; Kafesaki, M.; Kenanakis, G.; Economou, E.N. Meta-Atoms with Toroidal Topology for Strongly Resonant Responses. *Micromachines* **2023**, *14*, 468. https://doi.org/10.3390/mi14020468

Academic Editors: Luigi Sirleto and Giancarlo C. Righini

Received: 24 January 2023
Revised: 9 February 2023
Accepted: 15 February 2023
Published: 17 February 2023

1. Introduction

Toroidal multipoles are a class of fundamental electromagnetic excitations that complement the more familiar electric and magnetic multipole families [1–5]. The toroidal dipole, the lowest order member of the toroidal family, first considered by Zel'dovich [6], originates from conduction or displacement currents circulating on a torus along the meridians, producing a closed loop of magnetic field circulation. Although it is distinctly different in its construction from the electric dipole (**p**, a simple separation of electric charges), it emits radiation with the same angular momentum and parity properties as the electric dipole. As a result, the two cannot be easily discerned by observing the far field radiation.

Recently, metamaterials have provided fertile ground for the observation of the toroidal dipole and higher order multipoles through the ability to judiciously shape the meta-atom/meta-molecule geometry in the unit cell. The first demonstration was published in 2010 by Kaelberer et al. [7]. Since then, a broad range of structures have been studied, based on both dielectric materials (displacement currents) [8–12] and metals (conduction currents) [13–15]. In particular, planar and, ideally, single-metallization-layer structures are favorable for easier fabrication [16–20]. An important point of attention is the possibility to attain a quasi-dark (almost non-radiating) state, which is termed a "dynamic anapole", and results from the near-destructive interference of the toroidal dipole, **T**, with the electric dipole, **p** [21]. As perfect destructive interference is approached through $\mathbf{p} + ik\mathbf{T} \to 0$, one can obtain arbitrarily high radiation quality factors, Q_{rad} (provided that contribution from other multipoles is suppressed). The total quality factor, Q_{tot}, can be very high as well, provided that a material system with low absorption is used [22]. This concept is related to the topics of trapped [23], broken-symmetry [24,25] and BIC (bound states in the continuum) [26] resonances. In such cases, sharply resonant responses and narrow spectral linewidths can be achieved, which are particularly useful for functionalities requiring enhanced local fields. Examples of such applications with metasurfaces supporting anapole states are, for instance, nonlinear effects [27–29], lasers [30] and (bio)sensing [31,32].

In recent years, a broad range of metasurfaces (single-layer metamaterial structures) have been proposed for demonstrating toroidal dipole- and anapole-based phenomena [33]. In most cases, the designs are based on a toroidal topology. Here, we adopt such a design aimed at supporting an anapole state [13]; it consists of a unit cell geometry meant to enhance the toroidal character, it is planar and it requires a single metallization layer that can be free-standing (without a substrate), which are important advantageous traits regarding fabrication and practical applications. We show that, indeed, a quasi-dark resonant state can be supported, which leads to an arbitrarily narrow linewidth, limited only by the resistive quality factor. Varying the appropriate geometric parameters, both the linewidth *and* lineshape of the resonance can be controlled, offering valuable degrees of freedom in shaping the spectral response. By performing a rigorous multipole expansion, we find that the toroidal dipole moment in the structure is strong. However, the corresponding far-field scattering is cancelled exactly by the magnetic quadrupole, $\mathbf{Q}^m$. Thus, a quasi-dark resonance is achieved by controlling the residual electric dipole moment through asymmetry in the unit cell geometry (central vs. outer gaps) and not by satisfying the anapole condition. Subsequently, we experimentally verify our calculations through measurements within a rectangular metallic waveguide setup. More specifically, toroidal meta-atom samples, fabricated by a very low-cost technique based on 3D printing and subsequent metallization with a conductive paste, are loaded in the waveguide cross-section, and the S parameters (reflection and transmission coefficients) are measured with a vector network analyzer.

One goal of our work is to demonstrate a practical, easily realizable meta-atom geometry that provides freedom in shaping the electromagnetic response in terms of both linewidth and lineshape. Importantly, this study also clearly demonstrates that an interpretation suggested by the unit cell geometry (its topology) does not necessarily lead directly to the physical interpretation of the phenomenon taking place. Instead, in order to reveal the actual physical behavior, a careful analysis should be performed in all cases. Thus, our work can act as a reference for the mindful discussion of the physical mechanisms behind the resonant response of toroidal metamaterials (but not limited exclusively to this special class of structures). In this paper, the physical evidence is provided by the rigorous multipole expansion and is supported by the experimental verification of the spectral response.

2. Materials and Methods

2.1. Metasurface Full-Wave Simulations and Multipole Expansion

The metasurfaces studied in this work are composed of a unit cell made of conductive material. In order to study the effect of finite conductivity and increasing resistive loss, conductivities in the range of 10^3–10^7 S/m have been assumed. Full-wave simulations are conducted with the finite element method (FEM) using the commercial software COMSOL Multiphysics. They concern scattering and guided-wave simulations with a CW excitation and eigenvalue simulations without excitation.

Two types of metasurfaces are considered. The first is a free-space metasurface, where the excitation is a normally incident plane wave with E_y polarization (Figure 1a). In this case, a single unit cell is simulated with periodic boundary conditions in the x- and y-axes.

In the second case, two periods of the metasurface along the x-axis are inserted within a rectangular metallic waveguide (Figure 1b), since the aspect ratio of the waveguide cross-section ($a \times b$) is approximately 2:1. The structure is excited with the TE_{10} mode of the waveguide, exhibiting the characteristic field distribution $E_y(x) \sim \sin(\pi x/a)$. The waveguide walls are modeled as perfect electric conducting (PEC) walls.

The multipole expansion is performed for the free-space geometry by using the expressions found in ref. [3]. The field scattered by the terms of the multipole expansion can be used for reconstructing the reflected and transmitted field [3,34,35]. The correctness of the calculated multipole moments is checked through the agreement between the direct and reconstructed reflection/transmission. This way, we can also specify where to truncate the multipole expansion without sacrificing reconstruction accuracy.

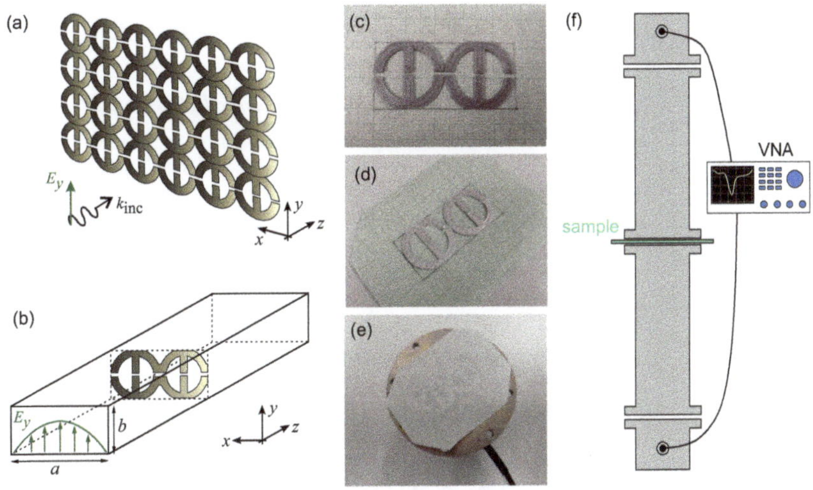

Figure 1. (**a**) Periodic metasurface illuminated with a normally incident plane wave of E_y polarization. (**b**) Two unit cells fitted in a rectangular metallic waveguide. (**c**) 3D-printed unit cells using a PLA filament. (**d**) Structure after coating with a conductive silver paste. (**e**) Test fitting of the structure under study on the flange of the waveguide-to-coax adapter. The unit cells are lightweight and can be simply glued on a piece of paper and inserted at any junction between waveguide segments. (**f**) Measurement setup including a vector network analyzer allowing to measure reflection (S_{11}) and transmission (S_{21}) coefficients.

2.2. Meta-Atom Fabrication with 3D Printing

For a fast and cost-effective fabrication of the metasurface under study, 3D printing technology was employed. A commercial fused filament fabrication (FFF) 3D printer was utilized (MakerBot Replicator 2x, New York, NY, USA). A common polylactic acid (PLA) filament was used as a spool material for building the unit cell geometry (Figure 1c). Details regarding the printing process and conditions can be found in refs. [36,37]. Free-standing meta-atoms, such as those depicted in Figure 1c, were successfully fabricated. The fabricated metasurfaces exhibit negligible electrical conductivity, since PLA is an insulator. Therefore, the meta-atoms were subsequently coated with a thin (~100 μm) layer of conductive silver paste, as shown in Figure 1d. The silver epoxy exhibits electrical conductivity as high as 10^4–10^5 S/m. We have deduced these values in earlier works [36,37] by comparing experimental results with simulations. Such a meta-atom configuration, in which an insulating core is coated with an adequately thick metallic paint, has been successfully employed at microwave frequencies [36,38].

2.3. Electromagnetic Characterization with Rectangular Waveguide Setup

The electromagnetic response of the investigated structure was experimentally confirmed through microwave measurements. To obtain a well-defined and highly reproducible measurement environment, which is not susceptible to external perturbations, we employed a rectangular metallic waveguide setup. More specifically, we fitted 2×1 unit cells inside the waveguide to fill the cross-section (aspect ratio 2:1). A test fitting of the two-unit-cell structure within the cross-section of a waveguide-to-coax adapter is shown in Figure 1e.

Measurements were conducted in the vicinity of 5 GHz by using an HP 8722ES vector network analyzer (Agilent Technologies Inc., Santa Clara, CA, USA). WR-187 rectangular waveguides (cross-section: 47.55 mm × 22.2 mm) were used, able to cover the frequency range of 3.95–5.85 GHz. The measurement setup is depicted in Figure 1f and allows for measuring both the reflection (S_{11}) and transmission (S_{21}) coefficients.

3. Results

3.1. Free-Space Metasurface with Controllable Strongly Resonant Response

The meta-atom considered in this work is depicted in Figure 2a. Its geometry is selected in such a way so as to lead to a circulation of induced magnetization giving rise to a toroidal dipole moment [13]; we will come back to this while discussing Figure 2c. The gap in the inner branch is denoted by g_1 and the gaps in the outer branches are denoted by g_2. For the inner and outer radii, $R_{\rm inn}$ and $R_{\rm out}$, it holds $R_{\rm inn} = R_{\rm out} - w$, where w is the width of the central and outer branches. The lattice constant (pitch) is denoted by a for both x and y axes (square periodicity). When $R_{\rm out} > a/2$, the neighboring meta-atoms are touching; in this case, the conductive meta-atoms can form an interconnected metasurface and no substrate is strictly required. This can be important for avoiding additional resistive loss from the substrate and avoiding excess thickness. Preserving the vertical symmetry of the structure has been also shown to lead to higher modulation depth in transmission [39].

First, we examined a periodic metasurface. The reflection/transmission/absorption power coefficients (R/T/A) for plane wave scattering under normal incidence (E_y polarization) are depicted in Figure 2b. The dimensions of the specific example are $a = 15$ mm, $R_{\rm out} = 7$ mm, $R_{\rm inn} = 5$ mm ($w = 2$ mm), $g_1 = 0.5$ mm and $g_2 = 0.5$ mm. The height of the conductive meta-atom is $h = 1$ mm. In this case, the meta-atoms are not touching. We observed a Fano lineshape associated with the excitation of a resonance at ~8.71 GHz and its interference with a non-resonant electric polarizability background. Even if the gaps are closed and the meta-atom is non-resonant, the conductive unit cell would still produce significant reflection. The conductivity adopted in these simulations is high ($\sigma = 10^7$ S/m), leading to low absorption.

We subsequently performed an eigenvalue analysis to identify the nature of the supported resonance. The mode profile is depicted in Figure 2c; the color corresponds to the E_y component (real part) and the arrows to the magnetic field. The magnetic field circles around the central branch and this should produce a toroidal dipole moment along the y axis (T_y). Looking at the fields in the gaps, one realizes that the central electric dipole moment is opposed by the two contributions of the outer gaps; when the opposing p_y contributions compensate each other exactly, no coupling via the electric dipole moment is possible. This balance can be used to tune the radiative strength of an electrically coupled meta-atom [40].

To further study the physics of the scattering process, we performed a multipole expansion based on the induced conduction currents on the meta-atom using the expressions found in ref. [3], and we focused on the multipole moments that produce scattered fields toward the z-axis with E_y (H_x) polarization. The eight leading terms of the expansion that satisfy the above criterion are p_y, m_x, T_y, Q^m_{xz}, Q^e_{yz}, Q^T_{yz}, O^m_{xzz} and O^e_{yzz}. The results are depicted in Figure 2d. It can be seen that the rationale behind the meta-atom geometry has indeed resulted in a strong toroidal dipole moment contribution in the spectral neighborhood of the resonance frequency. However, the magnetic quadrupole moment is equally dominant. In fact, their scattered fields cancel out, since on resonance it holds exactly $E_{\rm sca}^{T_y} = -E_{\rm sca}^{Q_{xz}^m}$. As a result, the response is dictated by the residual electric dipole moment, p_y, which is the third most dominant contribution. This can be further verified in Figure 2e. We first confirm the validity of the multipole expansion by reconstructing the reflection coefficient. The agreement is excellent and shows that using the eight leading terms up to the electric octupole is more than adequate in this case. Importantly, if we use only the field scattered by the electric dipole moment, we obtain a fairly accurate description of the scattering process, especially near the resonance, where there is mutual cancellation of the dominant T_y and Q_{xz}^m terms.

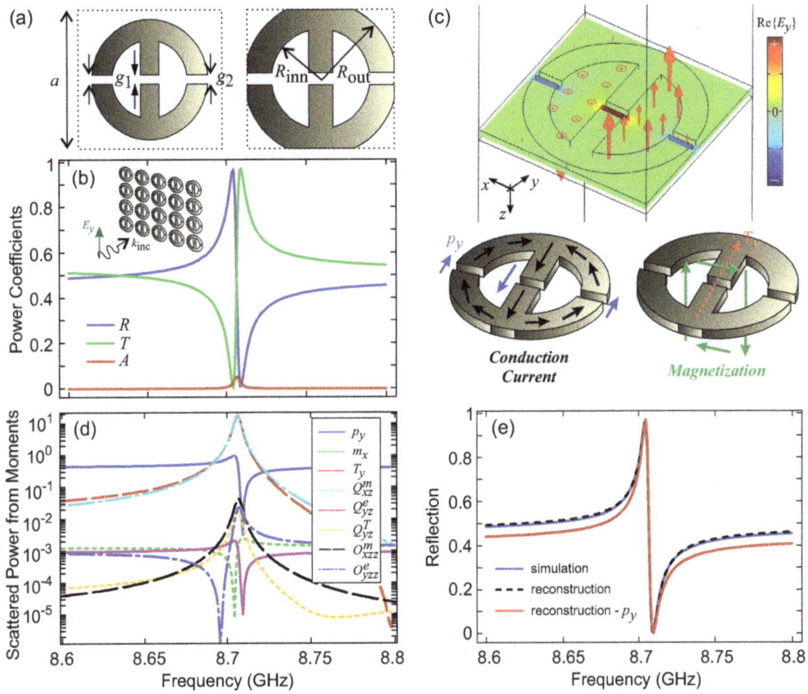

Figure 2. (**a**) Meta-atom geometry. When $R_{\rm out} > a/2$, the neighboring meta-atoms are touching; in this case, no substrate is necessary. (**b**) R/T/A power coefficients for plane-wave scattering by the periodic metasurface under normal incidence (E_y polarization). The dimensions are $a = 15$ mm, $R_{\rm out} = 7$ mm, $R_{\rm inn} = 5$ mm ($w = 2$ mm), $g_1 = 0.5$ mm and $g_2 = 0.5$ mm. The thickness of the conductive meta-atom is $h = 1$ mm. (**c**) Supported resonance associated with the spectral feature in panel (**b**). The color corresponds to the E_y component (real part) and the arrows correspond to the magnetic field distribution. The eigenmode is characterized by a residual electric dipole moment due to counteracting contributions from the inner gap vs. the outer gaps. The magnetic field circulation gives rise to a strong toroidal dipole moment, T_y. (**d**) Power scattered from each multipole. The toroidal dipole and magnetic quadrupole moments are dominant at the resonant frequency. However, their scattered fields cancel out, since $E_{\rm sca}^{T_y} = -E_{\rm sca}^{Q_{xz}^m}$ exactly. Thus, the response is dictated by the residual electric dipole moment, p_y. (**e**) Comparison of the reflection coefficient calculated from the full-wave simulation with those reconstructed from the multipole moments. The response can be described fairly accurately when only the electric dipole moment is considered.

Next, we investigated different geometric degrees of freedom for tuning the resonance linewidth and lineshape. In Figure 3a, we varied the central gap (g_1) while keeping the outer gaps (g_2) constant at 0.5 mm. This controls the residual electric dipole moment and consequently the radiative strength of the resonance. For $g_1 = 0.4$ mm, the two opposing contributions (see Figure 2c) cancel each other out almost perfectly. In this case, very little coupling via the electric dipole moment is possible and the resonance becomes quasi-dark, resulting in a very narrow linewidth (high quality factor). The non-negligible reflection that remains is due to the non-resonant electric polarizability background. For $g_1 > 0.4$ mm, the central p_y contribution dominates over those originating from the outer gaps, while the opposite occurs for $g_1 < 0.4$ mm. In both cases, the linewidth (and radiative strength) increases. The corresponding total quality factors are compiled in Table 1. They have been obtained from the solution of an eigenvalue problem using the complex eigenfrequency $\tilde{\omega} = \omega' + i\omega''$ (the eigenvalue) through the expression $Q = \omega'/(2\omega'')$. For additional details regarding approaches to calculating the quality factor we refer the interested reader to ref. [22]. We document the total quality factor, Q_{tot}, which is directly associated with the linewidth of the spectral feature, along with the corresponding resonant frequency. Note that $Q_{rad} \approx Q_{tot}$ holds, since $Q_{res} \to \infty$ ("res" stands for resistive) for such a high value of the conductivity ($\sigma = 10^7$ S/m).

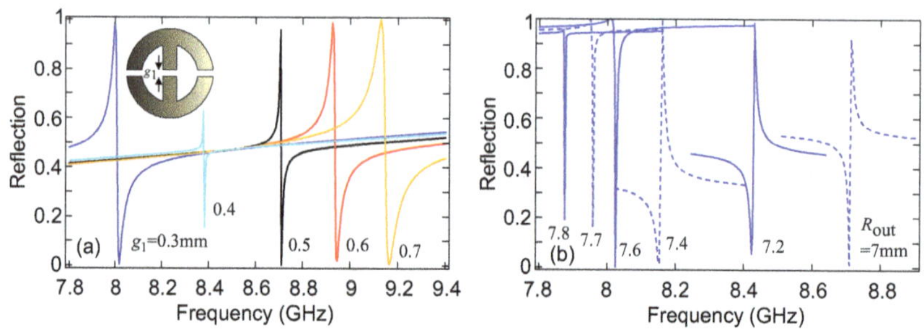

Figure 3. (**a**) Tuning the central gap (g_1), while keeping other gaps (g_2) constant at 0.5 mm. This controls the residual electric dipole moment and consequently the radiative strength of the resonance. (**b**) Varying the outer radius for $R_{inn} = R_{out} - w$ with $w = 2$ mm. As the radius increases, the non-resonant (background) electric dipole moment changes and with it the Fano lineshape. For R_{out} values of 7.6, 7.7, and 7.8 mm, the meta-atoms are touching; this modifies the residual electric dipole moment, as the outer gaps of each meta-atom begin to merge with those of the neighboring one.

Table 1. Quality factors for the cases depicted in Figure 3a. g_2 is held constant at 0.5 mm. They have been calculated using the complex eigenfrequency through $Q = \omega'/(2\omega'')$.

Central Gap g_1 (mm)	**Resonant Frequency (GHz)**	Q_{tot}
0.3	8.02	446
0.4	8.37	14,000
0.5	8.71	2,114
0.6	8.94	517
0.7	9.14	260

A different option is explored in Figure 3b, where we tune the outer radius. In all cases, the inner radius is also appropriately modified, so that the branch width is kept constant, i.e., $R_{inn} = R_{out} - w$, with $w = 2$ mm. This primarily modifies the non-resonant (background) electric dipole moment, leading to changes in the Fano lineshape from highly asymmetric to quasi-Lorentzian. For R_{out} values of 7.6, 7.7, and 7.8 mm, the meta-atoms are touching, this additionally modifies the residual electric dipole moment, as the outer gaps of each meta-atom begin to merge with those of the neighboring atom, producing high values of background reflectance. Importantly, by tuning R_{out} we are able to achieve very narrow linewidths and, at the same time, a large resonance depth (compare Figure 3b with Figure 3a). The total quality factors for $R_{out} = 7.7$ and 7.8 mm are ~5000 and ~30,000, respectively. Note that this quality factor corresponds to radiation losses and will drop if resistive losses increase.

3.2. Meta-Atoms in a Rectangular Waveguide Setup—Experimental Verification

In this Section, the unit cells under study are inserted inside a rectangular waveguide environment (see Figure 1). This is exercised in order to end up with a well-defined and highly reproducible measurement environment, which is not susceptible to external perturbations. Furthermore, the electromagnetic response is anticipated to be similar to the free-space periodic structure, since the top/bottom PEC walls emulate a periodic repetition along the y axis and the main difference is the $sin(\pi x/a)$ profile of the incident guided wave along the x axis [35,37,41].

We first performed simulations. We tuned the dimensions in order to bring the resonant frequency near to the center of the band of the WR-187 rectangular waveguide (3.95–5.85 GHz) and fit 2×1 meta-atoms in the cross-section ($a \times b = 47.55$ mm $\times$ 22.2 mm). The dimensions are $R_{\text{out}} = 12$ mm, $R_{\text{inn}} = 8.6$ mm, $w = 3.4$ mm, $g_1 = 0.7$ mm and $g_2 = 1$ mm. The thickness of the conductive meta-atom is $h = 3.4$ mm. The response is depicted in Figure 4, where the reflection, $R = |S_{11}|^2$, transmission, $T = |S_{21}|^2$, and absorption, $A = 1 - |S_{11}|^2 - |S_{21}|^2$, are plotted. The meta-atoms are touching and the resonance lineshape resembles the respective cases in Figure 3b. The meta-atoms will be fabricated via first 3D printing a dielectric (PLA) scaffold and then by coating with a silver paste, which is characterized by a limited conductivity of approximately $\sigma = 10^4$–10^5 S/m. Thus, in Figure 4 we examine the conductivities $\sigma = 10^5$ S/m, $\sigma = 5 \times 10^4$ S/m, $\sigma = 10^4$ S/m and $\sigma = 5 \times 10^3$ S/m. As the conductivity decreases, the spectral features become broader and the absorption increases up to a maximum of 0.5 (for an electrically polarizable structure) before starting to decrease again (under-coupling regime). The mode profile is depicted in the inset of Figure 4d. Notice that the gaps near the edges of the waveguide do not accommodate strong fields, since the incident TE_{01} waveguide mode is zeroed out at the side walls (cf. Figure 2).

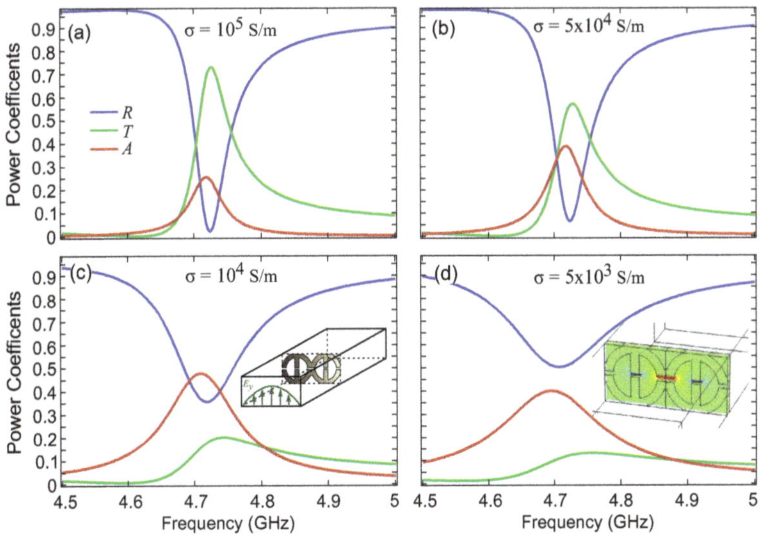

Figure 4. Simulations of the 3D-printed conductive unit cells within the WR-187 rectangular waveguide. The reflection ($R = |S_{11}|^2$), transmission ($T = |S_{21}|^2$) and absorption ($A = 1 - |S_{11}|^2 - |S_{21}|^2$) power coefficients are plotted for different conductivities: (**a**) $\sigma = 10^5$ S/m, (**b**) $\sigma = 5 \times 10^4$ S/m, (**c**) $\sigma = 10^4$ S/m and (**d**) $\sigma = 5 \times 10^3$ S/m. As the conductivity is decreased, the spectral feature becomes broader and the absorption on resonance increases up to a maximum of 0.5 before starting to decrease again.

In Figure 5, we depict the measurements of the 3D-printed conductive unit cells within the WR-187 rectangular waveguide. The reflection ($R = |S_{11}|^2$) and transmission ($T = |S_{21}|^2$) power coefficients are plotted in the frequency range of 4.5–5.5 GHz. The lineshape of the spectral feature and the linewidth (full-width half maximum of approximately 100 MHz, as extracted from the spectral feature in reflection) are in good qualitative agreement with those predicted by the simulations and specifically Figure 4c, which corresponds to a conductivity value we anticipate for the silver epoxy. The resonant frequency is slightly different; we attribute this discrepancy to the limited accuracy of the geometric dimensions achieved in the actual fabricated sample.

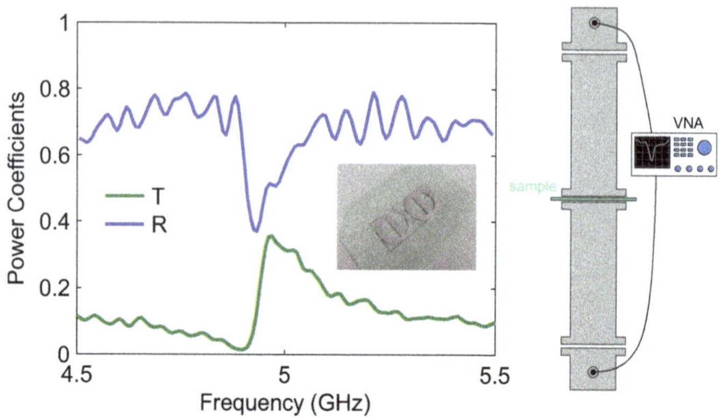

Figure 5. Measurement of the 3D-printed conductive unit cells within the WR-187 rectangular waveguide. The reflection ($R = |S_{11}|^2$) and transmission ($T = |S_{21}|^2$) power coefficients are plotted in the frequency range of 4.5–5.5 GHz.

4. Discussion and Conclusions

We have studied a conductive meta-atom of toroidal topology, meant to enhance the toroidal dipole moment. We have performed simulations of both a periodic free-space metasurface comprised of this meta-atom and, subsequently, we have inserted two such meta-atoms into a metallic, rectangular waveguide setup. The latter configuration has allowed for experimentally verifying the resonance characteristics (lineshape and linewidth) and thus supports the physical claims arising from the theoretical analysis and multipole expansion.

One goal of our work was to demonstrate a practical, easily realizable meta-atom geometry that provides freedom in shaping the electromagnetic response in terms of both linewidth and lineshape. Another important goal was to reveal and highlight the physical mechanisms dictating the resonant response. Importantly, despite the physical interpretation suggested by the unit cell geometry (its toroidal topology), a rigorous multipole expansion has revealed that, although the toroidal dipole moment is indeed strong, the corresponding scattered field is cancelled out exactly by that of the magnetic quadrupole. As a result, the scattering response can be described accurately by simply considering the electric dipole moment alone. Thus, we conclude that a multipole expansion must be performed in all cases, in order to reach safe conclusions regarding the multipole components that mediate the scattering process.

The studied meta-atom allows to control the residual electric dipole moment through the geometric asymmetry between the central and outer gaps. This is a very similar concept to "accidental BIC" or "trapped/broken-symmetry" resonances. It becomes possible to achieve a controllably sharp resonant response, reaching very high radiation quality factors (Q_{rad}). This is also possible in finite meta-atom arrays through spatially extended dark (sub-radiant) eigenmodes, as has been discussed in ref. [24]. Note, however, that ultimately the spectral linewidth is dictated by the total quality factor (Q_{tot}), which will be limited by resistive losses as well. As a result, pushing for very high Q_{rad} values which require slight asymmetry and precise fabrication is not necessary when it cannot be supported by low material losses. Interconnected meta-atoms that do not require a substrate can help to avoid excess losses stemming from the substrate material and excess radiation losses stemming from the breaking of the vertical symmetry.

Author Contributions: Conceptualization: E.N.E., O.T. and A.C.T.; methodology, O.T., A.C.T., D.C.Z. and E.N.E.; software: O.T. (implementation of multipole expansion and models); validation: O.T., Z.V. and G.K.; formal analysis, O.T.; investigation: O.T. and Z.V.; resources: M.K., D.C.Z., Z.V. and G.K.; data curation: O.T.; writing—original draft preparation, O.T.; writing—review and editing: Z.V., D.C.Z., M.K., A.C.T., G.K., E.N.E. and O.T.; visualization: O.T.; supervision: E.N.E., M.K. and G.K.; project administration: G.K.; funding acquisition: G.K. and D.C.Z. All authors have read and agreed to the published version of the manuscript.

Funding: This work was funded by the project "METAmaterial-based ENERGY autonomous systems (META-ENERGY) (Project ID 2936)" which is implemented under the 2nd Call for H.F.R.I. "Research Projects to Support Faculty Members & Researchers" funded by the Operational Programme "Competitiveness, Entrepreneurship and Innovation" (NSRF 2014–2020). D.C.Z. acknowledges the support of the project ECS00000024 "Ecosistemi dell'Innovazione"- Rome Technopole of the Italian Ministry of University and Research, public call n. 3277, PNRR-Mission 4, Component 2, Investment 1.5, financed by the European Union, Next Generation EU and of the CNR-FAPESP biennial (2022–2023) bilateral project StReAM "Strongly Resonant All-dielectric Metasurfaces based on quasi-dark and toroidal modes".

Institutional Review Board Statement: Not applicable.

Informed Consent Statement: Not applicable.

Data Availability Statement: The data presented in this study are available on request from the corresponding author.

Acknowledgments: The authors thank Varvara Mouzi for their help with silver paste coating of the 3D-printed meta-atoms.

Conflicts of Interest: The authors declare no conflict of interest.

References

1. Afanasiev, G.N.; Stepanovsky, Y.P. The electromagnetic field of elementary time-dependent toroidal sources. *J. Phys. A* **1995**, *28*, 4565–4580. [CrossRef]
2. Radescu, E.E.; Vaman, G. Exact calculation of the angular momentum loss, recoil force, and radiation intensity for an arbitrary source in terms of electric, magnetic, and toroid multipoles. *Phys. Rev. E* **2002**, *65*, 046609. [CrossRef]
3. Savinov, V.; Fedotov, V.A.; Zheludev, N.I. Toroidal dipolar excitation and macroscopic electromagnetic properties of metamaterials. *Phys. Rev. B* **2014**, *89*, 205112. [CrossRef]
4. Papasimakis, N.; Fedotov, V.A.; Savinov, V.; Raybould, T.A.; Zheludev, N.I. Electromagnetic toroidal excitations in matter and free space. *Nat. Mater.* **2016**, *15*, 263–271. [CrossRef]
5. Alaee, R.; Rockstuhl, C.; Fernandez-Corbaton, I. Exact Multipolar Decompositions with Applications in Nanophotonics. *Adv. Opt. Mater.* **2019**, *7*, 1800783. [CrossRef]
6. Zel'dovich, Y.B. Electromagnetic Interaction with Parity Violation. *Sov. Phys. JETP* **1958**, *6*, 1184.
7. Kaelberer, T.; Fedotov, V.A.; Papasimakis, N.; Tsai, D.P.; Zheludev, N.I. Toroidal dipolar response in a metamaterial. *Science* **2010**, *330*, 1510–1512. [CrossRef]
8. Basharin, A.A.; Kafesaki, M.; Economou, E.N.; Soukoulis, C.M.; Fedotov, V.A.; Savinov, V.; Zheludev, N.I. Dielectric Metamaterials with Toroidal Dipolar Response. *Phys. Rev. X* **2015**, *5*, 011036. [CrossRef]
9. Tasolamprou, A.C.; Tsilipakos, O.; Kafesaki, M.; Soukoulis, C.M.; Economou, E.N. Toroidal eigenmodes in all-dielectric metamolecules. *Phys. Rev. B* **2016**, *94*, 205433. [CrossRef]
10. Zografopoulos, D.C.; Ferraro, A.; Algorri, J.F.; Martín-Mateos, P.; García-Cámara, B.; Moreno-Oyervides, A.; Krozer, V.; Acedo, P.; Vergaz, R.; Sánchez-Pena, J.M.; et al. All-Dielectric Silicon Metasurface with Strong Subterahertz Toroidal Dipole Resonance. *Adv. Opt. Mater.* **2019**, *7*, 1900777. [CrossRef]
11. Jeong, J.; Goldflam, M.D.; Campione, S.; Briscoe, J.L.; Vabishchevich, P.P.; Nogan, J.; Sinclair, M.B.; Luk, T.S.; Brener, I. High Quality Factor Toroidal Resonances in Dielectric Metasurfaces. *ACS Photonics* **2020**, *7*, 1699–1707. . c00179. [CrossRef]

12. Zografopoulos, D.C.; Algorri, J.F.; Fuscaldo, W.; López-Higuera, J.M.; Vergaz, R.; Sánchez-Pena, J.M.; Karolos, I.A.; Beccherelli, R.; Tsioukas, V.E.; Yioultsis, T.V.; et al. All-Dielectric Toroidal Metasurfaces for Angular-Dependent Resonant Polarization Beam Splitting. *Adv. Opt. Mater.* **2021**, *9*, 2002143. [CrossRef]
13. Basharin, A.A.; Chuguevsky, V.; Volsky, N.; Kafesaki, M.; Economou, E.N. Extremely high *Q*-factor metamaterials due to anapole excitation. *Phys. Rev. B* **2017**, *95*, 035104. [CrossRef]
14. Liu, Z.; Du, S.; Cui, A.; Li, Z.; Fan, Y.; Chen, S.; Li, W.; Li, J.; Gu, C. High-Quality-Factor Mid-Infrared Toroidal Excitation in Folded 3D Metamaterials. *Adv. Mater.* **2017**, *29*, 1606298. [CrossRef]
15. Wu, P.C.; Liao, C.Y.; Savinov, V.; Chung, T.L.; Chen, W.T.; Huang, Y.; Wu, P.R.; Chen, Y.; Liu, A.; Zheludev, N.I.; et al. Optical Anapole Metamaterial. *ACS Nano* **2018**, *12*, 1920–1927. [CrossRef]
16. Fan, Y.; Wei, Z.; Li, H.; Chen, H.; Soukoulis, C.M. Low-loss and high-Q planar metamaterial with toroidal moment. *Phys. Rev. B* **2013**, *87*, 115417. [CrossRef]
17. Cong, L.; Srivastava, Y.K.; Singh, R. Tailoring the multipoles in THz toroidal metamaterials. *Appl. Phys. Lett.* **2017**, *111*, 081108. [CrossRef]
18. Gupta, M.; Savinov, V.; Xu, N.; Cong, L.; Dayal, G.; Wang, S.; Zhang, W.; Zheludev, N.I.; Singh, R. Sharp Toroidal Resonances in Planar Terahertz Metasurfaces. *Adv. Mater.* **2016**, *28*, 8206–8211. [CrossRef]
19. Fan, Y.; Zhang, F.; Shen, N.H.; Fu, Q.; Wei, Z.; Li, H.; Soukoulis, C.M. Achieving a high-*Q* response in metamaterials by manipulating the toroidal excitations. *Phys. Rev. A* **2018**, *97*, 033816. [CrossRef]
20. Han, S.; Gupta, M.; Cong, L.; Srivastava, Y.K.; Singh, R. Toroidal and magnetic Fano resonances in planar THz metamaterials. *J. Appl. Phys.* **2017**, *122*, 113105. [CrossRef]
21. Yang, Y.; Bozhevolnyi, S.I. Nonradiating anapole states in nanophotonics: From fundamentals to applications. *Nanotechnology* **2019**, *30*, 204001. [CrossRef]
22. Christopoulos, T.; Tsilipakos, O.; Sinatkas, G.; Kriezis, E.E. On the calculation of the quality factor in contemporary photonic resonant structures. *Opt. Express* **2019**, *27*, 14505–14522. [CrossRef]
23. Pu, M.; Hu, C.; Huang, C.; Wang, C.; Zhao, Z.; Wang, Y.; Luo, X. Investigation of Fano resonance in planar metamaterial with perturbed periodicity. *Opt. Express* **2013**, *21*, 992. [CrossRef]
24. Jenkins, S.D.; Ruostekoski, J.; Papasimakis, N.; Savo, S.; Zheludev, N.I. Many-Body Subradiant Excitations in Metamaterial Arrays: Experiment and Theory. *Phys. Rev. Lett.* **2017**, *119*, 053901. [CrossRef]
25. Tsilipakos, O.; Maiolo, L.; Maita, F.; Beccherelli, R.; Kafesaki, M.; Kriezis, E.E.; Yioultsis, T.V.; Zografopoulos, D.C. Experimental demonstration of ultrathin broken-symmetry metasurfaces with controllably sharp resonant response. *Appl. Phys. Lett.* **2021**, *119*, 231601. [CrossRef]
26. Zografopoulos, D.C.; Tsilipakos, O. Recent advances in strongly resonant and gradient all-dielectric metasurfaces. *Mater. Adv.* **2023**, 4, 11–34. [CrossRef]
27. Shi, L.; Evlyukhin, A.B.; Reinhardt, C.; Babushkin, I.; Zenin, V.A.; Burger, S.; Malureanu, R.; Chichkov, B.N.; Morgner, U.; Kovacev, M. Progressive Self-Boosting Anapole-Enhanced Deep-Ultraviolet Third Harmonic During Few-Cycle Laser Radiation. *ACS Photonics* **2020**, *7*, 1655–1661. [CrossRef]
28. Timofeeva, M.; Lang, L.; Timpu, F.; Renaut, C.; Bouravleuv, A.; Shtrom, I.; Cirlin, G.; Grange, R. Anapoles in Free-Standing III–V Nanodisks Enhancing Second-Harmonic Generation. *Nano Lett.* **2018**, *18*, 3695–3702. [CrossRef]
29. Ahmadivand, A.; Semmlinger, M.; Dong, L.; Gerislioglu, B.; Nordlander, P.; Halas, N.J. Toroidal Dipole-Enhanced Third Harmonic Generation of Deep Ultraviolet Light Using Plasmonic Meta-atoms. *Nano Lett.* **2018**, *19*, 605–611. [CrossRef]
30. Totero Gongora, J.S.; Miroshnichenko, A.E.; Kivshar, Y.S.; Fratalocchi, A. Anapole nanolasers for mode-locking and ultrafast pulse generation. *Nat. Commun.* **2017**, *8*, 15535. [CrossRef]
31. Ahmadivand, A.; Gerislioglu, B.; Manickam, P.; Kaushik, A.; Bhansali, S.; Nair, M.; Pala, N. Rapid Detection of Infectious Envelope Proteins by Magnetoplasmonic Toroidal Metasensors. *ACS Sensors* **2017**, *2*, 1359–1368. [CrossRef]
32. Ahmadivand, A.; Gerislioglu, B.; Ramezani, Z.; Ghoreishi, S.A. Attomolar Detection of Low-Molecular Weight Antibiotics Using Midinfrared-Resonant Toroidal Plasmonic Metachip Technology. *Phys. Rev. Appl.* **2019**, *12*, 034018. [CrossRef]
33. Gupta, M.; Singh, R. Toroidal metasurfaces in a 2D flatland. *Rev. Phys.* **2020**, *5*, 100040. [CrossRef]
34. Takou, E.; Tasolamprou, A.C.; Tsilipakos, O.; Economou, E.N. Dynamic anapole in metasurfaces made of sculptured cylinders. *Phys. Rev. B* **2019**, *100*, 085431. [CrossRef]
35. Takou, E.; Tasolamprou, A.C.; Tsilipakos, O.; Viskadourakis, Z.; Kafesaki, M.; Kenanakis, G.; Economou, E.N. Anapole Tolerance to Dissipation Losses in Thermally Tunable Water-Based Metasurfaces. *Phys. Rev. Appl.* **2021**, *15*, 014043. [CrossRef]
36. Viskadourakis, Z.; Tamiolakis, E.; Tsilipakos, O.; Tasolamprou, A.C.; Economou, E.N.; Kenanakis, G. 3D-Printed Metasurface Units for Potential Energy Harvesting Applications at the 2.4 GHz Frequency Band. *Crystals* **2021**, *11*, 1089. [CrossRef]
37. Tasolamprou, A.C.; Mentzaki, D.; Viskadourakis, Z.; Economou, E.N.; Kafesaki, M.; Kenanakis, G. Flexible 3D Printed Conductive Metamaterial Units for Electromagnetic Applications in Microwaves. *Materials* **2020**, *13*, 3879. [CrossRef]
38. Ishikawa, A.; Kato, T.; Takeyasu, N.; Fujimori, K.; Tsuruta, K. Selective electroless plating of 3D-printed plastic structures for three-dimensional microwave metamaterials. *Appl. Phys. Lett.* **2017**, *111*, 183102. [CrossRef]
39. Taghizadeh, A.; Mørk, J.; Chung, I.S. Ultracompact resonator with high quality-factor based on a hybrid grating structure. *Opt. Express* **2015**, *23*, 14913. [CrossRef]

40. Tsilipakos, O.; Zhang, L.; Kafesaki, M.; Soukoulis, C.M.; Koschny, T. Experimental Implementation of Achromatic Multiresonant Metasurface for Broadband Pulse Delay. *ACS Photonics* **2021**, *8*, 1649–1655. [CrossRef]
41. Zografopoulos, D.C.; Algorri, J.F.; Ferraro, A.; García-Cámara, B.; Sánchez-Pena, J.M.; Beccherelli, R. Toroidal metasurface resonances in microwave waveguides. *Sci. Rep.* **2019**, *9*, 7544. [CrossRef] [PubMed]

micromachines

Article

Bloch Surface Waves in Open Fabry–Perot Microcavities

Niccolò Marcucci [1], Tian-Long Guo [2], Ségolène Pélisset [2], Matthieu Roussey [2], Thierry Grosjean [3] and Emiliano Descrovi [1,*]

[1] Dipartimento di Scienza Applicata e Tecnologia, Politecnico di Torino, 10129 Torino, Italy
[2] Center for Photonics Sciences, Department of Physics and Mathematics, University of Eastern Finland, 80101 Joensuu, Finland
[3] Department of Optics, FEMTO-ST Institute, UMR CNRS 6174, 25030 Besançon, France
* Correspondence: emiliano.descrovi@polito.it

Abstract: Thanks to the increasing availability of technologies for thin film deposition, all-dielectric structures are becoming more and more attractive for integrated photonics. As light–matter interactions are involved, Bloch Surface Waves (BSWs) may represent a viable alternative to plasmonic platforms, allowing easy wavelength and polarization manipulation and reduced absorption losses. However, plasmon-based devices operating at an optical and near-infrared frequency have been demonstrated to reach extraordinary field confinement capabilities, with localized mode volumes of down to a few nanometers. Although such levels of energy localization are substantially unattainable with dielectrics, it is possible to operate subwavelength field confinement by employing high-refractive index materials with proper patterning such as, e.g., photonic crystals and metasurfaces. Here, we propose a computational study on the transverse localization of BSWs by means of quasi-flat Fabry–Perot microcavities, which have the advantage of being fully exposed toward the outer environment. These structures are constituted by defected periodic corrugations of a dielectric multilayer top surface. The dispersion and spatial distribution of BSWs' cavity mode are presented. In addition, the hybridization of BSWs with an A exciton in a 2D flake of tungsten disulfide (WS_2) is also addressed. We show evidence of strong coupling involving not only propagating BSWs but also localized BSWs, namely, band-edge and cavity modes.

Keywords: Bloch Surface Waves; strong coupling; TMD; 2D materials

Citation: Marcucci, N.; Guo, T.-L.; Pélisset, S.; Roussey, M.; Grosjean, T.; Descrovi, E. Bloch Surface Waves in Open Fabry–Perot Microcavities. *Micromachines* **2023**, *14*, 509. https://doi.org/10.3390/mi14030509

Academic Editors: Luigi Sirleto and Giancarlo C. Righini

Received: 11 January 2023
Revised: 17 February 2023
Accepted: 20 February 2023
Published: 22 February 2023

1. Introduction

The existence of optical modes strongly confined at the truncation interface of planar dielectric multilayers (one-dimensional photonic crystals, -1DPC) has been known for decades as a result of the pioneering studies by P. Yeh and A. Yariv [1,2] and several later works [3–5]. With the advent of the Plasmonics era, interest toward optical surface modes on planar dielectric stacks has seen a significant boost, mainly due to the possibility of overcoming some the inherent limitations of surface plasmons, especially in sensing applications [6–9]. In the last 20 years, optical surface modes (hereafter called Bloch Surface Waves, or BSW) on flat and patterned dielectric multilayers have been investigated in many frameworks, such as strong light–matter coupling [10–13], integrated [14–18], guided [19–22] and fiber [23,24] optics, label-free [25–28] and fluorescence-based [29–31] sensing, metrology, [32] photon management [33–36], light-driven particle manipulation, [37,38] emitting devices [39,40], microscopy imaging, and spectroscopy [41–43].

Besides the advantages of tuning the BSWs' spectral position, polarization [44], as well as propagation and penetration lengths [45], one of the most intriguing features offered by 1DPC-based platforms is the possibility to exploit the 1DPC surface to set up an accessible framework for controlling light–matter interaction. Furthermore, shallow patterning of the 1DPC surface allows the BSW to be manipulated and possibly confined along transverse directions [40,46], thus providing new degrees of freedom for exploring complex phenomena. Having a photonic mode confined both out-of-plane and in-plane on a surface is advantageous when alleviating some difficulties in technological tasks such as the integration of emitters/absorbers, which are typically buried/grown within the photonic structure itself, where the radiation energy of photonic modes is generally localized.

In this article, we present a computational work describing some peculiar features of transverse electric (TE)-polarized BSWs confined within linear resonant structures recalling Fabry–Perot microcavities. In addition, the interaction of cavity BSWs with the A exciton in a WS_2 monolayer laying on top of the multilayered structure is investigated and discussed. WS_2 monolayers display a narrow and intense excitonic resonance at 2.03 eV that is well suited to promote mode hybridization with BSWs. Extending previous results reporting on strong coupling between the WS_2 exciton and BSWs on flat dielectric multilayers [47], we anticipate here that strong coupling can be also observed with BSWs confined within planar cavities covered by WS_2 flakes of down to about 1 μm in transverse size.

2. Computational Model

The exemplary model for the 1D photonic crystal considered in this work is constituted by a stack of five alternating pairs of TiO_2 and SiO_2 layers on a semi-infinite glass half-space (refractive index $n_{glass} = 1.5$). The TiO_2 and SiO_2 layers have thickness 85 nm and 110 nm, respectively, so that a main forbidden band is crossing the light line at energies of about 2 eV. On top of the stack, an additional pair of SiO_2 (90 nm) and TiO_2 (40 nm) layers is introduced, leading to a total of $N_L = 12$ layers. These two layers are aimed at tailoring the position of the BSW across the forbidden band, in the energy–momentum space. In addition, the upmost TiO_2 layer is intended as a functional medium carrying a topographic modulation. In fact, a linear grating acting as a Distributed Bragg Reflector (DBR) is engraved therein, with a modulation height $h = 25$ nm, period $\Lambda_{DBR} = 275$ nm, and fill factor 0.5. A linear defect (hereafter called a "spacer") with a width W is eventually introduced within the DBR, thus resulting in a linear Fabry–Perot cavity. A schematic of the structure is presented in Figure 1a.

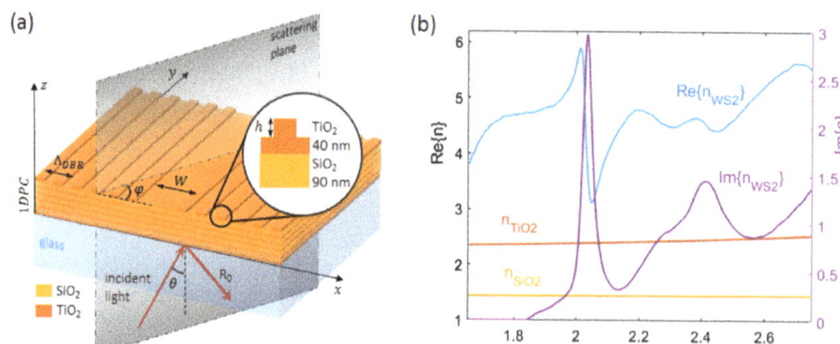

Figure 1. (**a**) Sketch of the patterned 1DPC in the conical diffraction mounting. The topographic modulation is oriented along the x-direction. Light is incident from the glass substrate at an angle of incidence θ; (**b**) refractive index dispersion of the materials considered in this work.

As will be apparent from the following, the DBR is aimed at perturbing the BSW dispersion by opening a bandgap at the crossing point of the BSW dispersion with the boundary of the first Brillouin zone defined by the DBR. The bandgap width is determined by the BSW (effective) refractive index contrast associated to the corrugation. Generally speaking, the larger the topographic height modulation, the wider the BSW bandgap. However, practical limitations to the maximum attainable width of the BSW bandgap exist. For example, if the modulation height or the top layer thickness are too large, the dielectric loading/unloading effect may cause BSWs to disappear in either the trenches (low effective refractive index) or the ridges (high effective refractive index), thus preventing a BSW bandgap to open. The geometry proposed here represents a (not unique) solution that avoids the occurrence of such an issue, while allowing the formation of a reasonably wide BSW bandgap for observing cavity BSWs and related hybridization with the WS_2 exciton.

This kind of dielectric multilayer can be fabricated by means of standard techniques, including RF sputtering [16], chemical vapor deposition (CVD) and plasma-enhanced CVD [48] plasma ion-assisted vacuum evaporation (PIAD) [49], and atomic layer deposition (ALD) [50]. In particular, the refractive indexes used to design the present stack are taken from ellipsometric data collected from ALD-deposited materials and plotted in Figure 1b. In addition, the refractive index of a WS_2 monolayer is plotted as well [51].

Calculations were performed by means of a freely available implementation of the Rigorous Coupled Wave Analysis (RCWA) RETICOLO [52,53]. The RCWA supercell includes the spacer surrounded by 180 DBR periods on each side. The periodic corrugation has an associated Bragg grating vector oriented along the x-axis, i.e., $K_{DBR} = \left(2\pi\cdot\Lambda_{DBR}^{-1}, 0, 0\right)$. The conical mounting configuration is considered so that the incoming radiation can be made incident onto the bottom interface of the multilayer at an incident angle θ with respect to the normal (z-axis) and at an angle φ with respect to the x-axis. At $\varphi = 0^o$, the incident light is parallel to the K_{DBR} and diffracted accordingly. Illumination is always provided from the glass substrate to reach momentum values beyond the light line in air, as required for BSW coupling. RCWA is a Fourier-based method involving a Rayleigh expansion of the diffracted field as well as a Fourier decomposition of the structure harmonics. For this reason, it is crucial to define the proper number of Fourier terms to be retained in the calculation. The case of the spacer surrounded by DBRs is addressed using N = 1200 Fourier terms. In the convergence plot in Figure 2, the spectral position of the BSW cavity mode and the corresponding reflectivity are shown as a function of the number of Fourier terms considered. Such a supercell contains a number of DBR periods large enough to avoid significant cross-talk effects between adjacent supercells occurring in the momentum–energy region wherein the cavity mode is located. Instead, the simpler case of a purely periodic corrugation (i.e., no spacer) is modeled as an elementary cell containing a single DBR period and N = 60 Fourier terms for assuring reliable results.

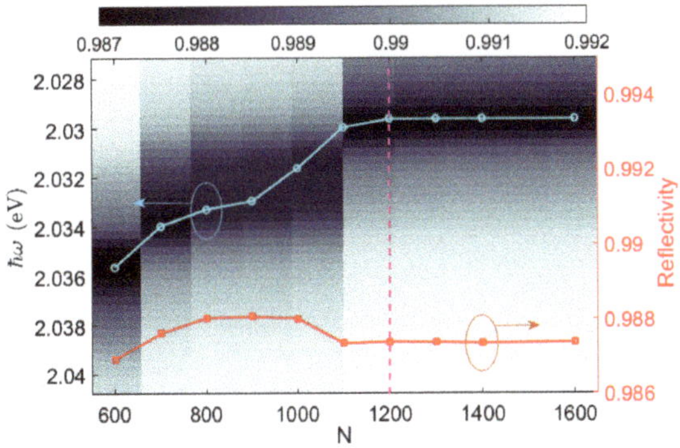

Figure 2. Reflectivity map $R(N, \hbar\omega)$ and spectral position of the BSW resonance (cyan circles) in an exemplary cavity with spacer width $W = 500$ nm and 360 DBR periods overall at $\varphi = 0^o$ incidence. N indicates the number of Fourier terms retained in the RCWA code. The reflectivity values at the resonance center (red squares) are also plotted. The dashed pink line indicates the number of harmonics used in this work.

The multilayer supports TE-polarized Bloch Surface Waves whose dispersion changes from the near-infrared to the visible spectrum. The dispersion of the BSWs on the flat multilayer can be inferred upon calculation of the reflectivity $R(\beta, \hbar\omega)$, wherein the effective refractive index is $\beta = n_{glass} \cdot \sin\theta$ and $\hbar\omega$ is the photon energy of the TE-polarized incident radiation. Without loss of generality, the BSW dispersion can be identified by the pairs (β_{BSW}, ω) corresponding to the bright line in Figure 3a. The TiO_2 and SiO_2 layers have been given a small imaginary refractive index $k_{SiO_2} = k_{TiO_2} = 10^{-3}$ to introduce enough losses to make the BSW dip visible on the calculated reflectivity maps. In Figure 3a,b, the $\log(1 - R(\beta, \hbar\omega))$ map and the normalized intensity profile of the BSW at $\hbar\omega = 2$ eV are respectively shown.

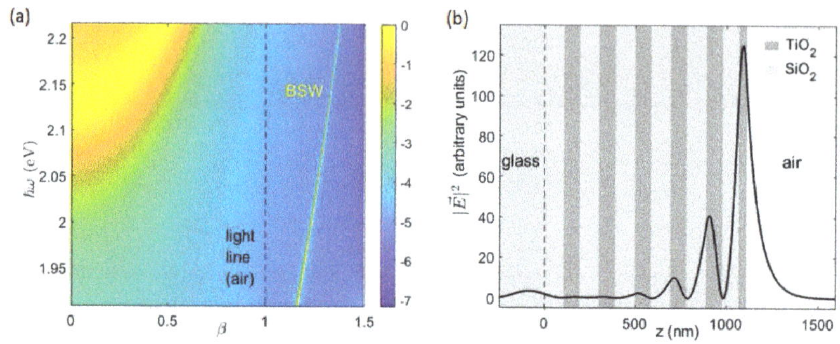

Figure 3. (**a**) Calculated TE−polarized $\log(1 - R(\beta, \hbar\omega))$ of the flat 1DPC, with illumination from the bottom glass substrate at an incidence angle θ such that $\beta = n_{glass} \cdot \sin\theta$; (**b**) intensity profile of the BSW at $\hbar\omega = 2$ eV, normalized to the amplitude of the incident wave.

3. Results

The case of the planar stack corrugated with a purely periodic surface modulation (no defect spacer) is considered first. In Figure 4a, the energy and angularly resolved full dispersion of the TE-polarized BSWs is shown as a 3D surface whose points $(\beta_x, \beta_y, \omega)$ are extracted from the calculated reflectivity $R(\beta, \varphi, \hbar\omega)$. There, $\beta_x = \beta \cdot \cos\varphi$ and $\beta_y = \beta \cdot \sin\varphi$, and they must match the BSW effective refractive index at any BSW energy $\hbar\omega$, i.e., $\sqrt{\beta_x^2 + \beta_y^2} = \beta_{BSW}(\hbar\omega)$. Reflectivity values refer to the diffraction efficiency of the 0th order at different illumination conditions.

The periodic modulation along the x-direction results in a folding of the BSW dispersion and the formation of a forbidden band that is dispersed in energy. To facilitate the interpretation, the BSW folding and the corresponding energy gap at $\beta_y = 0$ (corresponding to $\varphi = 0$, i.e., a direction parallel to the grating vector K_{DBR}) is presented in Figure 4b. The folding of the BSW dispersion due to diffraction at the -1st order results in a gap opening at about 2 eV. As pointed out elsewhere [40,54], the energy gap width depends on the effective refractive index contrast introduced by the periodic modulation. More interestingly, along directions different than K_{DBR}, the gap shifts to higher energies. Band edges are identified at the intersection of the surface $\sigma(\beta_x) = \pi\hbar c \cdot (\Lambda_{DBR} \cdot \beta_x)^{-1}$ (first Brillouin zone boundary, $K_{DBR}/2$) with the full BSW dispersion, as shown in Figure 4c. The forbidden band can be better appreciated by projecting said BSW dispersion onto the $(\beta_y, \hbar\omega)$ plane (Figure 4d). A maximum bandgap of about 150 meV at $\beta_y = 0$ is found. For larger values of $|\beta_y|$ and higher energies, the bandgap is narrowed down.

At each bandgap edge, two counterpropagating BSWs are produced, whose interference pattern has a spatial period equal to the DBR period Λ_{DBR}. In Figure 5a,b, the normalized intensity distributions of the BSWs at the lower and the upper band are respectively shown. The field localization on either the ridges (high refractive index regions) or trenches (low refractive index regions) reflects the different energies associated to the two band edge modes. Calculations are performed at $\hbar\omega_L = 1.953$ eV and $\hbar\omega_U = 2.1$ eV for the lower and the upper band edge modes, respectively. In both cases, β_x is taken on the boundary of first Brillouin zone, i.e., $\beta_x = \pi c \cdot (\Lambda_{DBR} \cdot \omega_{L,U})^{-1}$, while $\beta_y = 0$. Field amplitudes are normalized to the incident radiation amplitude.

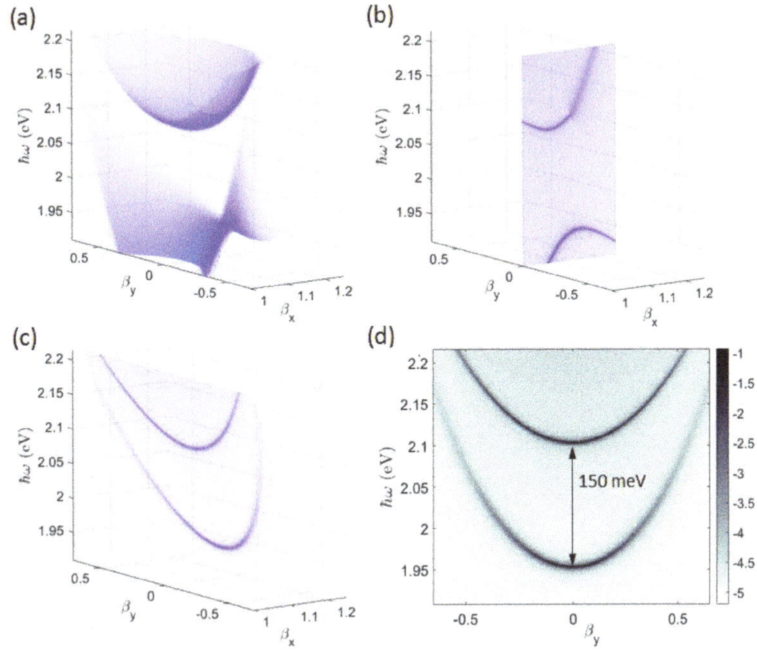

Figure 4. (**a**) Dispersion of a TE−polarized BSW represented as a 3D surface from the calculated reflectivity $R(\beta_x, \beta_y, \hbar\omega)$ of the 1DPC patterned with the periodic DBR. Only the region beyond the light-line in air $\sqrt{\beta_x^2 + \beta_y^2} \geq 1$ is considered; BSW dispersion on (**b**) the $\beta_y = 0$ plane and (**c**) the boundary of the first Brillouin zone defined by the shaded surface $\sigma = \pi\hbar c \cdot (\Lambda_{DBR} \cdot \beta_x)^{-1}$; (**d**) $\log(1 - R(\beta_x, \beta_y, \hbar\omega))$ map on the σ surface, where $\beta_x = \pi c \cdot (\Lambda_{DBR} \cdot \omega)^{-1}$.

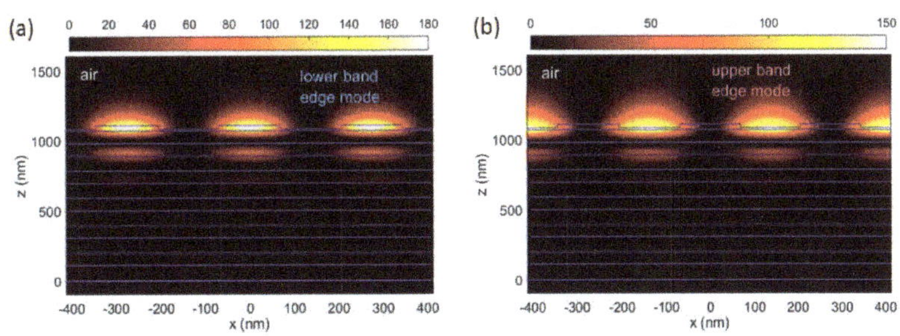

Figure 5. Intensity field distribution $\left| < \left| E_y(x,z) \right|^2 > \right|$ of (**a**) the lower band BSW, calculated at $\hbar\omega_L = 1.953$ eV, $\beta_x = \pi c \cdot (\Lambda_{DBR} \cdot \omega_L)^{-1}$, $\beta_y = 0$; (**b**) the upper band BSW, calculated at $\hbar\omega_U = 2.1$ eV, $\beta_x = \pi c \cdot (\Lambda_{DBR} \cdot \omega_U)^{-1}$, $\beta_y = 0$.

If a linear defect is introduced into the DBR, a cavity mode appears within the forbidden band created by the BSW folding. In Figure 6a, the BSW cavity mode corresponding to a spacer width $W = 500$ nm is visible as a narrow band starting at about 2.03 eV. In comparison to the spacer mode in a conventional planar microcavity, the BSW cavity mode follows a dispersion curve depending on the momentum component that is transverse to the direction of the refractive index modulation. In our case, since the pattern modulation is developing in the x-direction, the BSW cavity mode is energy dispersed as a function of β_y. It is worth noting that, for $|\beta_y| > 0.6$, the cavity mode tends to merge with the lower band edge mode, thus loosing spatial localization. Instead, the BSW cavity mode is dispersionless along the β_x direction (Figure 6b).

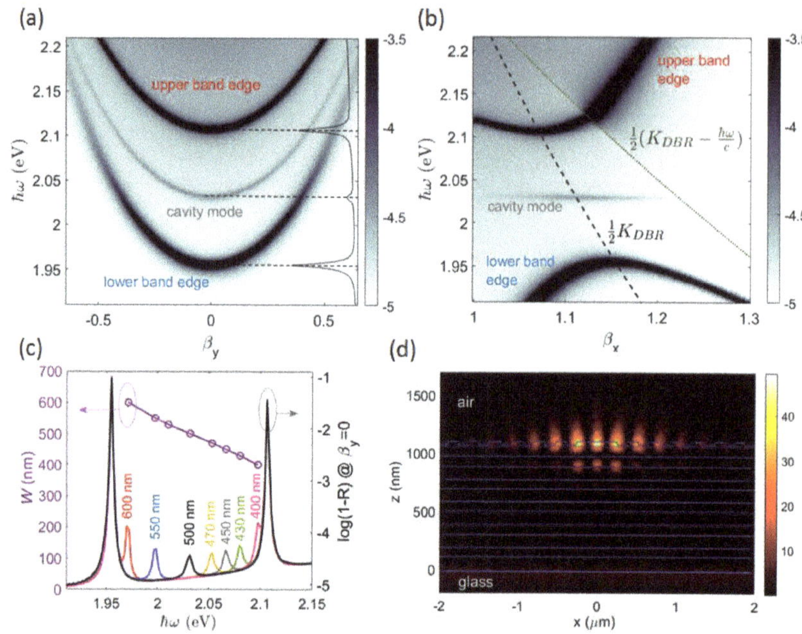

Figure 6. Dispersion of a TE−polarized BSW on the 1DPC patterned with a cavity spacer $W = 500$ nm surrounded by 180 DBR periods on each side, calculated on (**a**) the σ surface, (**b**) the $\beta_y = 0$ plane; (**c**) spectral position (purple circles) and corresponding spectra of the cavity mode across the forbidden band calculated at $(\beta_x, \beta_y) = \left(\pi c \cdot (\Lambda_{DBR} \cdot \omega)^{-1}, 0\right)$ for several spacer values W, i.e. 400 nm (pink line), 430 nm (green line), 450 nm (gray line), 470 nm (yellow line), 500 nm (black line), 550 nm (blue line), 600 nm (red line); (**d**) near-field intensity of the BSW cavity mode $|E_y(x,z)|^2$ for $W = 500$ nm, at $\hbar\omega = 2.032$ eV, $\beta_y = 0$.

The spectral position of the cavity mode can be modulated across the forbidden band by varying the spacer width W (Figure 6c). In Figure 6d, a cross-sectional view of the spatial near-field intensity of the BSW cavity mode at $\hbar\omega = 2.03$ eV is illustrated. As in a Fabry–Perot planar microcavity, the energy of the cavity mode is preferentially localized in the defect region. However, the very low refractive index contrast experienced by the BSW prevents a strong confinement within the spacer only, with a significant penetration of the mode profile in the surrounding periodic corrugations.

Very recently, multilayers sustaining BSWs coated with organic/inorganic thin layers exhibiting excitonic resonances have been exploited as novel platforms to enhance light–matter interactions [12,13,55]. In this framework, 2D materials (e.g., Transition Metal Dichalcogenides, orTMD monolayers) are particularly attractive because of the possibility to operate polariton manipulation at room temperature [56,57]. Note that the TE-polarization of BSW facilitates the interaction with excitons whose transition dipole lays on the plane, as occurs, for example, in WS_2 monolayers [58]. Strong coupling effects between BSWs and excitons have been experimentally studied on flat or periodically patterned multilayers, wherein the resulting polariton field exhibits a spatially delocalized distribution.

When our flat 1DPC model is coated with a tungsten disulfide (WS_2) monolayer (Figure 7a), the BSW dispersion exhibits the typical anti-crossing behavior illustrated in Figure 7b. The upper and lower BSW polariton branches repel each other via a vacuum Rabi splitting Ω that can be estimated by means of a simple coupled harmonic oscillator model. The system Hamiltonian can be expressed as a 2×2 matrix [12] with non-zero off-diagonal elements $H = \begin{pmatrix} E_{BSW} & \Omega/2 \\ \Omega/2 & E_X \end{pmatrix}$, where E_X is the exciton energy and $E_{BSW}(\beta_x)$ is the dispersion of the uncoupled BSW. Hamiltonian eigenvalues are: $E_{up} = \frac{1}{2}\left(E_{BSW} + E_X + \sqrt{(E_{BSW} - E_X)^2 + \Omega^2}\right)$; $E_{lp} = \frac{1}{2}\left(E_{BSW} + E_X - \sqrt{(E_{BSW} - E_X)^2 + \Omega^2}\right)$, which correspond to the dispersions of the Upper and the Lower Polariton Branch (UPB and LPB), respectively. A fitting procedure is used to estimate Ω and E_X. Worth specifying is that, to consider the dielectric loading effect introduced by the WS_2 monolayer, the values for $E_{BSW}(\beta_x)$ calculated for the bare 1DPC need be adjusted (redshift) by an additional energy term E_0 that is also obtained from the fit. In Figure 7b, the blue and the red dashed lines represent the lower and upper polariton dispersion, respectively, the yellow line represents the exciton energy E_X = 2.032 eV, and the green line represents the uncoupled BSW dispersion, redshift-corrected. From the fit, we obtain a Rabi splitting $\Omega = 52$ meV, which is similar to previously published results on a different BSW platform [13]

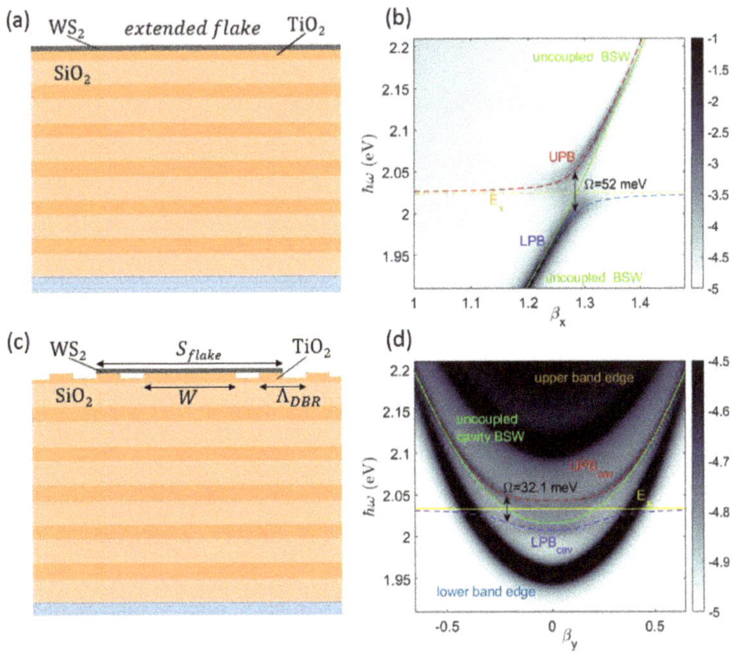

Figure 7. (**a**) Sketch of the z-cross section of the flat 1DPC topped with a uniform WS_2 monolayer with an indefinite extension; (**b**) corresponding dispersion of BSW polariton; (**c**) sketch of the z−cross section of the 1DPC patterned with the cavity topped with a WS_2 monolayer. The flake has a lateral size $S_{flake} = 1.6$ µm and a thickness 0.65 nm; (**d**) corresponding dispersion of the BSW cavity polariton branches (UPB_{cav} and LPB_{cav}).

When the BSW cavity is considered, the BSW–exciton interaction occurs over a spatial region that is substantially limited by the transverse size of the cavity mode. The sketch in Figure 7c shows the case of a WS_2 monolayer deposited on top of the BSW cavity, extending over the spacer and the first surrounding DBR period, with a total size $S_{flake} = 1.05$ µm and a thickness of 0.65 nm [59]. Similar to cavity modes in a planar microcavity, the cavity BSW hybridizes with the excitonic resonance, resulting in a mode dispersion splitting into upper and lower polariton branches (UPB_{cav} and LPB_{cav}), as shown in Figure 7d. The Rabi splitting estimated using the fitting procedure described above is found to be $\Omega = 32.1$ meV. Given the half-widths of the BSW cavity mode γ_{BSW} =6 meV and the A exciton of WS_2 γ_X =11 meV [60,61], the condition for the occurrence of strong coupling, i.e., $\Omega > \frac{1}{2}(\gamma_{BSW} + \gamma_X)$, is fulfilled.

Losses due to the WS_2 exciton resonance affect the mode quality of the two polariton branches. Figure 8a shows a calculated quality factor $Q = \omega_0/\delta\omega$ for the bare BSW cavity mode, UPB_{cav}, and LPB_{cav} in multilayers made of 8 to 16 layers. For all three modes, resonance frequencies ω_0 are recorded on the corresponding dispersion curve, at $\beta_y = 0$. While an increase in Q with the number of layers N_L is ascribed to a decrease in leaky power tunnelling through the 1DPC, the absorption by WS_2 is responsible for the lower quality of both polariton branches, as compared to the bare cavity mode. Furthermore, the upper branch polariton is more severely attenuated, as also observed experimentally in an analogous system involving strong coupling between ZnO excitons and BSWs [12]. Intensity distributions calculated for UPB_{cav} and LPB_{cav} at $\beta_y = 0$ are localized in the neighborhood of the spacer, similar to the bare cavity mode, with a substantially reduced enhancement due to losses (Figure 8b,c).

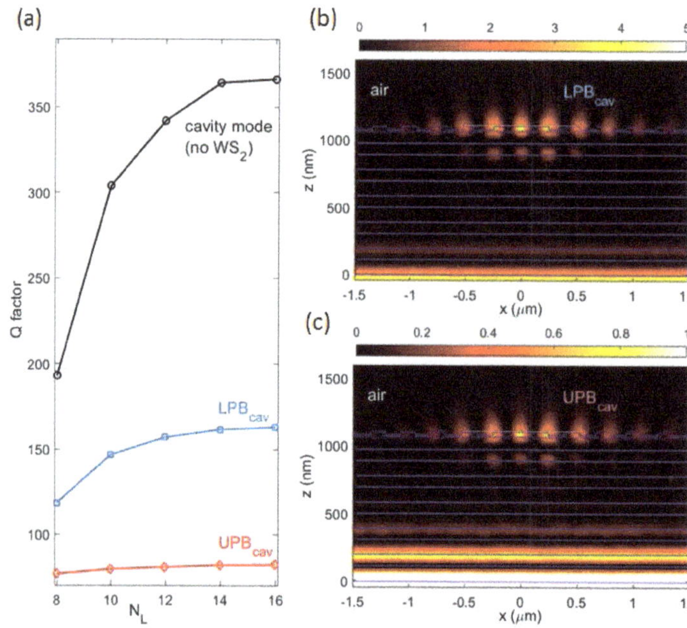

Figure 8. (**a**) Quality factor of upper (red diamonds) and lower (blue squares) BSW cavity polariton branches and the bare cavity mode (black circles) as a function of the overall number of layers N_L in the 1DPC; intensity field distribution $\left|< |E_y(x,z)|^2 >\right|$ of (**b**) the lower cavity polariton LPB_{cav}, calculated at $\hbar\omega_{LP} = 2$ eV, $\beta_x = \pi c \cdot (\Lambda_{DBR} \cdot \omega_{LP})^{-1}$, $\beta_y = 0$; (**c**) the upper cavity polariton UPB_{cav}, calculated at $\hbar\omega_{UP} = 2.04$ eV, $\beta_x = \pi c \cdot (\Lambda_{DBR} \cdot \omega_{UP})^{-1}$, $\beta_y = 0$. Both intensity distributions refer to an N_L = 12 1DPC.

It is interesting to note that the BSW hybridization presented above involves the cavity mode only. This is due to the limited size of the WS_2 monolayer, which overlaps mainly with the cavity spacer, where most of the energy is localized. Instead, if the whole patterned 1DPC is topped with an extended monolayer (Figure 9a), the BSW hybridization is found to occur also with the BSW band edge modes, as shown in Figure 9b,c, resulting in upper and lower band edge-polariton branches (UPB_{be} and LPB_{be}). Here, we find a Rabi splitting

$\Omega = 37$ meV for the BSW cavity polariton, and $\Omega = 37.6$ meV for the lower BSW band edge polariton. The different spatial distribution of the band edge and the cavity modes enables us to further modulate the photonic mode–exciton interaction by controlling the size and position of the WS_2 monolayer transferred onto the BSW platform.

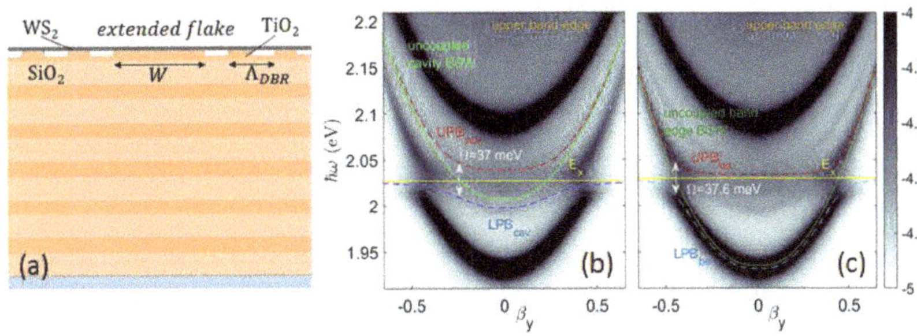

Figure 9. (**a**) Sketch of the z−cross section of the 1DPC patterned with the cavity topped with an indefinitely large WS_2 monolayer; corresponding dispersions of hybridized (**b**) BSW cavity polariton mode UPB_{cav}, LPB_{cav}, and (**c**) BSW band edge-polariton mode UPB_{be}, LPB_{be}.

4. Discussion

Fostered by an increasing interest toward surface waves on dielectric photonic platforms, we investigated some aspects of the transverse confinement of BSWs in open Fabry–Perot microcavities. Despite the low effective refractive index contrast produced by the ultra-shallow patterning/modulations of the multilayer surface, BSWs can be resonantly confined within sub-micrometric regions. As an extension of the findings presented here, an omnidirectional confinement on the multilayer plane can be obtained by introducing axis-symmetric corrugations [40] or dielectric ridges [46]. The design of the surface modulation is addressed in conjunction with the multilayer design, as dielectric loading/unloading effects can significantly influence the BSW dispersion, resulting in momentum mismatch with the pattern. In the exemplary cases presented here, we designed an open Fabry–Perot cavity by employing materials and geometries that can be realistically handled in clean-room fabrication processes. The cavity mode follows a parabolic energy dispersion as a function of the transverse in-plane momentum $(\propto \beta_y)$, thus suggesting a two-dimensional analogy to the case of 3D stacked planar microcavities. Cavity Q factors generally increase with the number of layers within the 1DPC due to reduced leakage losses. Although this feature may look advantageous when a BSW is near-field coupled from sources directly located on the 1DPC surface, it can be detrimental in cases of prism-coupling from external sources, because very little power is resonantly transferred from the glass substrate to the 1DPC top surface.

One of the main advantages in BSW-based platforms is the possibility of enhancing light–matter interaction on the surface of the photonic structure. This aspect is particularly useful in the framework of 2D materials such as TMD, that are often transferred on planar or structured surfaces upon exfoliation from bulk. Fostered by the potential applications of polariton handling at room temperature, we addressed the hybridization of a BSW cavity mode with the A exciton in a WS_2 monolayer deposited thereto. Strong coupling is found with both the cavity and the band-edge modes, depending on the spatial overlap with the interacting modal volumes, resulting in vacuum Rabi splitting in the range from 30 meV to 40 meV.

Our findings contribute to the scientific understanding of BSW-based platforms, particularly of polariton control at room temperature. We anticipate future opportunities for embedding TMDs on dielectric multilayers from the perspective of integrated quantum nanophotonic devices [62].

Author Contributions: N.M. has set up the computational model and run the calculations; T.-L.G. has performed deposition of TiO_2 and SiO_2 layers, S.P. and M.R. have conducted ellipsometric measurements and fitting procedures for the refractive index estimation; T.G. has provided the BSW cavity design; E.D. has coordinated the work. All authors have contributed to writing the manuscript. All authors have read and agreed to the published version of the manuscript.

Funding: This research was partially funded by the "Dipartimento di Eccellenza 2018–2022" program from the Italian Ministry of Education, University and Research.

Data Availability Statement: Not applicable.

Acknowledgments: Computational resources were provided by HPC@POLITO, a project of Academic Computing within the Department of Control and Computer Engineering at the Politecnico di Torino http://hpc.polito.it (accessed on 21 February 2023).

Conflicts of Interest: The authors declare no conflict of interest. The funders had no role in the design of the study; in the collection, analyses, or interpretation of data; in the writing of the manuscript; or in the decision to publish the results.

References

1. Yeh, P.; Yariv, A.; Hong, C.-S. Electromagnetic Propagation in Periodic Stratified Media I General Theory. *J. Opt. Soc. Am.* **1977**, *67*, 423–438. [CrossRef]
2. Yeh, P.; Yariv, A.; Cho, A.Y. Optical Surface Waves in Periodic Layered Media. *Appl. Phys. Lett.* **1978**, *32*, 104–105. [CrossRef]
3. Meade, R.D.; Brommer, K.D.; Rappe, A.M.; Joannopoulos, J.D. Electromagnetic Bloch Waves at the Surface of a Photonic Crystal. *Phys. Rev. B* **1991**, *44*, 10961–10964. [CrossRef]
4. Robertson, W.M.; Arjavalingam, G.; Meade, R.D.; Brommer, K.D.; Rappe, A.M.; Joannopoulos, J.D. Observation of Surface Photons on Periodic Dielectric Arrays. *Opt. Lett.* **1993**, *18*, 528. [CrossRef]
5. Robertson, W.M. Experimental Measurement of the Effect of Termination on Surface Electromagnetic Waves in One-Dimensional Photonic Bandgap Arrays. *J. Light. Technol.* **1999**, *17*, 2013–2017. [CrossRef]
6. Villa, F.; Regalado, L.E.; Ramos-Mendieta, F.; Gaspar-Armenta, J.; Lopez-Ríos, T. Photonic Crystal Sensor Based on Surface Waves for Thin-Film Characterization. *Opt. Lett.* **2002**, *27*, 646. [CrossRef]
7. Shinn, M.; Robertson, W.M. Surface Plasmon-like Sensor Based on Surface Electromagnetic Waves in a Photonic Band-Gap Material. *Sens. Actuators B Chem.* **2005**, *105*, 360–364. [CrossRef]
8. Descrovi, E.; Frascella, F.; Sciacca, B.; Geobaldo, F.; Dominici, L.; Michelotti, F. Coupling of Surface Waves in Highly Defined One-Dimensional Porous Silicon Photonic Crystals for Gas Sensing Applications. *Appl. Phys. Lett.* **2007**, *91*, 241109. [CrossRef]
9. Liscidini, M.; Sipe, J.E. Enhancement of Diffraction for Biosensing Applications via Bloch Surface Waves. *Appl. Phys. Lett.* **2007**, *91*, 253125. [CrossRef]
10. Liscidini, M.; Gerace, D.; Sanvitto, D.; Bajoni, D. Guided Bloch Surface Wave Polaritons. *Appl. Phys. Lett.* **2011**, *98*, 121118. [CrossRef]
11. Pirotta, S.; Patrini, M.; Liscidini, M.; Galli, M.; Dacarro, G.; Canazza, G.; Guizzetti, G.; Comoretto, D.; Bajoni, D. Strong Coupling between Excitons in Organic Semiconductors and Bloch Surface Waves. *Appl. Phys. Lett.* **2014**, *104*, 051111. [CrossRef]
12. Henn, S.; Grundmann, M.; Sturm, C. Strong Coupling of Bloch Surface Waves and Excitons in ZnO up to 430 K. *New J. Phys.* **2021**, *23*, 093031. [CrossRef]
13. Lerario, G.; Ballarini, D.; Fieramosca, A.; Cannavale, A.; Genco, A.; Mangione, F.; Gambino, S.; Dominici, L.; De Giorgi, M.; Gigli, G.; et al. High-Speed Flow of Interacting Organic Polaritons. *Light Sci. Appl.* **2017**, *6*, e16212. [CrossRef]
14. Dubey, R.; Vosoughi Lahijani, B.; Barakat, E.; Häyrinen, M.; Roussey, M.; Kuittinen, M.; Herzig, H.P. Near-Field Characterization of a Bloch-Surface-Wave-Based 2D Disk Resonator. *Opt. Lett.* **2016**, *41*, 4867. [CrossRef]
15. Rodriguez, G.A.; Aurelio, D.; Liscidini, M.; Weiss, S.M. Bloch Surface Wave Ring Resonator Based on Porous Silicon. *Appl. Phys. Lett.* **2019**, *115*, 011101. [CrossRef]
16. Deng, C.-Z.; Ho, Y.-L.; Yamahara, H.; Tabata, H.; Delaunay, J.-J. Near-Zero-Index Slabs on Bloch Surface Wave Platform for Long-Range Directional Couplers and Optical Logic Gates. *ACS Nano* **2022**, *16*, 2224–2232. [CrossRef]
17. Kim, M.-S.; Vosoughi Lahijani, B.; Descharmes, N.; Straubel, J.; Negredo, F.; Rockstuhl, C.; Häyrinen, M.; Kuittinen, M.; Roussey, M.; Herzig, H.P. Subwavelength Focusing of Bloch Surface Waves. *ACS Photonics* **2017**, *4*, 1477–1483. [CrossRef]
18. Augenstein, Y.; Roussey, M.; Grosjean, T.; Descrovi, E.; Rockstuhl, C. Inverse Design of Cavities for Bloch Surface Waves Interfaced to Integrated Waveguides. *Photonics Nanostructures Fundam. Appl.* **2022**, *52*, 101079. [CrossRef]
19. Descrovi, E.; Sfez, T.; Quaglio, M.; Brunazzo, D.; Dominici, L.; Michelotti, F.; Herzig, H.P.; Martin, O.J.F.; Giorgis, F. Guided Bloch Surface Waves on Ultrathin Polymeric Ridges. *Nano Lett.* **2010**, *10*, 2087–2091. [CrossRef]
20. Baghbadorani, H.K.; Aurelio, D.; Barvestani, J.; Liscidini, M. Guided Modes in Photonic Crystal Slabs Supporting Bloch Surface Waves. *J. Opt. Soc. Am. B* **2018**, *35*, 805. [CrossRef]
21. Yu, L.; Barakat, E.; Sfez, T.; Hvozdara, L.; Di Francesco, J.; Peter Herzig, H. Manipulating Bloch Surface Waves in 2D: A Platform Concept-Based Flat Lens. *Light Sci. Appl.* **2014**, *3*, e124. [CrossRef]

22. Dubey, R.; Barakat, E.; Häyrinen, M.; Roussey, M.; Honkanen, S.K.; Kuittinen, M.; Herzig, H.P. Experimental Investigation of the Propagation Properties of Bloch Surface Waves on Dielectric Multilayer Platform. *J. Eur. Opt. Soc. Rapid Publ.* **2017**, *13*, 5. [CrossRef]
23. Scaravilli, M.; Castaldi, G.; Cusano, A.; Galdi, V. Grating-Coupling-Based Excitation of Bloch Surface Waves for Lab-on-Fiber Optrodes. *Opt. Express* **2016**, *24*, 27771. [CrossRef]
24. Scaravilli, M.; Micco, A.; Castaldi, G.; Coppola, G.; Gioffrè, M.; Iodice, M.; La Ferrara, V.; Galdi, V.; Cusano, A. Excitation of Bloch Surface Waves on an Optical Fiber Tip. *Adv. Opt. Mater.* **2018**, *6*, 1800477. [CrossRef]
25. Niu, D.; Zerrad, M.; Lereu, A.; Moreau, A.; Lumeau, J.; Zapien, J.A.; Passian, A.; Aubry, V.; Amra, C. Excitation of Bloch Surface Waves in Zero-Admittance Multilayers for High-Sensitivity Sensor Applications. *Phys. Rev. Appl.* **2020**, *13*, 054064. [CrossRef]
26. Occhicone, A.; Polito, R.; Michelotti, F.; Ortolani, M.; Baldassarre, L.; Pea, M.; Sinibaldi, A.; Notargiacomo, A.; Cibella, S.; Mattioli, F.; et al. Low-Temperature Stability and Sensing Performance of Mid-Infrared Bloch Surface Waves on a One-Dimensional Photonic Crystal. *ACS Appl. Mater. Interfaces* **2022**, *14*, 43853–43860. [CrossRef]
27. Gryga, M.; Ciprian, D.; Gembalova, L.; Hlubina, P. Sensing Based on Bloch Surface Wave and Self-Referenced Guided Mode Resonances Employing a One-Dimensional Photonic Crystal. *Opt. Express* **2021**, *29*, 12996. [CrossRef]
28. Zhang, C.; Liu, Q.; Peng, X.; Ouyang, Z.; Shen, S. Sensitive THz Sensing Based on Fano Resonance in All-Polymeric Bloch Surface Wave Structure. *Nanophotonics* **2021**, *10*, 3879–3888. [CrossRef]
29. Sinibaldi, A.; Doricchi, A.; Pileri, T.; Allegretti, M.; Danz, N.; Munzert, P.; Giordani, E.; Giacomini, P.; Michelotti, F. Bioassay Engineering: A Combined Label-Free and Fluorescence Approach to Optimize HER2 Detection in Complex Biological Media. *Anal. Bioanal. Chem.* **2020**, *412*, 3509–3517. [CrossRef]
30. Sinibaldi, A.; Montaño-Machado, V.; Danz, N.; Munzert, P.; Chiavaioli, F.; Michelotti, F.; Mantovani, D. Real-Time Study of the Adsorption and Grafting Process of Biomolecules by Means of Bloch Surface Wave Biosensors. *ACS Appl. Mater. Interfaces* **2018**, *10*, 33611–33618. [CrossRef]
31. Bhaskar, S.; Das, P.; Srinivasan, V.; Bhaktha, S.; Ramamurthy, S.S. Bloch Surface Waves and Internal Optical Modes-Driven Photonic Crystal-Coupled Emission Platform for Femtomolar Detection of Aluminum Ions. *J. Phys. Chem. C* **2020**, *124*, 7341–7352. [CrossRef]
32. Wang, R.; Lei, X.; Jin, Y.; Wen, X.; Du, L.; Wu, A.; Yuan, X. Directional imbalance of Bloch surface waves for ultrasensitive displacement metrology. *Nanoscale* **2021**, *13*, 11041–11050. [CrossRef]
33. Stella, U.; Grosjean, T.; De Leo, N.; Boarino, L.; Munzert, P.; Lakowicz, J.R.; Descrovi, E. Vortex Beam Generation by Spin-Orbit Interaction with Bloch Surface Waves. *ACS Photonics* **2020**, *7*, 774–783. [CrossRef]
34. Bezus, E.A.; Bykov, D.A.; Doskolovich, L.L. Integrated Diffraction Gratings on the Bloch Surface Wave Platform Supporting Bound States in the Continuum. *Nanophotonics* **2021**, *10*, 4331–4340. [CrossRef]
35. Deng, C.-Z.; Ho, Y.-L.; Clark, J.K.; Yatsui, T.; Delaunay, J.-J. Light Switching with a Metal-Free Chiral-Sensitive Metasurface at Telecommunication Wavelengths. *ACS Photonics* **2020**, *7*, 2915–2922. [CrossRef]
36. Kovalevich, T.; Belharet, D.; Robert, L.; Kim, M.-S.; Herzig, H.P.; Grosjean, T.; Bernal, M.-P. Experimental Evidence of Bloch Surface Waves on Photonic Crystals with Thin-Film LiNbO_3 as a Top Layer. *Photonics Res.* **2017**, *5*, 649. [CrossRef]
37. Shilkin, D.A.; Lyubin, E.V.; Soboleva, I.V.; Fedyanin, A.A. Direct Measurements of Forces Induced by Bloch Surface Waves in a One-Dimensional Photonic Crystal. *Opt. Lett.* **2015**, *40*, 4883. [CrossRef]
38. Xiang, Y.; Tang, X.; Fu, Y.; Lu, F.; Kuai, Y.; Min, C.; Chen, J.; Wang, P.; Lakowicz, J.R.; Yuan, X.; et al. Trapping Metallic Particles Using Focused Bloch Surface Waves. *Nanoscale* **2020**, *12*, 1688–1696. [CrossRef]
39. Safronov, K.R.; Popkova, A.A.; Markina, D.I.; Pushkarev, A.P.; Makarov, S.V.; Bessonov, V.O.; Fedyanin, A.A. Efficient Emission Outcoupling from Perovskite Lasers into Highly Directional and Long-Propagation-Length Bloch Surface Waves. *Laser Photonics Rev.* **2022**, *16*, 2100728. [CrossRef]
40. Stella, U.; Boarino, L.; De Leo, N.; Munzert, P.; Descrovi, E. Enhanced Directional Light Emission Assisted by Resonant Bloch Surface Waves in Circular Cavities. *ACS Photonics* **2019**, *6*, 2073–2082. [CrossRef]
41. Pirotta, S.; Xu, X.G.; Delfan, A.; Mysore, S.; Maiti, S.; Dacarro, G.; Patrini, M.; Galli, M.; Guizzetti, G.; Bajoni, D.; et al. Surface-Enhanced Raman Scattering in Purely Dielectric Structures via Bloch Surface Waves. *J. Phys. Chem. C* **2013**, *117*, 6821–6825. [CrossRef]
42. Ray, K.; Badugu, R.; Lakowicz, J.R. Bloch Surface Wave-Coupled Emission from Quantum Dots by Ensemble and Single Molecule Spectroscopy. *RSC Adv.* **2015**, *5*, 54403–54411. [CrossRef] [PubMed]
43. Mouttou, A.; Lemarchand, F.; Koc, C.; Moreau, A.; Lumeau, J.; Favard, C.; Lereu, A.L. Optimization of Resonant Dielectric Multilayer for Enhanced Fluorescence Imaging. *Opt. Mater. X* **2022**, *17*, 100223. [CrossRef]
44. Mogni, E.; Pellegrini, G.; Gil-Rostra, J.; Yubero, F.; Simone, G.; Fossati, S.; Dostálek, J.; Vázquez, R.M.; Osellame, R.; Celebrano, M.; et al. One-Dimensional Photonic Crystal for Surface Mode Polarization Control. *Adv. Opt. Mater.* **2022**, *10*, 2200759. [CrossRef]
45. Vosoughi Lahijani, B.; Descharmes, N.; Barbey, R.; Osowiecki, G.D.; Wittwer, V.J.; Razskazovskaya, O.; Südmeyer, T.; Herzig, H.P. Centimeter-Scale Propagation of Optical Surface Waves at Visible Wavelengths. *Adv. Opt. Mater.* **2022**, *10*, 2102854. [CrossRef]
46. Perani, T.; Aurelio, D.; Liscidini, M. Bloch-Surface-Wave Photonic Crystal Nanobeam Cavity. *Opt. Lett.* **2019**, *44*, 5133. [CrossRef]
47. Barachati, F.; Fieramosca, A.; Hafezian, S.; Gu, J.; Chakraborty, B.; Ballarini, D.; Martinu, L.; Menon, V.; Sanvitto, D.; Kéna-Cohen, S. Interacting Polariton Fluids in a Monolayer of Tungsten Disulfide. *Nat. Nanotechnol.* **2018**, *13*, 906–909. [CrossRef]

48. Mandracci, P.; Frascella, F.; Rizzo, R.; Virga, A.; Rivolo, P.; Descrovi, E.; Giorgis, F. Optical and Structural Properties of Amorphous Silicon-Nitrides and Silicon-Oxycarbides: Application of Multilayer Structures for the Coupling of Bloch Surface Waves. *J. Non Cryst. Solids* **2016**, *453*, 113–117. [CrossRef]
49. Munzert, P.; Danz, N.; Sinibaldi, A.; Michelotti, F. Multilayer Coatings for Bloch Surface Wave Optical Biosensors. *Surf. Coat. Technol.* **2017**, *314*, 79–84. [CrossRef]
50. Kovalevich, T.; Ndao, A.; Suarez, M.; Tumenas, S.; Balevicius, Z.; Ramanavicius, A.; Baleviciute, I.; Häyrinen, M.; Roussey, M.; Kuittinen, M.; et al. Tunable Bloch Surface Waves in Anisotropic Photonic Crystals Based on Lithium Niobate Thin Films. *Opt. Lett.* **2016**, *41*, 5616. [CrossRef]
51. Hsu, C.; Frisenda, R.; Schmidt, R.; Arora, A.; de Vasconcellos, S.M.; Bratschitsch, R.; van der Zant, H.S.J.; Castellanos-Gomez, A. Thickness-Dependent Refractive Index of 1L, 2L, and 3L MoS2, MoSe2, WS2, and WSe2. *Adv. Opt. Mater.* **2019**, *7*, 1900239. [CrossRef]
52. Hugonin, J.P.; Lalanne, P. RETICOLO Software for Grating Analysis. *arXiv* **2021**, arXiv:2101.00901.
53. Lalanne, P.; Morris, G.M. Highly Improved Convergence of the Coupled-Wave Method for TM Polarization. *J. Opt. Soc. Am. A* **1996**, *13*, 779–784. [CrossRef]
54. Sfez, T.; Descrovi, E.; Dominici, L.; Nakagawa, W.; Michelotti, F.; Giorgis, F.; Herzig, H.-P. Near-Field Analysis of Surface Electromagnetic Waves in the Bandgap Region of a Polymeric Grating Written on a One-Dimensional Photonic Crystal. *Appl. Phys. Lett.* **2008**, *93*, 061108. [CrossRef]
55. Ardizzone, V.; De Marco, L.; De Giorgi, M.; Dominici, L.; Ballarini, D.; Sanvitto, D. Emerging 2D Materials for Room-Temperature Polaritonics. *Nanophotonics* **2019**, *8*, 1547–1558. [CrossRef]
56. Lerario, G.; Ballarini, D.; Dominici, L.; Fieramosca, A.; Cannavale, A.; Holwill, M.; Kozikov, A.; Novoselov, K.; Gigli, G. Bloch Surface Waves for MoS2 Emission Coupling and Polariton Systems. *Appl. Sci.* **2017**, *7*, 1217. [CrossRef]
57. Maggiolini, E.; Polimeno, L.; Todisco, F.; Renzo, A.D.; Giorgi, M.D.; Ardizzone, V.; Cannavale, A.; Pugliese, M.; Maiorano, V.; Gigli, G.; et al. Strongly Enhanced Light-Matter Coupling of a Monolayer WS2 from a Bound State in the Continuum. *arXiv* **2022**, arXiv:2209.00129.
58. Li, H.; Chen, B.; Qin, M.; Wang, L. Strong Plasmon-Exciton Coupling in MIM Waveguide-Resonator Systems with WS_2 Monolayer. *Opt. Express* **2020**, *28*, 205. [CrossRef]
59. Kim, H.-C.; Kim, H.; Lee, J.-U.; Lee, H.-B.; Choi, D.-H.; Lee, J.-H.; Lee, W.H.; Jhang, S.H.; Park, B.H.; Cheong, H.; et al. Engineering Optical and Electronic Properties of WS_2 by Varying the Number of Layers. *ACS Nano* **2015**, *9*, 6854–6860. [CrossRef]
60. Flatten, L.C.; He, Z.; Coles, D.M.; Trichet, A.A.P.; Powell, A.W.; Taylor, R.A.; Warner, J.H.; Smith, J.M. Room-Temperature Exciton-Polaritons with Two-Dimensional WS2. *Sci. Rep.* **2016**, *6*, 33134. [CrossRef]
61. Zhang, L.; Gogna, R.; Burg, W.; Tutuc, E.; Deng, H. Photonic-Crystal Exciton-Polaritons in Monolayer Semiconductors. *Nat. Commun.* **2018**, *9*, 713. [CrossRef]
62. Ballarini, D.; De Liberato, S. Polaritonics: From Microcavities to Sub-Wavelength Confinement. *Nanophotonics* **2019**, *8*, 641–654. [CrossRef]

micromachines

Article

Thermooptical PDMS-Single-Layer Graphene Axicon-like Device for Tunable Submicron Long Focus Beams

Giancarlo Margheri [1,*], André Nascimento Barbosa [2], Fernando Lazaro Freire, Jr. [2] and Tommaso Del Rosso [2]

[1] Institute for Complex Systems of National Council of Researches of Italy, Separate Location of Sesto Fiorentino, Via Madonna del Piano, Sesto, 50019 Florence, Italy
[2] Department of Physics, Pontifícia Universidade Católica do Rio de Janeiro, Rua Marques de São Vicente, Rio de Janeiro 22451-900, Brazil
* Correspondence: giancarlo.margheri@cnr.it; Tel.: +39-522-6619

Abstract: Submicron long focusing range beams are gaining attention due to their potential applications, such as in optical manipulation, high-resolution lithography and microscopy. Here, we report on the theoretical and experimental characterization of an elastomeric polydimethylsiloxane/single layer graphene (PDMS/SLG) axicon-like tunable device, able to generate diffraction-resistant submicrometric spots in a pump and probe configuration. The working principle is based on the phase change of an input Gaussian beam induced in the elastomer via the thermo-optical effect, while the heating power is produced by the optical absorption of the SLG. The phase-modified beam is transformed by an objective into a long focus with submicron diameter. Our foci reach an experimental full width at half maximum (FWHM) spot diameter of 0.59 μm at the wavelength of 405 nm, with the FWHM length of the focal line greater than 90 μm. Moreover, the length of the focal line and the diameter of the focus can be easily tuned by varying the pump power. The proposed thermo-optical device can thus be useful for the simple and cheap improvement of the spatial resolution on long focus lines.

Keywords: axicons; thermo-optical devices; beam shaping; long focusing devices; graphene–PDMS hetero interfaces

Citation: Margheri, G.; Barbosa, A.N.; Freire, F.L., Jr.; Del Rosso, T. Thermooptical PDMS-Single-Layer Graphene Axicon-like Device for Tunable Submicron Long Focus Beams. *Micromachines* **2022**, *13*, 2083. https://doi.org/10.3390/mi13122083

Academic Editor: Andrey V. Novitsky

Received: 12 October 2022
Accepted: 22 November 2022
Published: 26 November 2022

1. Introduction

Long focusing range beams (LFRB), also known as quasi-Bessel beams (QBBs), are appealing due to their non-diffractive properties. Their use in various important applications, including scanning microscopy [1–3], optical manipulation [4,5] and lithography [6–8], would be enhanced if the focal spot had a submicron full width at half maximum (FWHM) diameter. As is well known, the submicron requirement can be fulfilled only if the device producing LFRBs has a high numerical aperture (NA). This is not an easy task to achieve with refractive axicons (refraxicons) whose NA cannot exceed 0.75 [9]; the other widespread axicon devices use a high NA microscope objective with a central obscuration that strongly lowers the efficiency.

Moreover, refraxicons have two typical fabrication errors: (i) an elliptical cross-section resulting in a nonrotational symmetric beam, and (ii) imperfections of the tip, which is typically shaped as an aspherical micro lens [10], introducing strong fluctuations in the focus line. Several techniques have been developed to bypass the effects of imperfect tips, such as beam expansion, Fourier filtering [9], the use of liquid immersion [11], laser wet etching [12] or the use of photonic crystals and metasurfaces [9,13,14]. However, these solutions are often not efficient or are technologically difficult to implement.

An alternative is the use of positive reflective axicons, with benefits including the absence of chromatism, tolerance to high power densities and a very small cuspidal region [15]. These characteristics allow working with small laser beams and minimize the intensity fluctuations in the Bessel focus [10]. However, in a reflective geometry, the focus is close to the sample surface and on the same side of the input beam, with consequent obscuration effects. In order to avoid these drawbacks, the refraxicon must work in a tilted geometry, which is quite prone to aberrations, particularly for foci with submicrometric spot diameters. Thus, when a system based on a reflecting axicon is dedicated to submicrometric focusing, it usually presents a relevant opto-mechanical complexity [16], low stability and high cost.

Another desirable feature when working with LFRBs is the possibility to tune the light distribution without moving parts in the experimental set-up. In this case, the tuning function is generally obtained with complicated and high-cost experimental benches, or by the use of specialized components based on metasurfaces [17–19].

In this scenario, the formation of LFRBs with submicrometric cross-section is a field in which improvements are still demanded, especially in terms of optical efficiency, simplicity in the fabrication of the device and the tunability of both the focus diameter and extension.

Here, we propose an alternative solution in which LFRBs are formed by exploiting bilayers composed of a weakly absorbing film and a thicker layer possessing thermo-optical properties. Under proper illumination, the latter film develops a spatial variation of the refractive index, which modifies in an axicon-like way the phase of a probe beam. Many organic polymers exhibit negative thermo-optical coefficients (TOC) with a high absolute value (i.e., polymetylmetacrilate (PMMA) or polystyrene (PS)). Among these materials, we found that the most favorable performances concerning TOC, transparency, durability, stability, processability, cost and ease of supply are exhibited by the elastomer polydimethylsiloxane (PDMS). For PDMS, which is stable up to 200 °C, the absolute value of the TOC is around 4.5×10^{-4} °C^{-1}, approximately 4-fold higher than that of PMMA or PS.

The heating function is performed by the weakly absorbing film, which must also have high thermal conductivity. The choice of the proper optical absorbing layer is not a simple task. In fact, the ideal absorber should present the maximum transparency together with suitable heating behavior in the whole Vis–NIR region of the electromagnetic spectrum. A possible approach would be to consider absorbers based on metallic or semiconductors thin films, whose fabrication, however, is not an easy task using conventional techniques [20–23]. Indeed, the vapor condensation on the dielectric support may lead to a randomly nanostructured surface [21–23]. This is particularly problematic for noble metals thin films, where the nanostructured surface is characterized by a wavelength-dependent localized surface plasmon resonance (LSPR). While the metal nanostructures can be fruitfully used to perform numerous surface-enhanced spectroscopies [24,25], they represent a serious drawback for the performances of a wideband thermo-optical axicon. Metallic alloys based on nickel are used for the fabrication of commercial neutral optical density filters (see, for instance, Thorlabs neutral density reflective filters) up to a transmittance of 80% and an absorption of 10%, but they suffer from oxidation instability when the temperature rises up to 100 °C. Other nonmetallic absorbing substances, such as carbon black or graphite, could be used to heat the polymer, but they cannot be deposited in films thin enough to guarantee a transparency higher than 70% [26,27].

To summarize, the fabrication of ultrathin thin absorbers with (1) high flat spectral transmission, (2) sufficient absorption in the Vis–NIR spectral region, (3) high and uniform heat conduction and (4) high chemical and structural stability is currently a difficult challenge.

In this context, particular attention is focused on the emerging 2D materials, thanks to their intrinsic characteristic of small effective thickness [28] and interesting optical properties for photonics applications [29–32].

Among the low-dimensional materials, graphene was the first to be obtained experimentally [33,34], and now it can be synthesized by chemical vapor deposition (CVD) in a controlled number of layers and in foils having linear dimensions in the cm range [35]. These features greatly simplify the optical excitation of the 2D material, together with the precise control of the optical properties, which strictly depend on the number of single-layer graphene (SLG) foils. Indeed, in spite of its effective thickness of 0.34 nm, a typical SLG has an unusual optical absorption of 2.3%, almost flat in the Vis–NIR spectrum [36,37], which represents an unsurpassed advantage with respect to other light absorbers. Moreover, its thermal conduction is one of the highest known (4000 $Wm^{-1} K^{-1}$), and it is expected to efficiently drain the heating power at the graphene–PDMS interface.

In this work, we demonstrate that in spite of its low absorption, the SLG can significantly raise its temperature, even delivering moderate optical pump powers, heating the PDMS sufficiently to push its temperature close to the tolerance limit (~470 K) and generating a remarkable refractive index gradient. Moreover, the control of the transverse and longitudinal dimensions of the spot is performed optically by varying the pump power, allowing the fine-tuning of the focus diameter and its extension.

2. Thermooptical and Electromagnetic Modeling

2.1. Model of the PDMS/SLG Thermooptical Device

The working principle of the proposed thermooptical axicon-like device is reported in Figure 1.

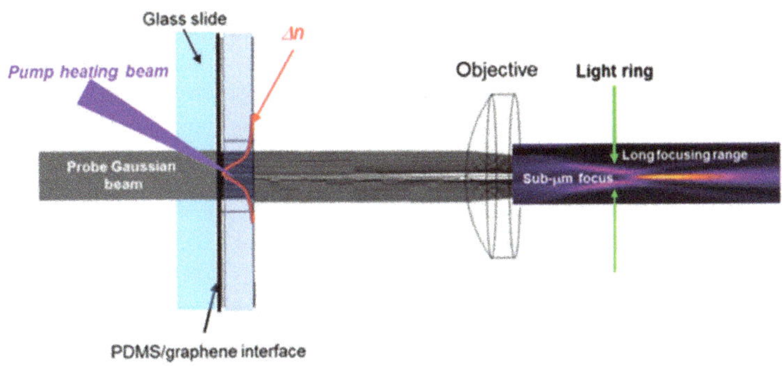

Figure 1. Working principle of the PDMS/SLG thermo-optical device. The absorbed pump heats up the PDMS locally, creating a refractive index profile Δn characteristic of a negative axicon. The input beam is focused into a light ring by an objective. The successive propagation produces the sub-micrometric LFRB.

The device is based on a PDMS/SLG bilayer used in a pump and probe configuration. The pump beam (wavelength: 405 nm) is tightly focused onto the SLG, which absorbs part of the light and heats the elastomer, generating a negative gradient index (GRIN) distribution Δn. A Gaussian probe beam is launched to the device from the back side and its phase is transformed by the GRIN distribution. As the heating of PDMS is pointy, the thermo-optical phase issued to the input beam resembles that produced by a negative axicon. An objective finally focuses the modified probe beam in a tight ring, with a long focusing range and a submicrometric cross-section.

The geometrical details of the graphene–PDMS thermo-optical device considered in the simulations are illustrated in Figure 2.

The diameter of the PDMS disk is 5 mm, and its thickness is 2 mm. The elastomer is in contact with an SLG, deposited on a BK7 glass disk with 1 mm thickness. The whole system is surrounded by air.

The heat source, located in the SLG, is simulated with a Gaussian heat surface power distribution (W/m^2), with a waist diameter $2w_0$ = 70 μm. The heating power density is $Q(r) = P_h/(\pi w_0^2)\exp(-(r/w_0)^2)$, where P_h is the heating power and r is the radial coordinate on the graphene surface. The maximum P_h is 20 mW, and is varied in steps of 5 mW. The calculations were performed with the COMSOL MULTIPHYSICS software 5.3a, whose database of materials provides the physical constants of PDMS and BK7, while those of graphene are taken from the available literature. As visible in Figure 2, the resulting temperature distributions $T(r,z)$ in PDMS exhibit a central absolute maximum $T(0,0)$ of 470 K corresponding to P_h = 20 mW, decreasing on the axis with a rate of 10 K/mW.

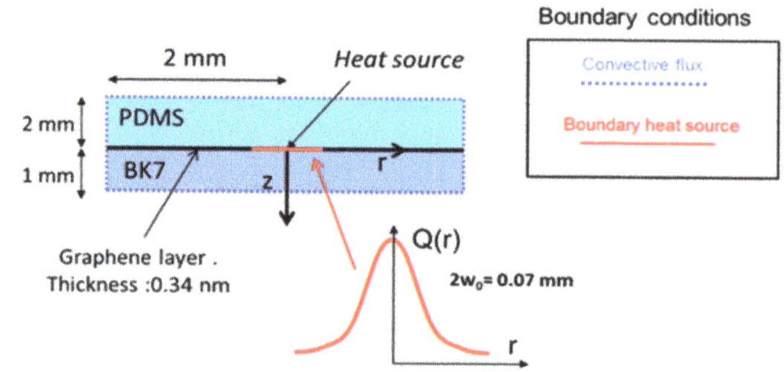

Figure 2. Geometry of the thermo-optical model, completed with the boundary conditions used in the COMSOL code. The convective parameter is $h = 10\ W/m^2K$.

Even if the data sheets of PDMS (specifically Sylgard 187) claim that the elastomer is stable until ~470 K, in order to guarantee a margin of 20 K for a safe operation of the device, we will consider as the maximum heating power the value $P_h = 16$ mW, corresponding to a temperature of 450 K (~150 Celsius). The FWHM of the radial temperature distribution is constant with P_h and equal to 0.086 mm, indicating a smearing with respect to the width of the heating source (0.07 mm), likely due to the high thermal conduction of graphene [37]. The temperature distribution $T(r,z)$ gives rise to the variation of the refractive index of Figure 3 that follows the relationship $\Delta n(r,z) = -4.5 \times 10^{-4}(T(r,z)-T_0)$, where $T_0 = 293.15$ K is the ambient temperature. For convenience reasons, in the following, we will consider the absolute value of $\Delta n(r,z)$ instead of the actual negative value. As anticipated, the distribution $|\Delta n(r,z)|$ resembles a tipped structure that, in analogy with classical axicons, is responsible for the generation of annular beams (see next section). The apex region, reported in the inset, has a smoothed apex with a diameter of about 106 μm, higher than the beam waist of the focused pump beam.

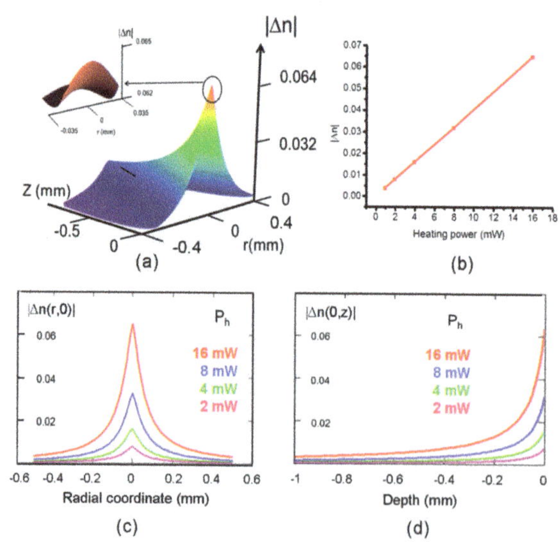

Figure 3. Results of the thermo−optical modeling. (**a**) Spatial variation of the refractive index due to the thermo-optical effect. In the inset, the apex distribution is shown. (**b**) Variation of $|\Delta n(0,0)|$ with the heating power. (**c**) Radial distribution of the modulus on the graphene layer surface for several heating powers and (**d**) its axial distribution for the same powers.

The variation of |Δn(0,0), that is, the absolute maximum of |Δn(*r*,*z*)|, is linear with P_h, as illustrated in Figure 3b, with a growth rate of 0.004 refractive index units (RIU)/mW, and reaches its maximum of 0.065 at P_h = 16 mW.

The radial behavior of |Δn| at *z* = 0 mm, namely |Δn(*r*,0)|, is shown in Figure 3c. Following the trend of the central maximum, at a given radius r, it decreases proportionally, with P_h halving at 0.067 mm from the center.

The axial behavior of |Δn| with *z* is represented in the plot of |Δn(0,*z*)| reported in Figure 3d. Its maximum |Δn (0,0)| halves at a depth *z* = 0.062 mm from the graphene layer. The behavior of |Δn(0,*z*)| is linear with P_h at a given *z*-coordinate, as it happens for the r-coordinate.

Given the linear relationship between |Δn(*r*,*z*)| and T(*r*,*z*), the temperature distribution resembles the distribution of |Δn(*r*,*z*)|, and it has not been reported.

2.2. *Electromagnetic Modeling*

As anticipated, the device exploits the phase variation of a probe beam that travels through the heated PDMS and experiences a thermo-optical phase variation. A further phase change is added by a focusing objective to generate the real LFRB. These variations of the phase of the probe electric field along its propagation are schematically reported in Figure 4.

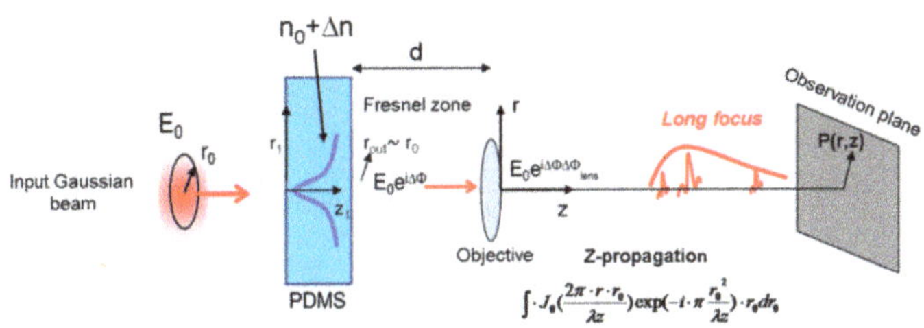

Figure 4. Propagation of the electromagnetic field through the optical chain. The input field E_0 acquires two phase delays, ΔΦ and $\Delta\Phi_{lens}$, due to the refractive index gradient in the PDMS and the lens, respectively. The objective is located at distance *d* from the sample of about 15 mm. The beam is finally transformed into a long focusing range beam. See text for further details.

We assume that the input electric field has a Gaussian distribution $E_0(r_0) = E_{amp} \exp(-(r_0/w_0)^2)$ where r_0 is the radial coordinate at the input plane and w_0 is the beam waist (w_0 = 0.5 mm, E_{amp} = 1 V/m). Concerning the probe wavelengths, we chose two values representing approximately the range of the visible spectrum, namely λ_1 = 633 nm and λ_2 = 405 nm, for red and violet radiation, respectively. Considering the propagation through the heated PDMS, the electric field experiences an overall phase change $\Delta\Phi(r_{out})$ given by:

$$\Delta\Phi(r_{out}) = \frac{2\pi}{\lambda}\int_0^L n(r_1, z_1)dl \tag{1}$$

where $dl = (dr_1^2 + dz_1^2)^{1/2}$ is the infinitesimal path on the ray trajectory, λ is the wavelength of the probe beam in vacuum, *L* is the overall length of the ray trajectory and r_{out} is the output radial coordinate. r_1 and z_1 are the radial and longitudinal coordinates inside the PDMS, and the refractive index of the heated elastomer is expressed as $n(r_1, z_1) \sim n_0 + \Delta n(r_1, z_1)$, where n_0 is the refractive index of PDMS at ambient temperature, and $\Delta n(r_1, z_1)$ is the refractive index change due to the thermo-optical effect. In the usual paraxial approximation, $dr/dz_1 << 1$, so that $dl \sim dz_1$, and the integral simplifies in:

$$\Delta\Phi(r_0) = \frac{2\pi}{\lambda}\int_0^s n(r_0, z)dz \quad (2)$$

where s is the sample thickness, r_0 is the input radial coordinate and r_{out} has been substituted by r_0.

The field at the PDMS output plane is thus given by $E_0(r_0)\exp(i\Delta\Phi(r_0))$. Then, the beam propagates towards a focusing objective whose focal length is f. As the objective is placed after the PDMS surface at a distance $d \sim 15$ mm, it is located in the Fresnel zone of the output pupil (approximately represented by the field beam waist), as the Fresnel number at wavelength λ_1 is $w_0^2/ld = 26.3 >> 1$ [38]. By exploiting the discussion on the Fresnel approximation, the field at the lens input plane is in good approximation of the geometrical projection of the field present at the PDMS exit face. The focusing objective issues a phase shift to the incoming field that, in the case of diffraction limited behavior, is quadratic with r_0 and expressed by the well-known relationship $\Delta\Phi_{lens}(r_0) = -\pi r_0^2/\lambda f$. After the lens, the field propagates towards the observation point P(r,z), where r is the distance from the optical axis and z is the distance from the lens. The resulting field at P is found with the application of the Fresnel integral and is calculated with the equation:

$$E(r,z) = A\int_0^\infty E_0(r_0)\cdot J_0\left(\frac{2\pi\cdot r\cdot r_0}{\lambda z}\right)\exp(i\cdot\Delta\Phi_{lens}(r_0))\exp(i\Delta\Phi(r_0))\exp\left(-i\cdot\pi\frac{r_0^2}{\lambda z}\right)\cdot dr_0 \quad (3)$$

where the term $\exp(-\pi r_0 \lambda z)$ represents the phase change due to the propagation towards the observation point, $A = A_0 \cdot \exp(ikz)/\lambda z$ and A_0 is the amplitude of the impinging electric field. The intensity is $\sim |E(r,z)|^2$ and is expressed in normalized or arbitrary units.

The electromagnetic modeling is used at first to calculate the effect due exclusively to the thermally induced refractive index distribution, obtained by dropping the phase contribution $\Delta\Phi_{lens}(r_0)$ in the integral (3). The phase term $\Delta\Phi(r_0)$ is evaluated by considering the refractive index distribution calculated in the thermo-optical model without approximations.

The theoretical far field intensities for the wavelength λ_1 are shown in Figure 5. The central region of the intensity pattern gradually depletes in response to the increase in P_h, and the peak to peak diameter increases as well. The axial intensity halves at $P_h = 2.4$ mW (Figure 3a) and vanishes at $P_h = 4.2$ mW (Figure 3b). For higher powers, the probe beam propagates into the far field as an empty annular beam (Figure 3c), reaching an angular semi-aperture of 0.0037 rads at $P_h = 16$ mW.

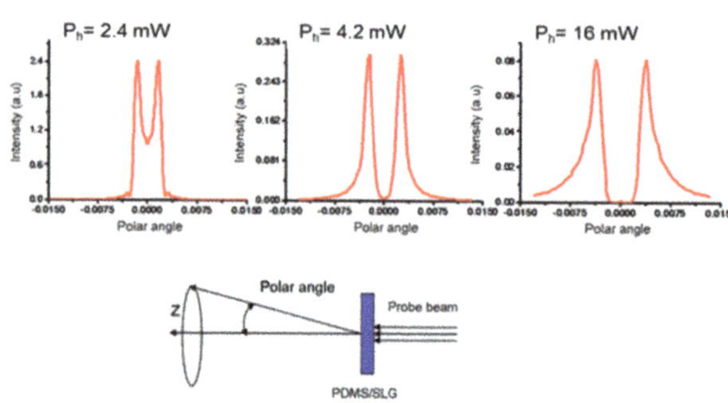

Figure 5. Upper part: Theoretical far field at three different heating powers P_h at the wavelength of 633 nm. Lower part: definition of the polar angle.

The results obtained for the probe wavelength λ_2 are similar, because the induced refractive index gradient does not have significant dispersion in the visible range, because the dominant effect is dependent on the temperature T [39].

The phase-modulated beam is in turn transformed by the focusing lens (objective), and the corresponding normalized axial and transversal intensity distributions are calculated by solving the integral of Equation (3) for the three pump powers P_h = 4 mW, 8 mW and 16 mW and considering the focal length f = 2.6 mm of a commercial objective (see Section 3).

The results obtained for the probe wavelength λ_1 are reported in Figure 6a,b, while Figure 6c,d reports the normalized intensities for the wavelength λ_2.

Table 1. Theoretical performances of the LFRB system based on PDMS/SLG interface. Objective focal length: 2.6 mm. Probe wavelengths: 633 nm (*405* nm).

P_h (mW)	$FWHM_{line}$ (μm)	D_F (μm)	$L_{\mu m}$* (μm)	R_G (μm)
4	11.5 (*9.3*)	1.32 (*0.6*)	ND (*ND*)	4.4 (*1.5*)
8	26.7 (*25.5*)	1.21 (*0.6*)	ND (*21.3*)	3.6 (*1.4*)
16	88.2 (*90.3*)	0.96 (*0.5*)	8.60 (*72*)	*2.3 (1.23)*

* Axial length within which D_F < 1 μm.

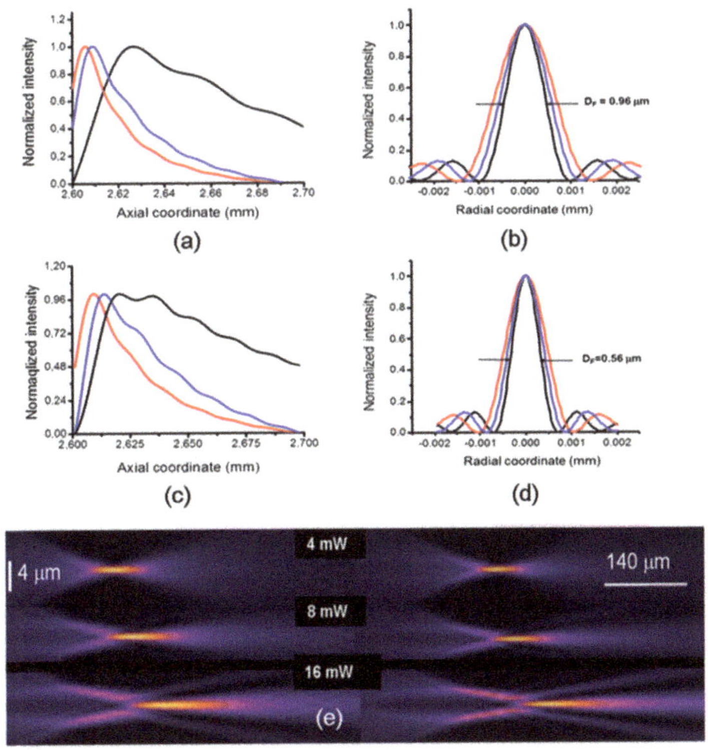

Figure 6. (**a–d**) Theoretical normalized light distributions beyond the focusing optics (focal length, origin of the x-axis = 2.6 mm) for three heating powers P_4 = 4 mW, 8 mW, 16 mW. Upper row (**a**,**b**): λ_1 = 633 nm; lower row (**c**,**d**): λ_2 = 405 nm. (**a**,**c**) Axial and (**b**,**d**) radial intensities. In (**b**,**d**), only the radial FWHMs relative to P_h = 16 mW are reported. The other parameters are listed in Table 1 for a simpler comparison. (**e**) Computed light intensities for the same heating powers for two wavelengths λ_1 (left side) and λ_2 (right side). The light propagates from left to right. Notice the converging ring on the left of each focal line that evolves in the long range focus.

Starting from the probe at λ_1, the axial distributions show fast rises followed by slower declines, reaching absolute intensity maxima at coordinates z_{max}. A weak low spatial frequency modulation is present, while the high-frequency oscillations near z_{max} are absent, demonstrating that the effect of the tip curvature (see Figure 3) is negligible [10]. The axial coordinate z_{max} has a right shift of about +20 μm passing from P_h = 4 mW to P_h = 16 mW. At P_h = 16 mW, the axial light distribution has a FWHM value (hereafter $FWHM_{line}$) of 88.2 μm, and lowers significantly at lower P_h, down to 11.5 μm (Table 1). The radial intensity distributions are calculated at the coordinate z_{max} corresponding to each heating power, and are plotted in Figure 6b. The curves show that their FWHMs (hereafter D_F) decrease at increasing P_h, and reach the minimum value D_F = 0.96 μm at P_h = 16 mW. Thus, the theoretical model predicts that for the red probe radiation at λ_1, D_F is submicrometric at the maximum P_h, while at lower heating powers, it enlarges by approximately 25% (see Table 1), reaching a value of 1.32 μm at the minimum P_h = 4 mW.

The long focusing feature is evident by comparing the $FWHM_{line}$ of LFRB with the depth of focus obtained with conventional Gaussian beams, which is routinely considered as the double of the Rayleigh range, $2\pi w_0^2/\lambda$ (hereafter R_G). For instance, considering a Gaussian beam with a waist equal to the half of the D_F calculated for P_h = 16 mW, 0.48 μm, it results in R_G = 2.3 μm, which is 24 times lower than $FWHM_{line}$ (see Table 1). Passing to the probe at wavelength λ_2, the trend of the axial light distributions (Figure 6c) follows approximately the same behavior of those calculated for λ_1, showing that the $FWHM_{line}$ is relatively wavelength-independent (Table 1), while D_F decreases with it. Considering, for instance, the maximum P_h = 16 mW, we have D_F = 0.56 μm, against D_F = 0.96 μm at λ_1. The Gaussian depth of focus corresponding to $D_F/2$ = 0.28 μm results in R_G = 1.23 μm, 73.4 times lower than $FWHM_{line}$.

For the sake of clarity, the behavior of $FWHM_{line}$ vs. P_h for λ_2, the more favorable condition, is resumed in the 2-Y plot of Figure 7, showing an almost linear trend with a rate of growth of 7.5 μm/mW.

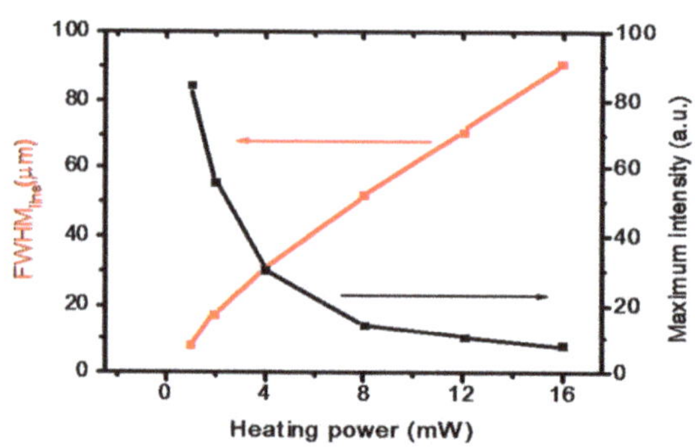

Figure 7. Focal line length ($FWHM_{line}$) and maximum on-axis intensity vs. the heating power P_h for λ_2 = 405 nm. The focal length of the focusing lens is 2.6 mm. The increase in $FWHM_{line}$ is accompanied by a reduction in the intensity. See text for details.

The 2D intensity plots of the focal regions are shown in the false color plots of Figure 6e. The focusing of the incoming light crowns into annular light distributions appears evident (see Figure 5), in agreement with the results of the ray tracing simulation presented in Figure 1.

It is worth noting that as the focus line lengthens at increasing P_h, the side lobes of the radial intensity shrink towards the center, subtracting power to the central lobe, whose D_F decreases at the same time. This last effect, however, is not sufficient to compensate for the loss of intensity due to the spread of the optical power in the side lobes, and as a consequence, the maximum intensity of the central spot decreases. This consideration is clearly evidenced in the plot of Figure 7, calculated for λ_2 = 405 nm. Here, the central intensity shows a decrease of approximately a decade passing from P_h = 1 mW to P_h = 16 mW, while the $FWHM_{line}$ expands approximately with the same ratio. As the PDMS/SLG behaves in a negative axicon-like way, similar trends are expected for a real axicon coupled to a positive lens [40]. The above result suggests that the operation with this device requires a preliminary tradeoff between the requirement of a large $FWHM_{line}$ and a proper power concentration in the focus. In this respect, the tunability of the presented device can be useful to finely adjust the proper operational conditions.

3. Experimental Section

3.1. Fabrication of the Device

The predicted behavior was experimentally verified using a sample of SLG deposited on glass and covered with PDMS. The SLG was synthesized by chemical vapor deposition (CVD) and deposited onto BK7 glass slides of 1 mm thickness by using the experimental procedures outlined in [41,42]. The details of the methods are reported briefly in the Supplementary Material, where we show in Figure S1 the Raman spectra of the graphene deposited on the glass, confirming the presence of a high-quality SLG. A small portion of the glass surface is not covered by graphene to serve as a reference for the absorption measurements.

The graphene monolayers were covered by a 2 mm thick PDMS layer prepared using a 10:1 mixture of monomer and curing liquid. After the mixture degassing, 1 mL of the liquid PDMS was poured onto the graphene layer (the viscosity of the liquid mixture was sufficient to avoid the overflow of the liquid out of the borders of the support) and heated at 50 °C for 20 min to efficiently activate the polymerization, completed overnight in air at ambient temperature.

3.2. Experimental Set-Up

We performed two different sets of measurements, the first aimed at demonstrating the formation of the predicted far field annular beams, prerequisite for the formation of LFRBs, and the second at measuring the diameter and length of the long focus line.

The scheme of the experimental apparatus is shown in Figure 8.

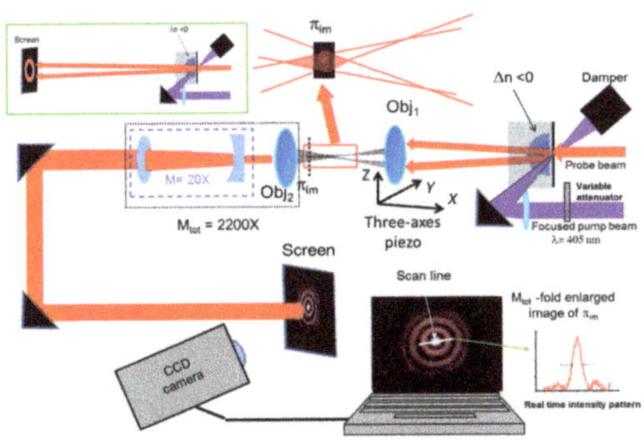

Figure 8. Experimental apparatus. The input probe beam travels through the heated PDMS, which modifies and transmits it to the focusing objective Obj_1 (focal length 2.6 mm), mounted on a three-axis piezo. The focusing action determines the formation of a superposition volume (evidenced in the red box and exploded in the upper part) where the long focus develops. The section of the focus located on the plane π_{im} is conjugated onto the observation screen by a 2-stage optical magnifier, composed of an objective Obj_2 (focal length 2.6 mm, measured magnification 110x, and a 20x Galilean telescope. The focus image, magnified 2200 times, is then projected onto the screen and imaged by a CCD camera. An image processing software permits checking the image intensity on a given line in real time, allowing the monitoring of the change in the spot dimensions in response to changes in P_h, obtained with a variable attenuator. The measure of the far field annular distributions is performed without Obj_1 and the 2200x magnification system (green frame in the upper left side).

In the former case, the PDMS/SLG was illuminated from the PDMS side with a focused pump beam generated by a diode pumped solid state laser (λ_2 = 405 nm), whose power can be adjusted by using an optical variable attenuator (Figure 8). At the same time, the sample was illuminated from the other side with a collimated probe beam coming from either a He-Ne laser (probe at λ_1) or a diode pumped solid-state laser (probe at λ_2). The diameter of the probe beam was adjusted to obtain an almost input Gaussian waist w_0 = 0.5 mm, corresponding to an intensity waist of 0.36 mm, while the focus of the pump beam was adjusted in a spot with an average diameter of 70 μm. The probe beam propagates through the PDMS and is transformed into an annular ring that is projected onto a screen placed at 350 mm distance, as illustrated in the inset of Figure 8.

The images were acquired with a CCD camera (Watec 902H) and elaborated by using the software IMAGE PRO Plus, providing in real time the light distribution on a chosen scan line during the pump power changes. After a calibration step, the measurement of the beam D_F can be easily performed.

In the latter case, the measurements on the submicron LFRB were performed using the complete optical system shown in Figure 8. While the generation of a phase-modulated beam at the PDMS output remains the same as in the previous paragraph, a high NA objective Obj_1 (Newport MVC-60x, 0.85 NA, 60x, focal length 2.6 mm), whose first interface is placed at a ~15 mm distance from the sample, focuses the light coming from the output surface of PDMS. The objective Obj_1 is mounted on a three-axis piezo translator (Piezojena Tritor 100 piezo stage working in closed loop operation, with 80 μm translation range and 0.53 μm/V resolution), and moved in steps of 5.8 μm.

The focal spot produced by Obj_1 is magnified by a two-stage optical system, formed by a second Newport MVC-60x objective Obj_2 (0.85 NA, 60x, focal length: 2.6 mm) with 110x magnification, and a further 20x Galilean telescope. Thus, the transverse light distribution is magnified 2200 times, and then it is projected onto a screen placed at a fixed position, where it can be reliably observed by a CCD camera. The image was then acquired and elaborated with the same procedure used for the analysis of the annular patterns.

The position of best focus is obtained by moving Obj_1 axially with the piezo translator until the intensity pattern on a chosen diametric line reaches the minimum radius, and the contrast with the side lobes is maximum. A similar measurement was performed at P_h = 16 mW with the other probe beam at wavelength λ_2.

In order to test the long focusing performance of the device, we measured the maximum light intensity on the focal line at P_h = 16 mW for both radiations at λ_1 and λ_2. This operation was performed by moving horizontally Obj_1, causing the axial shift of the focus line in front of the magnifying system. The screen collected the magnified image of the field intensity present on the conjugated plane π_{im} (see Figure 6). We acquired the image of the spot focus for different displacements of Obj_1, and recorded the central maximum intensity for each displacement.

4. Results and Discussion

The measured absorption of the PDMS/SLG bilayer was 3.4%, higher than that reported for a free-standing SLG (2.3%) at the wavelength λ_1. As this film has sufficient quality (see Section 3.1), this effect is mainly attributable to the multireflections in the bilayer, even if a minor contribution can come from some polymer residuals. To demonstrate this, we performed a simulation using a graphene refractive index of 2.7 + 1.33i [42], a monolayer thickness of 0.34 nm and a PDMS refractive index of 1.43. The calculations, carried out with the multilayer evaluation software WINSPALL, give at 45° incidence angle a reflection R = 2.3% and a transmission T = 0.94%, with a resulting absorption $A = 1 - R - T$ = 3.6%, close to the measured value.

This weak absorption of SLG is nevertheless high enough to generate the predicted transformation of the input beam into an annular beam in the far field, as shown in Figure 9. The images of the rings recorded at a distance of 350 mm from the sample (inset of Figure 8) show the gradual depletion of the central region with the increase in P_h. At

P_h = 5 mW, the center appears dark, in agreement with the expected theoretical value of 4.2 mW. Figure 9b reports the linear increase in the circles' diameter (distance between the maxima) with P_h. The diameter subtends a full angle whose rate of growth is approximately 5.3×10^{-4} rads/mW. This value is close for the two wavelengths, confirming the weak dispersion of the thermo-optical coefficient of PDMS in the visible range. It is worth noting that the capability to shape annular rings could be exploited to perform important functions [43], but the optimization of the working conditions would require a more detailed investigation that is beyond the scope of the present work.

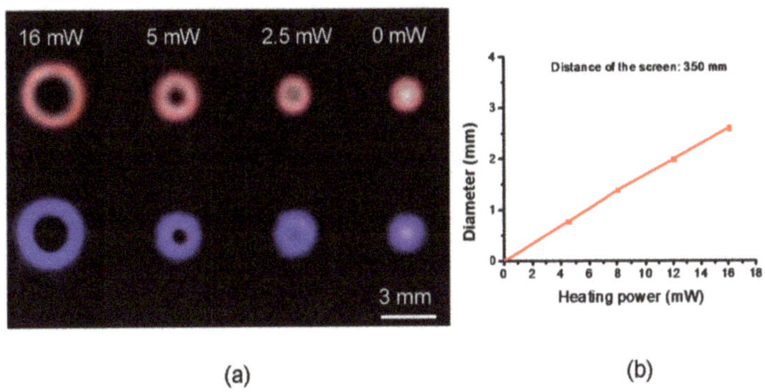

Figure 9. (**a**) Annular rings produced at 350 mm distance from the sample at λ_1 = 633 nm (upper part of the photo) and λ_2 = 405 nm (lower part) for different heating powers. (**b**) The ring diameter, intended as the distance between the maxima directly measured on a diametric scanning line (see caption of Figure 7), vs. the heating power P_h.

The magnified (2200x) radial light distributions of the focused probe beam at λ_1 are illustrated in Figure 10. The resulting spot diameters are D_F = 0.97 μm, 1.25 μm and 1.4 μm for the three heating powers of 4 mW, 8 mW and 16 mW, respectively. In the case of the probe beam at λ_2 (Figure 11), the measured spot diameters at the same P_h decrease as theoretically predicted, and their values are D_F = 0.59 μm, 0.63 μm and 0.73 μm for the three powers, respectively, in agreement with the theoretical findings reported in Table 1.

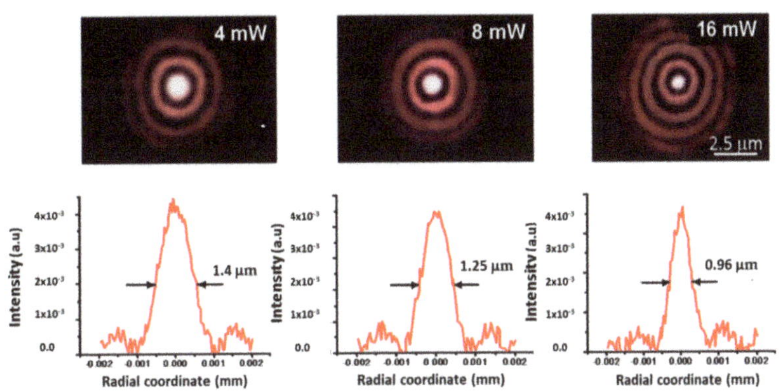

Figure 10. Upper part: photos of the focal LFRB spots magnified 2200 times at different heating powers at λ_1 = 633 nm. The scale bar indicates the effective dimensions of the focal cross−section. Lower part: Corresponding measured diametric light distributions.

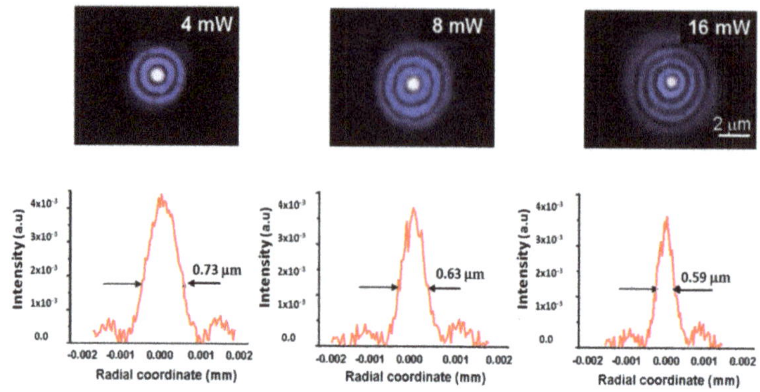

Figure 11. Upper part: Photos of the focal LFRB spots magnified 2200 times at different heating powers at λ_2 = 405 nm. The scale bar represents the actual dimensions of the focal cross-section. Lower part: Corresponding measured diametric light distributions.

The experimental axial intensity distributions measured at P_h = 16 mW for the two probes at wavelengths λ_1 and λ_2 are shown in Figure 12a,b, respectively. As shown, the theoretical predictions and the experimental findings agree well. In order to check the foci robustness to diffraction, we further measured the spot diameters at the extremes of the $FWHM_{line}$, axially displacing Obj_1 until the two positions P_1 and P_2, where D_F is ~1 μm, are locked. At the probe wavelength λ_1, this happens when the distance P_1P_2 is equal to 7.8 μm, while at λ_2 nm, this distance is equal to 68 μm, not far from the theoretical values (8.6 μm and 75 μm, respectively).

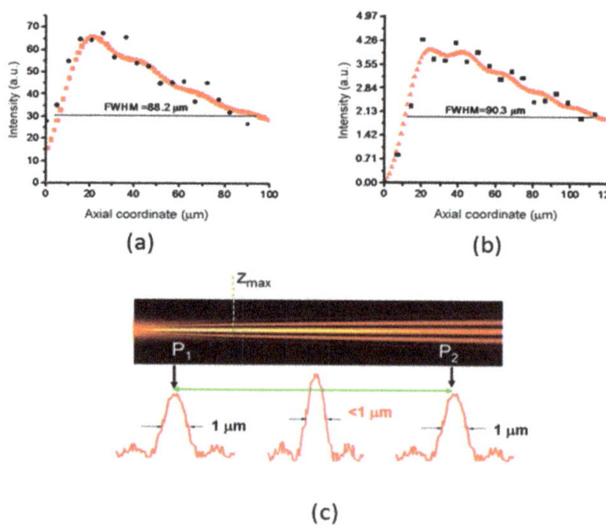

Figure 12. Upper part: Measured (black dots) and theoretical (red dots) light distributions in the focal lines for (**a**) λ_1 = 633 nm and (**b**) λ_2 = 405 nm. (**c**) Definition of the focal LFRB line P_1P_2, within which the FWHM spot diameter is submicrometric.

The literature on the topic of long focusing with submicron focal spots is surprisingly not hefty, and our results can be favorably compared with other ones obtained with more sophisticated tools. For instance, in [15], the use of a refraxicon fabricated with an ablation femtosecond laser allowed the theoretical formation of a long focus with a FWHM diameter down to 0.51 μm and a FWHM depth of focus of 95 μm, at a wavelength of 780 nm. The research shown in [9] reports on the performance of a metasurface-based axicon, with a FWHM focus diameter of 0.163 μm at the wavelength of 405 nm, maintained up to a measurable distance of 160 μm, while from the data reported herein, a FWHM of the focal length of ~40 μm can be evinced. Moreover, in this latter case, the transmission efficiency is strongly wavelength-dependent. Indeed, while at the design wavelength of 532 nm, the maximum experimental transmission of the device is 50%, the efficiency can become as low as 5% at red wavelengths. A similar drawback is reported in the metasurface-based axicon reported in [14], where it is demonstrated that the efficiency loss is mainly due to the geometrical structuration of the single unit cell of the metasurface, and thus inherently affects all the devices based on this approach. In that paper, a wavelength-independent spot size of $D_F = 0.25$ μm is reported, but the propagation length within which the spot diameter is unchanged is limited to 7 μm, while a FWHM axial line focus length of ~11 μm can be inferred from the reported data.

In our case, the transmission efficiency is ~85% in the whole visible spectrum, being limited only by the transmission of the objective. Our best focusing performances, experimentally verified at a heating power of 16 mW, are a FWHM spot and depth of focus of 0.56 μm and 90.3 μm, respectively. Noticeably, as shown in Figure S2, performances similar to those reported in [13] can be reached using the focal length of 0.9 mm of a commercial GRIN rod lens.

Moreover, it is worth noting that the value of D_F may not be so decisive, depending on the application. For instance, in [16], it was experimentally demonstrated that submicron features about 5-fold smaller than the FWHM spot diameter can be obtained by pulsed laser ablation, provided that the on-axis laser fluence exceeds a certain threshold. As a consequence, submicrometric features can be fabricated using spot sizes even higher than 1 μm. In this perspective, the depth of focus can play a more significant role, and our results are aligned with other more cutting-edge long focus forming systems. Moreover, when the interplay between the focus dimension and the laser fluence can make it difficult to precisely define the parameters of the fabrication process, the availability of a tunable focus may be an important tool to refine the performances of a working facility.

The most important drawback of the presented device is the high loss of pump power when the transmitted part is not used as a probe. However, this problem can be reduced by using multilayer graphene. Indeed, thanks to the progress of the fabrication procedures, a deposition of up to eight layers has been reported [34]. The amount of absorbed optical power could be multiplied, maintaining a high transmissivity at the same time, strongly relaxing the concern of optical pump loss.

5. Conclusions

In this paper, we have reported on a proof-of-concept of a single-layer graphene–PDMS long focusing thermo-optical device producing foci with submicrometric diameters. Thanks to the high concentration of the pump power in the subnanometric thickness of graphene, significant heating can be produced even with the small amount of absorbed optical pump power, causing temperature variations that give rise to a huge thermo-optical effect in PDMS, while maintaining a high optical transmission. The generated negative gradient index distribution has a spatial cuspidal-like shape, as the apex can be tightly reduced by properly focusing the heating optical pump. An input probe beam is phase-modulated in the heated PDMS and focused by an objective with short focal length, obtaining a long focus line where the spot size has submicrometric dimensions. The focusing performances (length of focus line, diameter of the spot focus) are easily modulable by changing the pumping power. This capability can be used to find the best tradeoff between the long focusing action and the achievement of on-axis high light intensity.

Calculations and experiments have shown that with the proper choice of the focusing lens, our device can reach performances close to those exhibited by other more refined tools (for instance, reflective axicons or metasurfaces), with the advantage of an easier and inexpensive fabrication process. Indeed, at the probe wavelength of 405 nm, we were able to produce a beam with a radial FWHM of 0.59 μm, and a FWHM depth of focus of about 90 μm. The present proof-of-concept can be designed in a more compact structure by substituting the bulky objective with a micro lens, such as a GRIN rod lens, available on the market even in sub-millimeter lengths.

Supplementary Materials: Refs. [44–47] are cited in Supplementary Materials. The following supporting information can be downloaded at: https://www.mdpi.com/article/10.3390/mi13122083/s1, Figure S1: Raman spectra of transferred single layer grapheme. Figure S2: The focusing performances of a PDMS/SLG device coupled to a GRIN lens with 0.9 mm focal length.

Author Contributions: Conceptualization: G.M. and T.D.R.; methodology: G.M., T.D.R., F.L.F.J. and A.N.B.; software: G.M.; validation: G.M., A.N.B., F.L.F.J. and T.D.R.; resources: T.D.R.; data curation: G.M. and T.D.R.; writing—original draft preparation: G.M.; writing review and editing: G.M., A.N.B., F.L.F.J. and T.D.R.; resources: T.D.R. and F.L.F.J.; supervision: G.M. and T.D.R.; funding acquisition: T.D.R. and F.L.F.J. All authors have read and agreed to the published version of the manuscript.

Funding: This study was financed in part by the Coordenação de Aperfeiçoamento de Pessoal de Nível Superior—Brasil (CAPES)—Finance Code 001. Funding from FAPERJ for processes E-26/010.000980/2019, E-26/211.180/2019 and E-26/211.540/2021 is acknowledged. We also acknowledge CNPq and the Instituto Nacional de Engenharia de Superfícies (INCT-INES) (processes 423349/2018-0 and 465423/2014-0).

Institutional Review Board Statement: Not applicable.

Informed Consent Statement: Not applicable.

Data Availability Statement: Not applicable.

Conflicts of Interest: The authors declare no conflict of interest.

References

1. Planchon, T.A.; Gao, L.; Milkie, D.E.; Davidson, M.W.; Galbraith, J.A.; Galbraith, C.G.; Betzig, E. Rapid threedimensional isotropic imaging of living cells using Bessel beam plane illumination. *Nat. Methods* **2011**, *5*, 417–423. [CrossRef] [PubMed]
2. Gao, L.; Shao, L.; Chen, B.-C.; Betzig, E. 3D live fluorescence imaging of cellular dynamics using Bessel beam plane illumination microscopy. *Nat Protoc.* **2014**, *9*, 1083–1101. [CrossRef] [PubMed]
3. Fahrbach, F.O.; Rohrbach, S.P.; Rohrbach, A. Microscopy with self-reconstructing beams. *Nat. Photonics* **2010**, *4*, 780–785. [CrossRef]
4. Brzobohatý, O.; Karásek, V.; Šiler, M.; Chvátal, L.; Čižmár, T.; Zemanek, P. Experimental demonstration of optical transport, sorting and self-arrangement using a 'tractor beam'. *Nat Photonics* **2013**, *7*, 123–127. [CrossRef]
5. Novitsky, A.; Qiu, C.-W.; Wang, H.F. Single gradientless light beam drags particles as tractor beams. *Phys. Rev. Lett.* **2011**, *107*, 203601. [CrossRef]
6. Yalizay, B.; Ersoy, T.; Soylu, B.; Akturk, S. Fabrication of nanometer-size structures in metal thin films using femtosecond laser Bessel beams. *Appl. Phys. Lett.* **2012**, *100*, 31104. [CrossRef]
7. Tseng, M.L.; Wu, P.C.; Sun, S.; Chang, C.M.; Chen, W.T.; Chu, C.H.; Chen, P.L.; Zhou, L.; Huang, D.W.; Yen, T.J.; et al. Fabrication of multilayer metamaterials by femtosecond laser-induced forward-transfer technique. *Laser Photonics Rev.* **2012**, *6*, 702–707. [CrossRef]
8. Tahir, P.O.G.; Zaman, Q.; Concas, G.C.; Gisbert, M.; Cremona, M.; Freire, F.L., Jr.; Carvalho, I.C.S.; Bevilaqua, P.H.C.; Santos de Sá, D.; Canellas, A.P.; et al. Thermoelastic pulsed laser ablation of silver thin films with organic metal–SiO_2 adhesion layer in water: Application to the sustainable regeneration of glass microfluidic reactors for silver nanoparticles. *J. Phys. Commun.* **2022**, *6*, 55005. [CrossRef]
9. Chen, W.T.; Khorasaninejad, M.; Zhu, A.Y.; Oh, J.; Devlin, R.C.; Zaidi, A.; Capasso, F. Generation of wavelength-independent subwavelength Bessel beams using metasurfaces. *Light Sci. Appl.* **2017**, *6*, e16259. [CrossRef] [PubMed]
10. Brzobohatý, O.; Čižmár, T.; Zemánek, P. High quality quasi-Bessel beam generated by round-tip axicon. *Optics Exp.* **2008**, *18*, 12688–12700. [CrossRef]
11. Akturk, S.; Arnold, C.L.; Prade, B.; Mysyrowicz, A. Generation of high quality tunable Bessel beams using a liquid-immersion axicon. *Opt. Commun.* **2009**, *282*, 3206. [CrossRef]
12. Skora, J.-L.; Gaiffe, O.; Bargiel, S.; Cote, J.-M.; Tavernier, L.; De Labachelerie, M.; Passilly, N. High-fidelity glass micro-axicons fabricated by laser-assisted wet etching. *Opt. Exp.* **2022**, *30*, 3749–3759. [CrossRef] [PubMed]
13. Kurt, H. Limited-diffraction light propagation with axicon-shape photonic crystals. *J. Opt. Soc. Am. B* **2009**, *26*, 981–986. [CrossRef]

14. Li, D.; Wang, X.; Ling, J.; Yuan, Y. Planar efficient metasurface for generation of Bessel beam and super-resolution focusing. *Opt. Quantum Elect.* **2021**, *53*, 143. [CrossRef]
15. Boucher, P.; Del Hoyo, J.; Billet, C.; Pinel, O.; Labroille, G.; Courvoisier, F. Generation of high conical angle Bessel–Gauss beams with reflective axicons. *Appl.Opt.* **2018**, *57*, 6725–6728. [CrossRef]
16. Gerhorst, A.; Sivankutty, S.; Pallier, G.; Labroille, G. Submicrometer surface structuring with a Bessel beam generated by a reflective axicon. *J. Laser Appl.* **2021**, *33*, 42013.
17. Milne, G.; Jeffries, G.D.M.; Chiu, D.T. Tunable generation of Bessel beams with a fluidic axicon. *Appl. Phys. Lett.* **2008**, *92*, 261101. [CrossRef]
18. Butt, M.A.; Savelyev, D. Bessel beams produced by axicon and spatial light modulator: A brief analysis. In Proceedings of the 2021 International Conference on Information Technology and Nanotechnology, Samara, Russia, 20–24 September 2021; pp. 1–6.
19. Dorrah, A.H.; Capasso, F. Tunable structured light with flat optics. *Science* **2022**, *376*, 367. [CrossRef]
20. Manyiara, R.A.; Rodrigo, D.; Yu, R.; Canet-Ferrer, J.; Ghosh, D.S.; Yongsunthon, R.; Baker, D.E.; Rezikyan, A.; García de Abajo, F.J.; Pruneri, V. Tunable plasmons in ultrathin metal films. *Nat. Photonics* **2019**, *13*, 328–333. [CrossRef]
21. Huang, W.; Wang, M.; Hu, L.; Wang, C.; Xie, Z.; Zhang, H. Recent Advances in Semiconducting Monoelemental Selenium Nanostructures for Device Applications. *Adv. Funct. Mater.* **2020**, *30*, 2003301. [CrossRef]
22. Huang, W.; Zhu, Y.; Wang, M.; Hu, L.; Tang, Y.; Shu, Y.; Xie, Z.; Zhan, H. Emerging Mono-Elemental Bismuth Nanostructures: Controlled Synthesis and Their Versatile Applications. *Adv. Funct. Mater.* **2021**, *31*, 2007584. [CrossRef]
23. He, Z.; Zhang, C.; Wang, Z.; Chen, Y.; Liu, J.; Fan, D. Real-time observation of Q-switched mode-locking in a tin selenide modulated ultrafast fiber laser. *App. Phys. Exp.* **2021**, *14*, 42009. [CrossRef]
24. Zhang, J.; Zhang, L. Nanostructures for surface plasmons. *Adv. Opt. Photonics* **2012**, *4*, 157–321. [CrossRef]
25. Nishijima, Y.; Rosa, L.; Juodkazis, S. Surface plasmon resonances in periodic and random patterns of gold nano-disks for broadband light harvesting. *Opt. Exp.* **2012**, *20*, 11466–11477. [CrossRef] [PubMed]
26. Jan, C.J.; Walton, M.D.; McConnell, E.P.; Jang, W.S.; Kim, Y.S.; Grunlan, J.C. Carbon black thin films with tunable resistance and optical transparency. *Carbon* **2006**, *44*, 1974–1981. [CrossRef]
27. Rahman, D.Y.; Sutisna; Utami, F.D.; Margaretta, D.O.; Sustini, E.; Abdullah, M. Highly Easy and Low Cost Fabrication of Graphite based Flexible Transparent Conducting Film. *J. Phys.* **2019**, *1204*, 12084.
28. Zeng, M.; Xiao, Y.; Liu, J.; Yang, K.; Lei, F. Exploring Two-Dimensional Materials toward the Next-Generation Circuits: From Monomer Design to Assembly Control. *Chem. Rev.* **2018**, *118*, 6236–6239. [CrossRef] [PubMed]
29. Huang, W.; Hu, L.; Tang, Y.; Xie, Z.; Zhang, H. Recent Advances in Functional 2D MXene-Based Nanostructures for Next-Generation Devices. *Adv. Funct. Mater.* **2020**, *30*, 2005223. [CrossRef]
30. Wu, L.; Xie, Z.; Lu, L.; Zhao, J.; Wang, Y.; Jiang, X.; Ge, Y.; Zhang, F.; Lu, S.; Guo, Z.; et al. Few-Layer Tin Sulfide: A Promising Black-Phosphorus Analogue 2D Material with Exceptionally Large Nonlinear Optical Response, High Stability, and Applications in All-Optical Switching and Wavelength Conversion. *Adv. Opt. Mater.* **2018**, *6*, 1700985. [CrossRef]
31. Zu, Y.; Guo, J.; Hao, Q.; Zhang, F.; Wang, C.; Liu, J.; Wang, B. Graphdiyne as a saturable absorber for 2-μm all-solid-state Q-switched laser. *Sci. China Mater.* **2021**, *64*, 683–690. [CrossRef]
32. Zaman, Q.; Tahir, F.L., Jr.; Shtepliuk, I.; Barbosa, A.N.; Maia da Costa Marcelo, E.H.; Diaz Mendoza, C.A.; Araujo, J.F.D.F.; Concas, C.G.; Cremona, M.; Zubahir, A.; et al. Water Diffusion Effects at Gold–Graphene Interfaces Supporting Surface Plasmon Polaritons. *J. Phys. Chem. C* **2022**, *126*, 13905–13919. [CrossRef]
33. Novoselov, K.S.; Geim, A.K.; Morozov, S.V.; Jiang, D.; Zhang, Y.; Dubonos, S.V.; Grigorieva, I.V.; Firsov, A.A. Electric field effect in atomically thin carbon films. *Science* **2004**, *306*, 666–669. [CrossRef] [PubMed]
34. Yoo, M.S.; Lee, H.C.; Wolf, C.N.; Nguyen, N.; Park, D.H.; Kim, J.; Lee, E.; Chung, H.J.; Cho, K. Growth of multilayer graphene with a built-in vertical electric field. *Chem. Mater.* **2020**, *32*, 5142–5152. [CrossRef]
35. Schrade, M.; Xing, W.; Thorsaugh, K.; Belle, B.D. Centimeter-Sized Monolayer CVD Graphene with High Power Factor for Scalable Thermoelectric Applications. *ACS Appl. Electron. Mater.* **2022**, *4*, 1506–1510. [CrossRef]
36. Docherty, C.J.; Johnston, M.B. Terahertz Properties of Graphene. *J. Infrared Milli Terahertz Waves* **2012**, *33*, 797–815. [CrossRef]
37. Pop, E.; Varshney, V.; Roy, A.K. Thermal properties of graphene: Fundamentals and applications. *MRS Bull.* **2012**, *37*, 1273–1281. [CrossRef]
38. Goodman, J.W. *Introduction to Fourier Optics*; Mc Graw Hill Book Company: New York, NY, USA, 1988; pp. 71–74.
39. Zhu, Z.; Liu, L.; Liu, Z.; Zhang, Y.; Zhang, Y. Surface-plasmon-resonance-based optical-fiber temperature sensor with high sensitivity and high figure of merit. *Opt. Lett.* **2017**, *42*, 2948–2951. [CrossRef]
40. Parigger, C.; Tang, Y.; Plemmons, D.H.; Lewis, J.W.L. Spherical aberration effects in lens–axicon doublets: Theoretical study. *Appl. Opt.* **1997**, *36*, 8214–8221. [CrossRef]
41. Del Rosso, T.; Zaman, Q.; Romani, E.C.; Pandoli, O.; Aucelio, R.Q.; Melo de Lima, L.; Cremona, M.; Dmitriev, V.; Queiroz da Costa, K.; Lazaro Freire, F., Jr.; et al. Enhanced stability of plasmonic metal thin films by CVD grown graphene transfer. *Thin Solid Film* **2017**, *644*, 65–70. [CrossRef]
42. Wang, X.; Chen, Y.P.; Nolte, D.D. Strong anomalous optical dispersion of graphene: Complex refractive index measured by Picometrology. *Opt. Exp.* **2008**, *16*, 22105–22112. [CrossRef] [PubMed]
43. Simo, W.; Jiangyong, Z.; Fanxing, L.; Wei, Y. Generation of soft annular beams with high uniformity, low ring width increment, and a smooth edge. *Opt. Exp.* **2022**, *30*, 27439–27452.

44. Do Nascimento Barbosa, A.; Figueroa, J.S.; Mendoza, C.D.; Pinto, A.L.; Freire, J.L., Jr. Characterization of graphene synthesized by low-pressure chemical vapordeposition using N-Octane as precursor. *Mater. Chem. Phys.* **2018**, *219*, 189–195. [CrossRef]
45. Gil, M.C.; Freire, F.L. Plasma-treated CVD graphene gas sensor performance in environmental condition: The role of defects on sensitivity. *J. Sens.* **2019**, *2019*, 5492583.
46. Barbosa, A.N.; Ptak, F.; Mendoza, C.D.; Maia da Costa, M.E.H.; Freire, F.L., Jr. Direct synthesis of bilayer graphene on silicon dioxide substrates. *Diam. Relat. Mater.* **2019**, *95*, 71–76. [CrossRef]
47. Ferrari, A.C.; Meyer, J.C.; Scardaci, V.; Casiraghi, C.; Lazzeri, M.; Mauri, F.; Piscanec, S.; Jiang, D.; Novoselov, K.S.; Roth, S. Raman spectrum of graphene and graphene layers. *Phys. Rev. Lett.* **2006**, *97*, 187401. [CrossRef]

micromachines

MDPI

Article

A Miniaturized Device Coupled with Digital Image Correlation for Mechanical Testing

Daniel J. Cruz [1], Jose Xavier [2,3,*], Rui L. Amaral [1] and Abel D. Santos [1,4]

1 INEGI, Institute of Science and Innovation in Mechanical and Industrial Engineering, Campus da FEUP, R. Dr. Roberto Frias 400, 4200-465 Porto, Portugal
2 UNIDEMI, Department of Mechanical and Industrial Engineering, NOVA School of Science and Technology, Universidade NOVA de Lisboa, 2829-516 Caparica, Portugal
3 LASI, Intelligent Systems Associate Laboratory, 4800-058 Guimarães, Portugal
4 FEUP, Faculty of Engineering, University of Porto, R. Dr. Roberto Frias, 4200-465 Porto, Portugal
* Correspondence: jmc.xavier@fct.unl.pt

Abstract: Miniaturized mechanical testing based on small sample testing technology is a powerful technique to characterize the mechanical properties of different materials, and it is being used in different application fields. However, the small size of the specimens poses several challenges because the results are highly sensitive to measurement accuracy and the corresponding mechanical properties can change substantially due to the so-called specimen size effect. In this work, a novel testing device based on miniaturized specimens is presented. The equipment is designed to test materials in tensile and compressive loadings, but it is also capable of performing reverse-loading tests. Buckling of the specimen is an inherent phenomenon in compression loadings, especially for thin materials. Therefore, specimen geometry is properly studied and optimized to mitigate this effect. To evaluate the deformation of the specimen, the digital image correlation (DIC) technique is used to capture the full-field strain in the central gauge section of the sample. A sensitivity analysis of the DIC setting parameters was performed for this application. To evaluate the performance of the developed system, experimental results of monotonic tests and tests with reverse loadings (tension-compression) are presented, considering two high-strength steels (DP500 and DP780).

Keywords: miniaturized specimen; digital image correlation (DIC); Small Specimen Test Techniques (SSTT); Bauschinger effect; sheet metal characterization

Citation: Cruz, D.J.; Xavier, J.; Amaral, R.L.; Santos, A.D. A Miniaturized Device Coupled with Digital Image Correlation for Mechanical Testing. *Micromachines* **2022**, *13*, 2027. https://doi.org/10.3390/mi13112027

Academic Editors: Luigi Sirleto and Giancarlo C. Righini

Received: 4 November 2022
Accepted: 17 November 2022
Published: 19 November 2022

1. Introduction

The mechanical behavior of engineering materials is typically determined by experimental tests using the so-called "macro specimens", with dimensions on the scale of millimeters. However, in some applications, the material to be characterized is not available in the form of large blocks [1], which has recently motivated an increasing interest in evaluating the mechanical properties of materials with "miniaturized" specimens, also known by other names such as "small-size", "sub-size", or "mini" specimens [2,3]. Miniaturized specimens were first developed in the 1970s to characterize materials for in-service nuclear reactor pressure vessels [4]; this is an application in which surveillance capsules are used to periodically monitor the degradation of vessel materials. These capsules typically have limited storage space, and for this reason, a limited specimen size is required [5]. As a result, Small Specimen Testing Technology (SSTT) has been developed and applied in different application fields such as residual life assessment of in-service components [6], local measurement of properties in the welding industry [7,8], the additive manufacturing process [9,10], and the evaluation of mechanical properties of miniature devices such as microelectromechanical systems (MEMS) [11,12].

Several non-standard test techniques are currently available to evaluate the mechanical behavior of materials using miniaturized specimens [3]. One of the most used SSTT is the miniaturized tensile test (MTT) [13], which uses specimens with dog-bone geometry with nominal dimensions ranging from several millimeters to hundreds of microns. No standards have yet been established for MTT, which has resulted in different specimen designs. The design of miniaturized specimens currently uses one of two approaches: (1) simply scaling down from standard specimens, or (2) scaling down from standard specimens, with a subsequent adjustment of some dimensions. A comprehensive review of this topic can be found in the recent work of [2].

Miniaturized tensile testing encompasses several challenges related to specimen manufacturing, handling, and procedural issues. In addition, the miniaturization may cause the so-called "specimen size effect", which leads to different behaviors at the miniaturized and macro scales. Specimen size effects can be categorized into two different levels. At the first level, the test results obtained from miniaturized specimens reflect the bulk mechanical properties described by continuum mechanics. However, for some materials, the results may deviate from standard specimens, and additional data processing techniques, e.g., the inverse Finite Element Method (FEM), are required [14–16]. On the other hand, at the second level, the specimen can no longer reflects the bulk properties as the thickness or diameter is further decreased. Typically, the critical thickness value is defined by the ratio t/d between thickness, t, and grain size, d, of the material. Generally, miniaturized specimens with a t/d ratio between 6 and 10 can achieve standard bulk behavior.

Sheet metal forming is an important manufacturing process in the production of metal components, applied in a wide variety of industrial areas such as automotive, naval, electrical, aeronautics, and aerospace [17]. Sheet metal forming includes simple processes such as bending, stretch forming and spinning, as well as more complex ones such as roll forming and deep drawing. In these processes, material behavior assessment is crucial for the correct calibration of material constitutive models. Therefore, a wide variety of tests are established in the literature to obtain reliable experimental data used in the identification of constitutive parameters.

The mechanical characterization of metallic sheets has standardized test procedures, by using the tensile test (uniaxial loading), according to ISO 6892-1 [18] and the hydraulic bulge test (biaxial loading), according to ISO 16808 [19]. On the other hand, no standardized test setup has yet been established to characterize the compressive loading of sheet metal materials [20]. Moreover, an uniaxial compressive stress state may cause buckling in flat sheet metal materials for small compressive strains. Therefore, the test method becomes invalid due to the change on the stress state. Over the past few years, different approaches have been developed to minimize buckling during compression tests. These approaches can be categorized into three groups: (i) the pack method, which involves packing a group of samples together in the thickness direction; (ii) a single sample with a lateral support system; and (iii) miniaturized test samples with a low gauge length/thickness ratio.

The pack method, developed by Aitchison et al. [21] and later improved by Jackman [22], involves joining together a group of samples along the thickness direction to form a unique specimen. With this strategy, the length/thickness ratio is reduced, and therefore the buckling instability under compression is minimized. A variation in the pack method was used by Yoshida et al. [23], in which five specimens were glued together and then covered with clamping plates. Because the pack method requires a large number of specimens and complex preparation, some authors prefer to use a single sheet with a lateral support system. Boger et al. [24] sandwiched a single flat specimen between two solid support plates using a constant restraining side force. To prevent buckling in the thickness direction (T-buckling), width direction (W-buckling), and unsupported gap (L-buckling), an oversized dog-bone specimen was designed. A device with two sets of fork-shaped supports was proposed in the same year by Sekine and Kuwabara [25] to eliminate the uncovered area in tensile-compressive relative movements. Because the male and female dies slide past one another as the sample is compressed, this design made it possible to support the entire length of the specimen.

The gauge-length/thickness ratio of the specimen has the biggest impact on the buckling phenomenon in compression. In order to effectively study the compression behavior of different sheet metal materials, test methods using miniaturized specimens have recently been suggested. Tritschler et al. [26], tested a titanium alloy with a thickness of 1.2 mm in compression using specimens with dimensions of 2.5 mm in length by 2 mm in width, achieving about 2% compressive strains. Hußnätter [27] used a specimen geometry of 2 mm length by 2 mm width to characterize the compressive behavior of magnesium alloy AZ31, achieving about 5% compressive strains before the specimen buckles.

Cruz et al. [28] developed an experimental equipment to perform uniaxial mechanical tests on miniaturized metallic specimens. The current work is a step forward in these developments, exploring the applicability of this equipment to describe the mechanical behavior of sheet metal materials, not only for monotonic uniaxial tests, i.e., tension and compression, but also for reverse loading tests, i.e., tension followed by compression or vice versa. Moreover, the digital image correlation technique was coupled with the miniaturized test device in order to quantify the quality of the test method regarding the specimen deformation at the scale of observation, in a balance between spatial resolution and accuracy [29].

2. Materials and Methods

2.1. Materials

In the present work, advanced high-strength steels (AHSS) produced by SSAB company were used, namely two grades of dual-phase (DP) steel sheets (DP500 and DP780) with an initial thickness of 0.8 mm. The dual-phase (DP) steel is an advanced high-strength steel (AHSS), with a good balance of strength and ductility due to its martensite islands in a ferrite matrix, as represented in Figure 1. Table 1 lists the percentage of alloying present in each material chemical composition.

Figure 1. Initial microstructure obtained from SEM of (**a**) DP500 and (**b**) DP780 dual-phase steel.

Table 1. Chemical composition of the DP500 and DP780 dual-phase steels.

Element [%]	C	Si	Mn	P	S	Cr	Ni	V	Cu	Al	Nb	B	N	EC [1]
DP500	0.079	0.31	0.65	0.003	0.003	0.03	0.03	0.01	0.01	0.038	0.0	0.0003	0.003	0.20
DP780	0.138	0.20	1.52	0.011	0.002	0.03	0.03	0.02	0.01	0.038	0.014	0.0002	0.003	0.40

[1] $EC = C + \frac{Mn}{6} + \frac{Ni+Cu}{15} + \frac{Cr+Mo+V}{5}$

2.2. Design of an Uniaxial Test Equipment for Miniaturized Specimens

The experimental test procedure for miniaturized specimens followed a similar procedure to the one typically carried out for conventional standard specimens, which is based on moving grips and results of force as a function of elongation/strain data. However, a very accurate procedure is required because results can be significantly affected by the testing protocol, including sample preparation, test setup, experimental operation, measurements, and data processing [2]. Several authors have emphasized different aspects to be taken into account when testing miniaturized specimens [3]: (a) the importance of the specimen-fabrication technique, which can have consequences for the measured properties, particularly on the yield strength; (b) good control and accuracy of the specimen dimensions, which directly relate to the reliability and reproducibility of the test data; (c) high-quality measuring instruments; and (d) the importance of a precise and well-aligned test frame, in order to minimize the extraneous bending strains.

Based on these considerations, a new type of equipment was developed, specifically designed to test miniaturized specimens on a reduced scale, and we refer to it as a Miniaturized Specimen Tester Device (MSTD) [28]. As represented in Figure 2, MSTD consists of three groups of components with their well-defined function. Firstly, there is the "motor group", which includes a ball screw, driven by a stepper motor, which guarantees continuous and precise movement of the specimen, making it possible to guarantee a minimum displacement of 2 µm for each step of the motor. Secondly, there is the "frame group", which ensures the layout and guidance of the various components, consisting of four linear bearings that slide on two ground shafts. Finally, there is the instrumentation group, consisting of a 5 kN "S"-type load cell and two extension measurement components: a linear ruler for measuring and controlling the positioning of the moving table and a digital image correlation (DIC) system. The MSTD's layout allows all components to be on a common plane, which is advantageous for a precise alignment of the specimen during the test. Consequently, the appearance of undesirable bending moments is minimized, especially in compression, which is also essential to prevent the buckling of the specimen. The equipment, designed to test miniaturized specimens, is capable of withstanding a maximum axial force of 2.5 kN.

The preparation of the specimen is crucial in MTT because miniaturized specimens are more susceptible to large-dimensional deviations. Accordingly, in this work the specimen profile was obtained by wire electrical discharge machining (EDM) using a 0.25 mm wire diameter. This process is widely used for miniaturized specimen manufacturing, resulting in good dimensional accuracy and surface finishing [2].

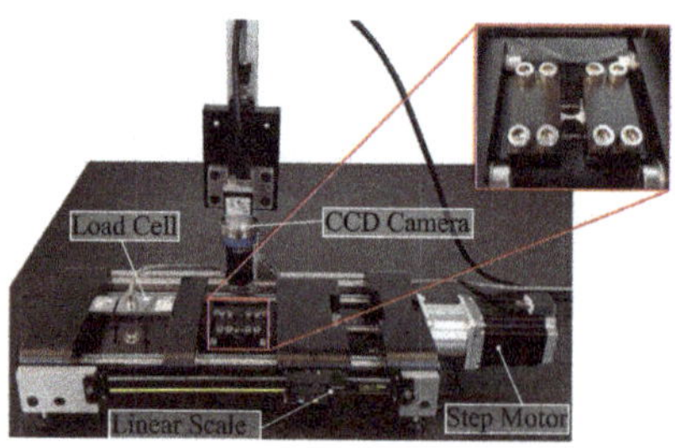

(a)

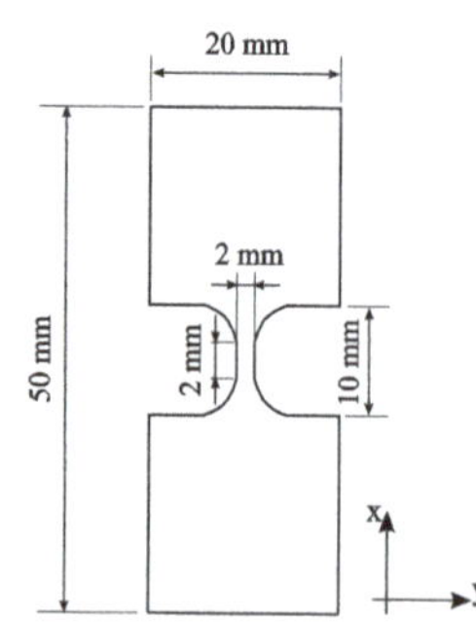

(b)

Figure 2. Miniaturized Specimen Tester Device (MSTD): (**a**) setup, and; (**b**) geometry of the miniaturized specimen.

2.3. Testing Conditions

In order to evaluate the monotonic behavior of DP500 and DP780 steels, tests were performed with tensile and compressive loading using miniaturized samples and the geometry shown in Figure 2b. Additionally, the same materials were tested on macro-tensile test pieces, in accordance with ISO 6892-1 [18]. These macro standard tensile tests were performed on a 300 kN Instron 5900 R testing machine. The experimental conditions for both geometries are shown in Table 2. The specimens were tested at a constant test speed, resulting in an initial strain rate of approximately 10^{-3} s^{-1}, until the material failed. To ensure the repeatability of the results, two samples were systematically tested for each geometry.

Table 2. Experimental conditions for macro and miniaturized samples.

	Macro Sample	Miniaturized Sample
Number of samples	2	2
Gauge length (L_0)	50 mm	2 mm
Crosshead speed	5 mm/min	0.5 mm/min
Frequency of data acquisition	20 Hz	20 Hz
Temperature	23 °C	23 °C
Humidity	53%	53%

2.4. *Digital Image Correlation*

2.4.1. Choice of the Optical Technique

The DIC technique was coupled with the mechanical tests to quantify the deformation of the material across the central region of interest for both macro and miniaturized specimens, with a suitable spatial resolution and accuracy at the scale of observation. Traditionally, in carrying out experimental tests (e.g., macro tensile test [18]), the strain components are measured by means of a strain gauge attached to the sample. However, for miniaturized specimens, these instrumentation devices may not be suitable due to the reduced sample size. Nevertheless, a non-contact optical technique can be conveniently used to access full-field strains over the target region. For that purpose, several white-light and interferometric optical methods have been proposed in the literature [30]. Among them, DIC has been increasingly used in recent years in the field of experimental solid mechanics [31,32]. This technique was introduced in the 1980s [33,34] and allows the measurement of full-field kinematic quantities by tracking unique features on images taken at different deformation stages, typically by means of a subset-based approach [35–38]. The DIC technique uses a camera-lens optical setup, and it allows variable sensitivity and resolution in strain measurements. Since the DIC method uses digital images, several high spatial-resolution digital devices can be used for image acquisition. For example, DIC can be coupled with Optical Microscopy (OM) [39,40], Scanning Electron Microscopy (SEM) [41,42], and Scanning Tunneling Microscope (STM) [43] to realize microscale-to-nanoscale deformation measurements. In the 2D DIC set-up, the planar surface of a specimen is imaged stringently parallel to the camera sensor and the surface must have a random pattern of a contrasted speckle pattern. The creations of a quality pattern for DIC at reduced scales can be challenging [44], and the accuracy of the measurements depends significantly on the spatial resolution of the imaging system [35].

2.4.2. Optical System and Speckle Pattern

A 2D DIC set-up using a 5 MP CMOS digital camera was selected in this work. The mounted lens was selected with regard to the specimen size. The details of the camera-lens system used for the miniaturized tests are summarized in Table 3. In this case, conversely, to a standard 50 mm focal lens (Fujinon HF50HB-1B 50 mm lens, f/2.3 C–Mount) used in macroscopic tensile tests, a telecentric lens was selected. It had the advantage of keeping the image magnification unchanged during the mechanical test, avoiding artifacts due to, for instance, Poisson effects or slightly out-of-plane movements during the test. In this case, the undeformed image was focused by fixing the working distance at 63.3 mm with a magnification factor of 1×. Images were recorded during the tests at an acquisition frequency of 5 Hz. The central part of the specimens was initially painted by spraying a thin layer of white paint in order to make the diffuse reflectivity of the surface uniform. The final speckle pattern was then created by spreading dark paint using an airbrush (Harder & Steenbeck—Evolution Silverline Solo) with a nozzle set of 0.2 mm. Considering the region of interest and camera-lens system, an average speckle size of about 6 pixels (21 µm) was obtained in the speckle quality evaluation. According to international guidelines [45], this means that a subset size larger than 21 pixels (0.075 mm) or 31 pixels (0.111 mm) must be used in the DIC measurements for accuracy.

Table 3. Camera-lens optical system for the miniaturized tests.

Camera	Basler acA2440-75um, 5 MPixel, CMOS sensor
Pixel resolution	2448 px × 2048 px
Lens	Opto Engineering TC 23 09
Field of view	8.8 mm × 6.6 mm
Magnification factor	1×
Working distance	63.3 mm
Image-conversion factor	3.5 µm/px
Image-acquisition frequency	5 Hz
Speckle pattern technique	Airbrush (nozzle set of 0.2 mm)
Average speckle size	6 px | 21.5 µm

2.4.3. Selecting DIC Setting Parameters

In performing DIC measurements, a convergence analysis must be carried out for the inherent parameters in the numerical imaging method [29,46]. The set of DIC parameters plays an important role in the spatial resolution and accuracy associated with both displacement and strain measurements [47]. A parametric analysis was therefore carried out in this study, using the Performance Analysis Tool of the MatchID subset-based 2D DIC software [48]. The parametric tool allows one to carry out several DIC analyses over the same set of images by considering several combinations of parameters, as reported in Table 4. Each relevant parameter was considered over a suitable design space by defining the minimum and maximum values and a step of an integer increment (i.e., a subset size with a minimum of 21 × 21 pixels^2, a maximum of 91 × 91 pixels^2, and increment of 10 pixels, Table 4). Results in terms of signal evaluation at the center of the specimen can then be discussed in light of the preselected DIC setting parameters. The linear strain component in the x direction (ε_{xx}) was used as a reference, since uniaxial tests were carried out along this axis. Each set of DIC parameters can be associated with a corresponding spatial resolution. To quantify this metric, the following Virtual Strain Gauge (VSG) measure was used [45,48]:

$$\mathrm{VSG} = [(\mathrm{SW} - 1) \times \mathrm{ST}] + \mathrm{SS} \quad [\text{pixels}] \tag{1}$$

where SW stands for a strain window, i.e., the number of displacement data points used to fit a local polynomial in the neighbor of the subset centroid under evaluation; ST is the subset step; and SS is the subset size. The VSG can be expressed in physical units of millimeters if the image conversion factor of the optical system is taken into account (Table 3).

In order to evaluate the ε_{xx} strain signal with regard to the DIC settings as a function of the VSG, three points of reference at the centre of the specimens were systematically analyzed for both tensile and compression tests, for the DP500 steel, as schematically shown in Figure 3.

Figure 4 shows the evolution of ε_{xx} as function of points A, B, and C (Figure 3) for both tensile and compression tests at two different deformation stages corresponding to low and high average strains, respectively. As can be observed, the DIC setting parameters converge systematically to an average reconstruction of the linear strain. In a more local analysis, Figure 5 shows the results of the reconstructed ε_{xx} value at point B in the center of the specimen (see Figure 3) as a function of VSG, for the different DIC setting parameters for both tensile and compression tests and for two stages at low and high strain values, respectively. A good balance is to be found with regard to the spatial resolution in reconstructing the strain field. As can be seen in all scenarios, the signal is rather stable with regard to a set of parameters. This may be expected for a homogeneous material and uniaxial loading configuration.

Table 4. Parameters used in the Performance Analysis Tool of MatchID 2D DIC.

Subset-Based Settings	
Subset size	SS $\in \{21, 31, 41, 51, 61, 71, 81, 91\}$ px
Subset step	ST = 10 px (fixed)
Shape function	{Affine, Quadratic}
Strain reconstruction-based settings	
Strain window	SW $\in \{3, 5, 7, 11, 13, 15, 17, 19\}$ data points
Polynomial order *	Bilinear (Q4), Biquadratic (Q8)
Strain convention	Green–Lagrange

* Local least-squares fitting approach for the strain evaluation.

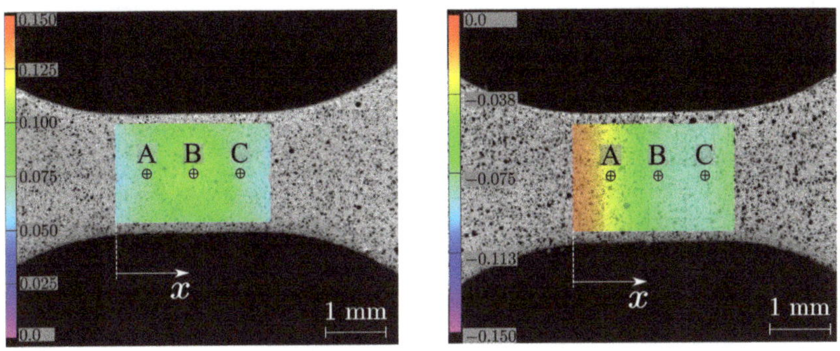

(**a**) (**b**)

Figure 3. Points of reference used in evaluating the reconstruction of the ε_{xx} strain signal with regard to the DIC settings and as a function of the VSG for both (**a**) tension and (**b**) compression tests.

Figure 6 shows a complementary analysis plotting the ε_{xx} value along the coordinate x at the center of both the tensile and the compression tests for several VSG values. As can be seen, the DIC setting corresponding to higher values of VSGshows a smooth signal evaluation, as it may be expected, in contrast with lower VSG DIC settings. In conclusion, from the analysis in this work, the DIC settings reported in Table 5 were finally selected in a suitable balance between spatial resolution and accuracy of results for the current application.

Table 5. DIC and strain settings selected in the full-field measurements.

DIC Settings	
Correlation criterion	ZNSSD
Interpolant	Bicubic spline
Subset shape function	Affine
Subset size	71 px
Step size	10 px
Image pre-filtering	Gaussian, 5 px kernel
Strain Settings	
Strain window size	9 data points
Strain interpolation	Bilinear Q4
Strain convention	Green–Lagrange

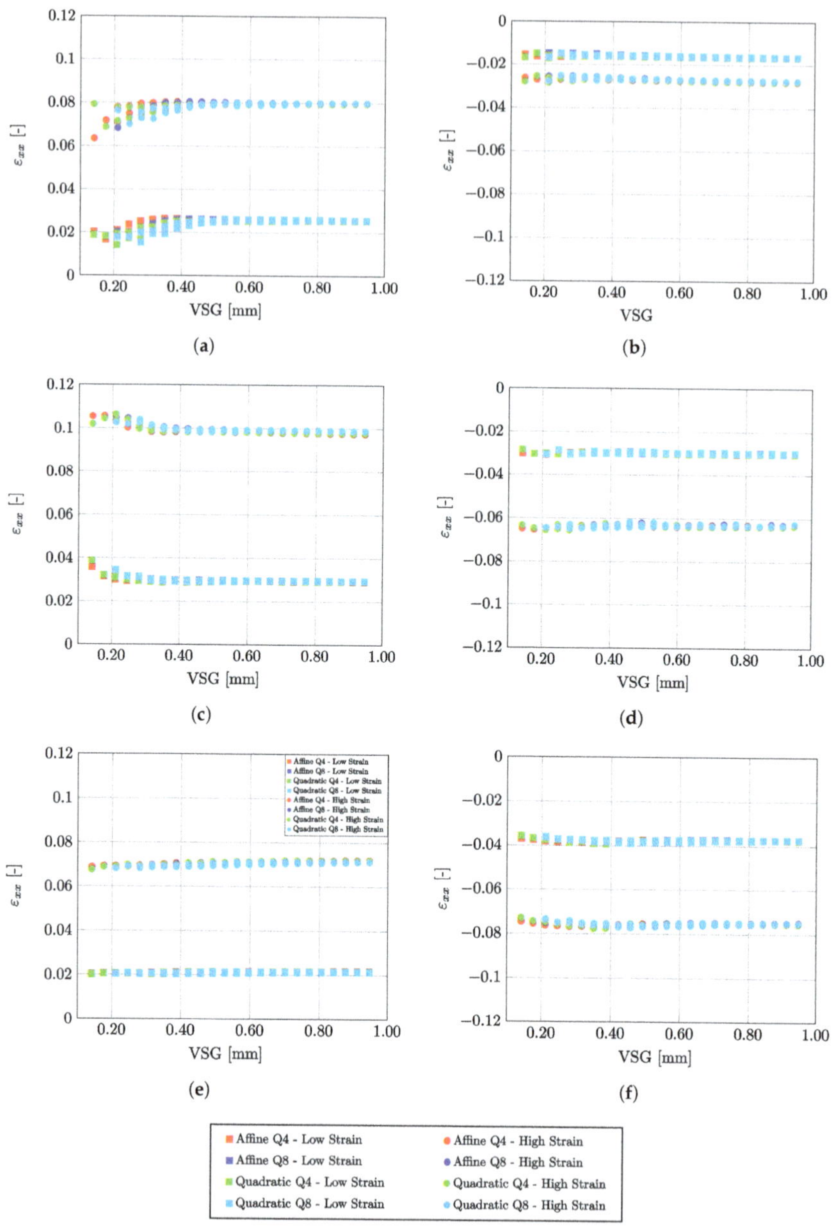

Figure 4. Evaluation of ε_{xx} for points A, B, and C (Figure 3) as a function of DIC parameters expressed with regard to Virtual Strain Gauge (VSG) for both tension and compression tests, at two stages of deformation corresponding low and high strains. (**a**) Point A—Tension; (**b**) Point A—Compression; (**c**) Point B—Tension; (**d**) Point B—Compression; (**e**) Point C—Tension; (**f**) Point C—Compression.

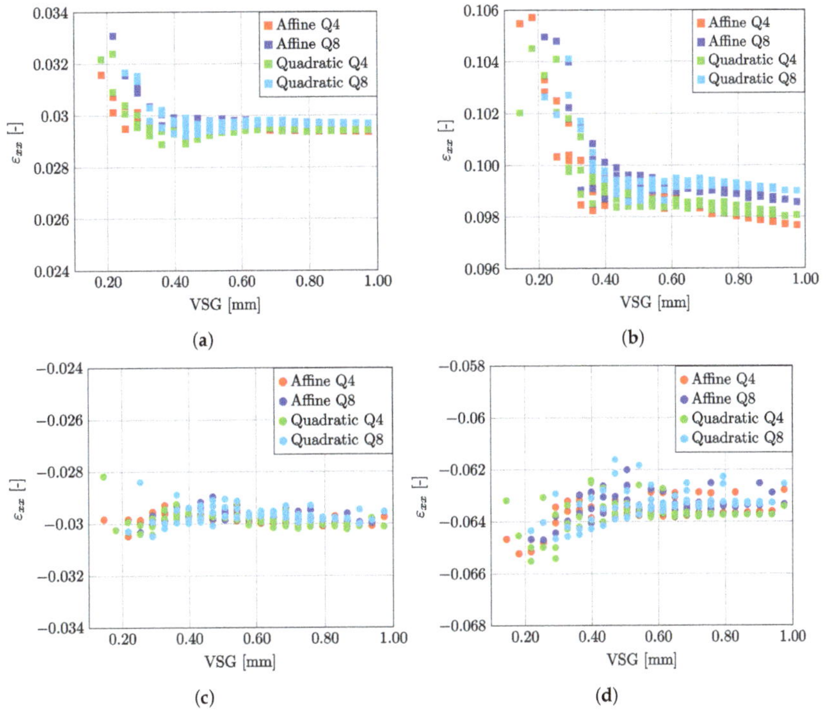

Figure 5. Evaluation of ε_{xx} at point B (Figure 3) as a function of DIC parameters expressed with regard to Virtual Strain Gauge (VSG). (**a**) Tension-Low Strain; (**b**) Tension—High Strain; (**c**) Compression—Low Strain; (**d**) Compression—High Strain.

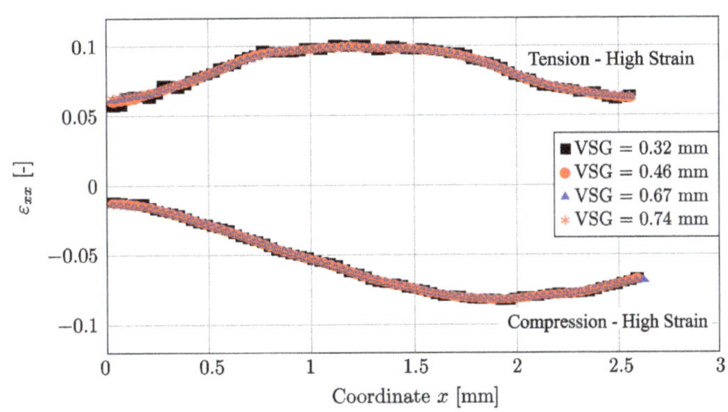

Figure 6. Evaluation of ε_{xx} as a function of the coordinate x for several Virtual Strain Gauge (VVSG) values for both tension and compression tests.

3. Results and Discussion

3.1. Monotonic Tension and Compression Tests

For the reconstruction of the strain signal during the mechanical tests, an average value was systematically calculated at the centre of the specimen corresponding to a region of 2 mm $\times$ 2 mm. Figure 7 shows a representative stress–strain curve for both DP500 and DP780 steel materials. The engineering stress–strain curves, shown in Figure 7a, indicate that the results obtained from both macro and miniaturized specimens are quite comparable. In fact, for both materials, the work hardening is very similar for the mini and macro geometry, as seen in the true stress–true strain curve ($\sigma - \varepsilon$) shown in Figure 7b. However, it can be seen that after the point of ultimate tensile strength, the miniaturized geometry presents a superior total elongation when compared to the macro geometry. This difference can be justified by the local necking that occurs after reaching the point of maximum stress, which translates into a localized strain zone whose size is independent of the gauge length. Therefore, for smaller gauge lengths, the localized strain has a higher impact on the total elongation value. Additionally, the variation in necking behavior should be taken into account as well. As reported by [49], total elongation and post-necking elongation tend to increase with a decreasing ratio, w/t, between specimen width and thickness. In fact, when w/t decreases, the shear deformation may occur in both the upper (lower) and side planes, resulting in complex stress distribution, and the localized necking changes to diffuse necking, which leads to higher post-necking elongation.

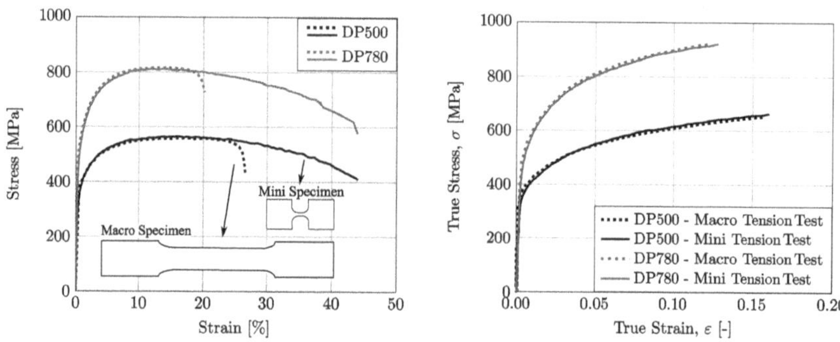

(**a**) (**b**)

Figure 7. Experimental tensile stress/strain behavior for DP500 and DP780 steels: (**a**) engineering stress–strain curves; (**b**) true stress–strain curves.

The hardening behavior of the two dual-phase steels under compressive loading was also evaluated using the miniaturized specimens. The true stress–true strain curve ($\sigma - \varepsilon$) for each material is shown in Figure 8a. Additionally, this figure represents, for each material, the tensile hardening curve obtained using miniaturized samples (curve defined by points). For an easy comparison with these two tensile curves, the compression results were reflected for the first quadrant and represented as a dashed line. When analyzing results, it was verified that the work hardening was similar in the tension and compression for both materials; however, in terms of compression, the maximum strain values in the compression test were lower than those reached in tension. As seen in Figure 8b, for a strain value close to 7%, the force reaches a maximum value. At this point, the test becomes non-planar and the buckling phenomenon becomes evident. This instability is translated in the decay of the stress, which means that the test is not longer valid.

3.2. *Reverse Loading Tests: Tension–Compression*

As represented in Figure 9, a typical hardening curve of a metallic material during reverse loading can be divided into four interconnected phenomena [50]: (a) the Bauschinger effect, which describes a reduction in the yield stress after the occurrence of plastic deformation during the initial loading; (b) the transient behavior, which is given by a smooth elasto-plastic transition characterized by a change in the work-hardening rate; (c) the work hardening stagnation, given by a transient behavior in which a stress–strain plateau is observed; (d) permanent softening, characterized by lower levels of stress after the transient period when compared to monotonic loading at a given accumulated plastic strain. These phenomena will hereafter be referred to as Bauschinger behavior. The experimental characterization of Bauschinger behavior is crucial to calibrate advanced constitutive models, e.g., the multi-surface Yoshida–Uemori (Y–U) [51] kinematic hardening model and the

Homogeneous anisotropic hardening (HAH) model proposed by [52], and consequently improves the accuracy of numerical simulation results of sheet-metal-forming processes. Different authors have shown that the Bauschinger behaviour has a strong influence on the prediction of the springback [53–55]. This phenomenon, which refers to the elastic distortion of a part when forming tools are removed, is a major issue in sheet metal forming processes and its mitigation is crucial.

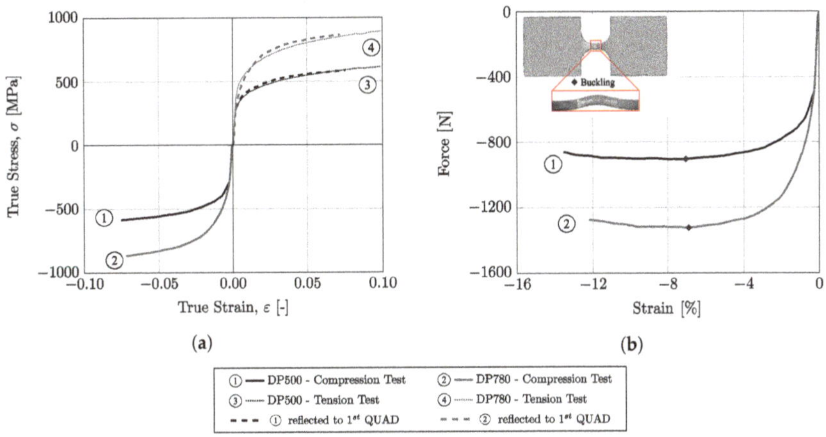

Figure 8. True stress–strain curves from monotonic compression tests and comparison with tension tests; (**b**) buckling point at the compression test.

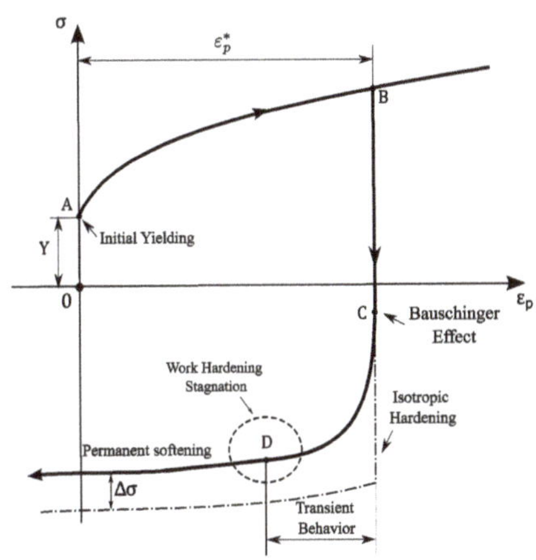

Figure 9. Schematic illustration of stress–strain response during forward and reverse deformation.

In this work, the Bauschinger behavior of the DP500 and DP780 materials was evaluated through uniaxial tests with reverse loading (tension–compression) using miniaturized specimens. The test includes a step of tensile loading, followed by a compression step. Initially, the tensile step is performed up to a defined pre-strain, ε_p^*. In this sense, three pre-strain values were selected, 0.025, 0.050, and 0.075. After this pre-strain in tensile loading, the compressive step was carried out until the buckling phenomenon began to be evident, which is shown, as already mentioned, by a decay of the stress value in the stress–strain curve. The results for DP500 and DP780 steels are shown in Figure 10a,b, respectively. For each material, the monotonic uniaxial tensile curve is also represented for comparison purposes. It can be seen that the tension step of tension–compression test matches the uniaxial tensile curve very well, for both materials, which confirms that the equipment provide excellent repeatability of results. The compression step curves for the three different pre-strain values show a similar changing trend across the entire strain range, indicating that the device can provide reliable results for the studied materials. To analyze the Bauschinger behavior of the dual-phase steels used, the compression-step-hardening curves were reflected to the first quadrant, as shown in the dashed line in Figure 10. The yield point for the uniaxial tensile curve and the reflected reverse loading curves, represented in round points, were taken at the 0.2% plastic strain and are summarized in Figure 11a. It can be observed that the studied materials present a strong Bauschinger effect, since the reverse yield stress is significantly reduced during reverse loading. A strong permanent softening effect is also visible, since the curves do not coincide with the monotonic tensile curve and tend to be approximately parallel.

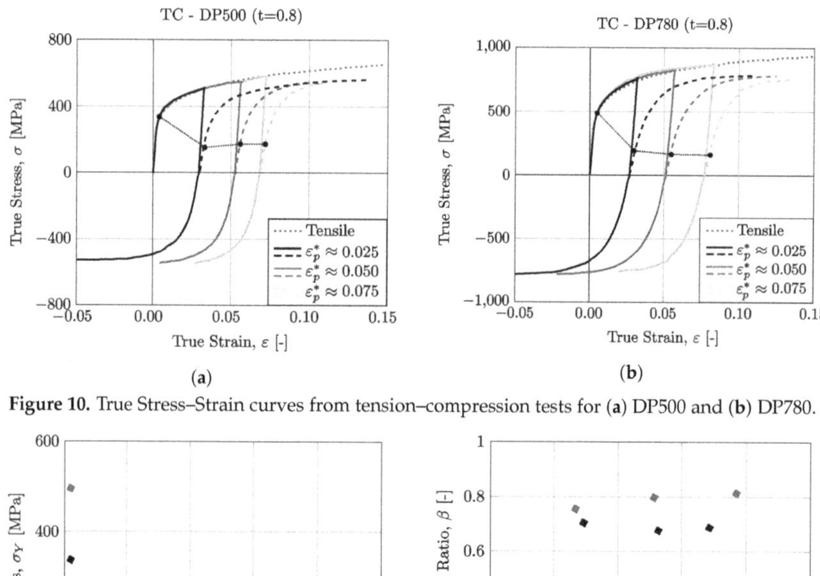

(a) (b)

Figure 10. True Stress–Strain curves from tension–compression tests for (**a**) DP500 and (**b**) DP780.

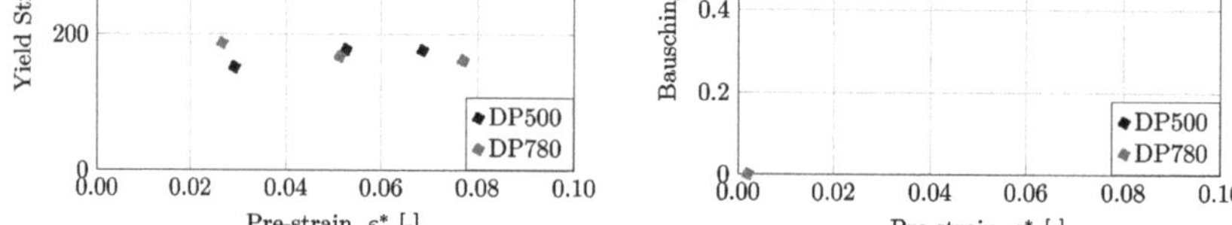

(a) (b)

Figure 11. (**a**) Yield stress, σ_Y, and (**b**) Bauschinger Ratio, β, values for different pre-strains, ε_p^*.

To quantify the Bauschinger effect, several indicators have been proposed over the years. One of the most used, the so-called Bauschinger ratio, β, is based on stress values and is expressed mathematically by Equation (2):

$$\beta = \frac{\sigma_{max}^T + \sigma_{Yield}^C}{\sigma_{max}^T} \quad (2)$$

where σ_{max}^T represents the maximum stress in the forward load (tensile step) and σ_{Yield}^C represents the yield stress in the backwards or reverse load (compression step)-point B and C, respectively, in Figure 9. The Bauschinger ratio quantifies the magnitude of the Bauschinger effect. The higher the parameter, the greater the Bauschinger effect of the material. In the case of isotropic hardening, the Bauschinger ratio is equal to zero. The Bauschinger ratio for the materials under this study is represented in Figure 11b, which is dependent on the applied pre-strain. From these results, it is possible to observe that both materials have a strong Bauschinger effect, since whenever the loading direction is reversed, the yield stress is reduced by about 70–80% compared to the value of the stress in the opposite direction.

4. Conclusions

This work presents the development of equipment dedicated to performing uniaxial, tensile–compression tests on miniaturized metallic specimens. The experimental equipment, called MSTD (Miniaturized Specimen Tester Device), allows the testing of specimens on a miniaturized scale, which opens the field of applications, one of them being compressive loading tests, having the advantage of overcoming buckling effects. In addition to uniaxial tensile, compression, and cyclic tests, it is also possible and expected that equipment will adapt to other fundamental tests for metallic sheets, such as shear and ductile damage tests.

The digital image correlation (DIC) technique was used to capture the full-field strain in the gauge section of the sample. A sensitivity analysis of the DIC setting parameters permitted to quantify the influence on results for subset size, subset step, strain window, shape function, and polynomial order. These three first parameters are included in the Virtual Strain Gage variable (VSG) and this variable was tested for different scenarios in tensile and compressive loading, giving a rather stable signal and a fast convergence of results. A good balance was found using a VSG of 151 px, which corresponds to 0.53 mm, thus giving the best ratio for efficiency and accuracy of results.

In order to validate the developed solution, several experimental tests were carried out using two dual-phase steels: DP500 and DP780. The tests performed on the developed equipment allowed the comparison of results for different types of loadings. Initially, monotonic tensile tests were carried out, and the results obtained with the miniaturized specimens were shown to be quite similar to those obtained by using the standardized geometry of tensile tests at the macro scale. The results of the compression tests showed that it is possible, with the developed equipment, to characterize materials up to 7.5% strain in compressive loading, without buckling, thus creating the possibility of sheet metal characterization for stress differential and Bauschinger behavior. Additionally, it was seen that the work-hardening of tested dual-phase steels (DP500, DP780) in compressive loading is very similar to their work hardening in tensile loading. Finally, uniaxial tests with reverse loadings showed that both DP500 and DP780 present a strong Bauschinger effect, since the yield stress undergoes a considerable decrease with reverse loading.

Author Contributions: Conceptualization, D.J.C., R.L.A. and A.D.S.; methodology, D.J.C., J.X., R.L.A. and A.D.S.; software, D.J.C., J.X., R.L.A. and A.D.S.; validation, D.J.C., J.X., R.L.A. and A.D.S.; formal analysis, D.J.C., J.X., R.L.A. and A.D.S.; investigation, D.J.C., J.X., R.L.A. and A.D.S.; resources, R.L.A. and A.D.S.; data curation, D.J.C., J.X., R.L.A. and A.D.S.; writing—original draft preparation, D.J.C., J.X., R.L.A. and A.D.S.; writing—review and editing, D.J.C., J.X., R.L.A. and A.D.S.; visualization, D.J.C., J.X., R.L.A. and A.D.S.; supervision, R.L.A. and A.D.S.; project administration, R.L.A. and A.D.S.; funding acquisition, R.L.A. and A.D.S. All authors have read and agreed to the published version of the manuscript.

Funding: Authors gratefully acknowledge the funding of Project POCI-01-0247-FEDER-069933-AMOB CUT3D, POCI-01-0247-FEDER-068492-AARM 4.0, POCI-01-0247-FEDER-047038-EcoForG4.0, cofinanced by Programa Operacional Competitividade e Internacionalização (COMPETE 2020), through Fundo Europeu de Desenvolvimento Regional (FEDER). The first author is also grateful to the FCT (Portuguese Foundation for Science and Technology) for the Doctoral grant 2022.12657.BD under the program POCH, cofinanced by the European Social Fund (FSE) and Portuguese National Funds from MCTES. The second author is also grateful to FCT by the support under the project UIDB/00667/2020 (UNIDEMI).

Data Availability Statement: Not applicable.

Conflicts of Interest: The authors declare no conflict of interest.

Abbreviations

The following abbreviations are used in this manuscript:

CCD	Charge-Coupled Device
DIC	Digital Image Correlation
EDM	Electrical Discharge Machining
ESPI	Electronic Speckle-Pattern Interferometry
FEM	Finite Element Method
HAH	Homogeneous Anisotropic Hardening
MEMS	Microelectromechanical systems
MSTD	Miniaturized Specimen Tester Device
MTT	Miniaturized Tensile Test
OM	Optical Microscopy
SEM	Scanning Electron Microscopy
SS	Subset Size
SSTT	Small Specimen Test Technology
ST	Subset Step
STM	Scanning Tunneling Microscope
SW	Strain Window
VSG	Virtual Strain Gauge

References

1. Gorji, M.B.; Furmanski, J.; Mohr, D. From macro- to micro-experiments: Specimen-size independent identification of plasticity and fracture properties. *Int. J. Mech. Sci.* **2021**, *199*, 106389. [CrossRef]
2. Zheng, P.; Chen, R.; Liu, H.; Chen, J.; Zhang, Z.; Liu, X.; Shen, Y. On the standards and practices for miniaturized tensile test—A review. *Fusion Eng. Des.* **2020**, *161*, 112006. [CrossRef]
3. Lucon, E. 1.08-Testing of Small-Sized Specimens. In *Comprehensive Materials Processing*; Hashmi, S.; Batalha, G.F., Van Tyne, C.J., Yilbas, B., Eds.; Elsevier: Oxford, UK, 2014; pp. 135–163. [CrossRef]
4. Klueh, R. Miniature tensile test specimens for fusion reactor irradiation studies. *Nucl. Eng. Des. Fusion* **1985**, *2*, 407–416. [CrossRef]
5. Şahin, S.; Saeed, A. Experimental evaluation of surveillance capsule assemblies for life assessment of CHASNUPP Unit-1 reactor pressure vessel. *Ann. Nucl. Energy* **2016**, *89*, 90–98. [CrossRef]
6. Lucon, E. Material damage evaluation and residual life assessment of primary power plant components using specimens of non–standard dimensions. *Mater. Sci. Technol.* **2001**, *17*, 777–785. [CrossRef]
7. Oluwasegun, K.; Cooper, C.; Chiu, Y.; Jones, I.; Li, H.; Baxter, G. Micro-tensile strength of a welded turbine disc superalloy. *Mater. Sci. Eng. A* **2014**, *596*, 229–235. [CrossRef]
8. Liu, P.; Bao, J.; Bao, Y. Mechanical Properties and Fracture Behavior of an EBW T2 Copper–45 Steel Joint. *Materials* **2019**, *12*, 1714. [CrossRef]
9. Gil, J.; Seca, R.; Amaral, R.; Emadinia, O.; Reis, A.; Jesus, A. 18Ni300 Maraging Steel Produced via Direct energy Deposition on H13 Tool Steel and DIN CK45. In *Key Engineering Materials*; Trans Tech Publications Ltd.: Wollerau, Switzerland, 2022; Volume 926, pp. 194–205. [CrossRef]
10. Azinpour, E.; Cruz, D.J.; de Sa, J.M.A.C.; Santos, A. Phase-field approach in elastoplastic solids: application of an iterative staggered scheme and its experimental validation. *Comput. Mech.* **2021**, *68*, 255–269. [CrossRef]
11. Spearing, S. Materials issues in microelectromechanical systems (MEMS). *Acta Mater.* **2000**, *48*, 179–196. [CrossRef]
12. Rajagopalan, J, Microelectromechanical Systems (MEMS)-Based Testing of Materials. In *Handbook of Mechanics of Materials*; Schmauder, S., Chen, C.S., Chawla, K.K., Chawla, N., Chen, W., Kagawa, Y., Eds.; Springer: Singapore, 2019; pp. 1955–1979. [CrossRef]
13. Rund, M.; Procházka, R.; Konopík, P.; Džugan, J.; Folgar, H. Investigation of Sample-size Influence on Tensile Test Results at Different Strain Rates. *Procedia Eng.* **2015**, *114*, 410–415. [CrossRef]
14. Read, D.T.; Liew, L.A.; White, R.M.; Barbosa, N.; Geaney, J. Measuring flow curve and failure conditions for a MEMS-scale electrodeposited nickel alloy. *Mater. Res. Express* **2019**, *6*, 0950a6. [CrossRef]
15. Hyun, H.C.; Kim, M.; Bang, S.; Lee, H. On acquiring true stress–strain curves for sheet specimens using tensile test and FE analysis based on a local necking criterion. *J. Mater. Res.* **2014**, *29*, 695–707. [CrossRef]
16. Kamaya, M.; Kitsunai, Y.; Koshiishi, M. True stress–strain curve acquisition for irradiated stainless steel including the range exceeding necking strain. *J. Nucl. Mater.* **2015**, *465*, 316–325. [CrossRef]

17. Hattalli, V.L.; Srivatsa, S.R. Sheet Metal Forming Processes–Recent Technological Advances. *Mater. Today Proc.* **2018**, *5*, 2564–2574. [CrossRef]
18. *ISO 6892-1:2019*; Metallic Materials—Tensile Testing—Part 1: Method of Test at Room Temperature. International Organization for Standardization: Geneva, Switzerland, 2019.
19. *ISO 16808:2022*; Metallic Materials—Sheet and Strip—Determination of Biaxial Stress-Strain Curve by Means of Bulge Test with Optical Measuring Systems. International Organization for Standardization: Geneva, Switzerland, 2022.
20. Hetz, P.; Kraus, M.; Merklein, M. Characterization of sheet metal components by using an upsetting test with miniaturized cylindrical specimen. *CIRP Ann.* **2022**, *71*, 233–236. [CrossRef]
21. Aitchison, C.; Tuckerman, L. *The Pack Method for Compressive Tests of Thin Specimens of Materials Used in Thin-Wall Structures*; NASA Technical Reports Server (NTRS): Washington, DC, USA, 1939.
22. Jackman, K. Improved methods for determining the compression properties of sheet metal. *Automot. Aviat. Ind.* **1944**, *11*, 36–38.
23. Yoshida, F.; Uemori, T.; Fujiwara, K. Elastic–plastic behavior of steel sheets under in-plane cyclic tension–compression at large strain. *Int. J. Plast.* **2002**, *18*, 633–659. [CrossRef]
24. Boger, R.K.; Wagoner, R.H.; Barlat, F.; Lee, M.G.; Chung, K. Continuous, large strain, tension/compression testing of sheet material. *Int. J. Plast.* **2005**, *21*, 2319–2343. [CrossRef]
25. Sekine, A.; Kuwabara, T. 616 Development of In-Plane Reverse Loading Test Apparatus and Measurement of the Bauschinger Effect of Sheet Metals. *Proc. Autumn Conf. Tohoku Branch* **2005**, *41*, 251–252. [CrossRef]
26. Tritschler, M.; Butz, A.; Helm, D.; Falkinger, G.; Kiese, J. Experimental analysis and modeling of the anisotropic response of titanium alloy Ti-X for quasi-static loading at room temperature. *Int. J. Mater. Form.* **2014**, *7*, 259–273. [CrossRef]
27. Hußnätter, W. Yielding of magnesium alloy AZ31. In Proceedings of the ICTP2008, Gyeongju, Republic of Korea, 7–11 September 2008; p. 109.
28. Cruz, D.J.; Shamchi, S.P.; Santos, A.D.; Amaral, R.L.; Tavares, P.J.; Moreira, P. Development of a mini-tensile approach for sheet metal testing using Digital Image Correlation. *Proc. Struct. Integr.* **2020**, *25*, 316–323. [CrossRef]
29. Pereira, J.; Xavier, J.; Ghiassi, B.; Lousada, J.; Morais, J. On the identification of earlywood and latewood radial elastic modulus of Pinus pinaster by digital image correlation: A parametric analysis. *J. Strain Anal. Eng. Des.* **2018**, *53*, 566–574. [CrossRef]
30. Grédiac, M.; Hild, F. (Eds.) *Full-Field Measurements and Identification in Solid Mechanics*; John Wiley and Sons: Hoboken, NJ, USA, 2012.
31. Pan, B. Digital image correlation for surface deformation measurement: historical developments, recent advances and future goals. *Meas. Sci. Technol.* **2018**, *29*, 082001. [CrossRef]
32. Cunha, F.; Santos, T.; Xavier, J. In Situ Monitoring of Additive Manufacturing Using Digital Image Correlation: A Review. *Materials* **2021**, *14*, 1511. [CrossRef] [PubMed]
33. Sutton, M.; Mingqi, C.; Peters, W.; Chao, Y.; McNeill, S. Application of an optimized digital correlation method to planar deformation analysis. *Image Vis. Comput.* **1986**, *4*, 143–150. [CrossRef]
34. Peters, W.H.; Ranson, W.F. Digital Imaging Techniques in Experimental Stress Analysis. *Opt. Eng.* **1982**, *21*, 427–431. [CrossRef]
35. Pan, B.; Qian, K.; Xie, H.; Asundi, A. Two-dimensional digital image correlation for in-plane displacement and strain measurement: A review. *Meas. Sci. Technol.* **2009**, *20*, 062001. [CrossRef]
36. Xavier, J.; de Jesus, A.; Morais, J.; Pinto, J. Stereovision measurements on evaluating the modulus of elasticity of wood by compression tests parallel to the grain. *Constr. Build. Mater.* **2012**, *26*, 207–215. [CrossRef]
37. Sousa, A.M.R.; Xavier, J.; Vaz, M.; Morais, J.J.L.; Filipe, V.M.J. Cross-Correlation and Differential Technique Combination to Determine Displacement Fields. *Strain* **2011**, *47*, 87–98. [CrossRef]
38. Sendrowicz, A.; Myhre, A.O.; Wierdak, S.W.; Vinogradov, A. Challenges and Accomplishments in Mechanical Testing Instrumented by In Situ Techniques: Infrared Thermography, Digital Image Correlation, and Acoustic Emission. *Appl. Sci.* **2021**, *11*, 6718. [CrossRef]
39. Zhang, D.; Luo, M.; Arola, D. Displacement/strain measurements using an optical microscope and digital image correlation. *OPT Eng.* **2006**, *45*, 2182108. [CrossRef]
40. Sun, Z.; Lyons, J.S.; McNeill, S.R. Measuring Microscopic Deformations with Digital Image Correlation. *Opt. Lasers Eng.* **1997**, *27*, 409–428. [CrossRef]
41. Sabaté, N.; Vogel, D.; Gollhardt, A.; Keller, J.; Michel, B.; Cané, C.; Gràcia, I.; Morante, J.R. Measurement of residual stresses in micromachined structures in a microregion. *Appl. Phys. Lett.* **2006**, *88*, 071910. [CrossRef]
42. Touchard, F.; Bridier, F.; Villechaise, P.; Brillaud, J. In-plane strain measurements on a microscopic scale by coupling digital image correlation and an in situ SEM technique. *Mater. Charact.* **2006**, *56*, 10–18. [CrossRef]
43. Vendroux, G.; Knauss, W. Submicron deformation field measurements: Part 1. Developing a digital scanning tunneling microscope. *Exp. Mech.* **1998**, *38*, 18–23. [CrossRef]
44. Dong, Y.; Pan, B. A Review of Speckle Pattern Fabrication and Assessment for Digital Image Correlation. *Exp. Mech.* **2017**, *57*, 1161–1181. [CrossRef]
45. Jones, E.; Iadicola, M. (Eds.) *A Good Practices Guide for Digital Image Correlation*; International Digital Image Correlation Society: Boston, MA, USA, 2018. [CrossRef]
46. Henriques, J.; Xavier, J.; Andrade-Campos, A. Identification of Orthotropic Elastic Properties of Wood by a Synthetic Image Approach Based on Digital Image Correlation. *Materials* **2022**, *15*, 625. [CrossRef]

47. Lava, P.; Jones, E.; Wittevrongel, L.; Pierron, F. Validation of finite-element models using full-field experimental data: Levelling finite-element analysis data through a digital image correlation engine. *Strain* **2020**, *56*, e12350. [CrossRef]
48. MatchID. *MatchID Manual*; MatchID: Metrology Beyond Colors: Gent, Belgium, 2017.
49. Yuan, W.; Zhang, Z.; Su, Y.; Qiao, L.; Chu, W. Influence of specimen thickness with rectangular cross-section on the tensile properties of structural steels. *Mater. Sci. Eng. A* **2012**, *532*, 601–605. [CrossRef]
50. Bauschinger, J. On the change of the position of the elastic limit of iron and steel under cyclic variations of stress. *Mitt. Mech. Tech. Lab. Munich* **1886**, *13*, 2.
51. Yoshida, F.; Uemori, T. A model of large-strain cyclic plasticity describing the Bauschinger effect and workhardening stagnation. *Int. J. Plast.* **2002**, *18*, 661–686. [CrossRef]
52. Barlat, F.; Gracio, J.J.; Lee, M.G.; Rauch, E.F.; Vincze, G. An alternative to kinematic hardening in classical plasticity. *Int. J. Plast.* **2011**, *27*, 1309–1327. [CrossRef]
53. Lee, M.G.; Kim, D.; Kim, C.; Wenner, M.L.; Chung, K. Spring-back evaluation of automotive sheets based on isotropic–kinematic hardening laws and non-quadratic anisotropic yield functions, part III: Applications. *Int. J. Plast.* **2005**, *21*, 915–953. [CrossRef]
54. Oliveira, M.; Alves, J.; Chaparro, B.; Menezes, L. Study on the influence of work-hardening modeling in springback prediction. *Int. J. Plast.* **2007**, *23*, 516–543. [CrossRef]
55. Yoshida, F. Material models for accurate simulation of sheet metal forming and springback. In Proceedings of the AIP Conference Proceedings, American Institute of Physics, Penang, Malaysia, 21–23 December 2010; Volume 1252, pp. 71–78.

micromachines

MDPI

Article

Electrically Tunable Lens (ETL)-Based Variable Focus Imaging System for Parametric Surface Texture Analysis of Materials

Jorabar Singh Nirwan [1], Shan Lou [2], Saqib Hussain [3], Muhammad Nauman [4], Tariq Hussain [5,6], Barbara R. Conway [1] and Muhammad Usman Ghori [1,*]

1 Department of Pharmacy, School of Applied Sciences, University of Huddersfield, Huddersfield HD1 3DH, UK; jorabar.nirwan@hud.ac.uk (J.S.N.); b.r.conway@hud.ac.uk (B.R.C.)
2 EPSRC Future Metrology Hub, University of Huddersfield, Queensgate, Huddersfield HD1 3DH, UK; s.lou@hud.ac.uk
3 Department of Mathematics and Physics, Texas A&M International University, Laredo, TX 78041, USA; saqib.hussain@tamiu.edu
4 Division of Mathematical and Physical Sciences, Institute of Science and Technology, 3400 Klosterneuburg, Austria; nouman.kakakhail@gmail.com
5 System Engineering Department, Military Technological College, Muscat 111, Oman; tariq.hussain@mtc.edu.om
6 The Wolfson Centre for Bulk Solid Handling Technology, University of Greenwich, London SE10 9LS, UK
* Correspondence: M.Ghori@hud.ac.uk; Tel.: +44-(0)-1484-4732-95

Abstract: Electrically tunable lenses (ETLs) are those with the ability to alter their optical power in response to an electric signal. This feature allows such systems to not only image the areas of interest but also obtain spatial depth perception (depth of field, DOF). The aim of the present study was to develop an ETL-based imaging system for quantitative surface analysis. Firstly, the system was calibrated to achieve high depth resolution, warranting the accurate measurement of the depth and to account for and correct any influences from external factors on the ETL. This was completed using the Tenengrad operator which effectively identified the plane of best focus as demonstrated by the linear relationship between the control current applied to the ETL and the height at which the optical system focuses. The system was then employed to measure amplitude, spatial, hybrid, and volume surface texture parameters of a model material (pharmaceutical dosage form) which were validated against the parameters obtained using a previously validated surface texture analysis technique, optical profilometry. There were no statistically significant differences between the surface texture parameters measured by the techniques, highlighting the potential application of ETL-based imaging systems as an easily adaptable and low-cost alternative surface texture analysis technique to conventional microscopy techniques.

Keywords: surface texture; electrically tunable lens; materials; hypromellose; surface topography; surface roughness; pharmaceutical tablet; variable focus imaging

Citation: Nirwan, J.S.; Lou, S.; Hussain, S.; Nauman, M.; Hussain, T.; Conway, B.R.; Ghori, M.U. Electrically Tunable Lens (ETL)-Based Variable Focus Imaging System for Parametric Surface Texture Analysis of Materials. *Micromachines* **2022**, *13*, 17. https://doi.org/10.3390/mi13010017

Academic Editors: Luigi Sirleto and Giancarlo C. Righini

Received: 29 October 2021
Accepted: 18 December 2021
Published: 23 December 2021

1. Introduction

Surface texture may be defined as the repetitive or random deviation from the nominal surface that forms the three-dimensional topography of the surface and includes roughness, waviness, lay, and flaws [1]. In general, roughness is caused by fluctuations in the surface of short wavelengths and is characterised by peaks and valleys of varying amplitudes and spacings. Waviness (macroroughness) is the surface irregularity of longer wavelengths and includes all irregularities whose spacing is greater than the roughness wavelengths. The principal direction of the surface pattern is referred to as lay and is predominately determined by the fabrication method, while flaws are unexpected, unwanted, and unintentional interruptions in the texture [1,2]. These surface components play a vital role in determining the functionality and behaviour of materials used in multiple applications including mechanical components, construction material, surgical implants, and solid pharmaceutical formulations; hence, the measurement and control of these properties is necessary for safe and effective functioning [3,4].

Current surface texture analysis techniques can be divided into two broad categories: a contact type in which during measurement a component of the measurement instrument contacts the surface to be measured and includes contact profilometers and atomic force microscopy (AFM); and a non-contact type which includes coherence scanning interferometry, confocal microscopy, and scanning laser microscope [2]. These techniques have been employed extensively in industry and academia with AFM currently considered the most powerful and versatile technique for nanoscale imaging and multiple studies have successfully employed this technique, either alone or in combination with another imaging or analytical technique, to quantitatively assess the surface attributes [3–7]. However, along with other contact-type techniques, this may cause damage to the sample surfaces and precise attention must be given for the normal loads to be low enough so that the contact stresses do not exceed the hardness of the surface to be measured [2]. Furthermore, measurement of samples can be time consuming, and although this may be improved with the use of non-contact techniques, these systems are generally large, expensive, and require high power consumption. Another drawback of currently employed surface texture analysis techniques is their inability to capture the full surface of a sample; thus, leaving users restricted to being able to analyse only a limited area. Though, to an extent digital holographic microscopy has mended this gap [8–12]; however, this imaging technique is expert and resource intensive with heavy computational load, which necessitates development of an accurate and sensitive numerical reconstruction algorithm for meaningful data processing and results [13]. Moreover, the coexistence of twin images is a continuous challenge with these imaging systems [14–16].

A unique solution to the aforementioned drawbacks of conventional surface texture analysis techniques may be found in electrically tunable lenses (ETL), which possess the ability to alter their optical power in response to an electric signal and have an imaging system integrated with an advanced image stitching algorithm [3,17,18]. The mechanism of ETL is achieved via two main approaches: local variations in refractive index which include liquid crystals, and controlling the shape of the lens with the two main techniques including electrowetting and shape-changing polymers [19]. In general, liquid crystal lenses display larger focal length changes with lower applied voltage, although they suffer from chromatic aberration. Electrowetting lenses have the characteristics of large power range and fast response speed, while requiring a high driving voltage and possessing an aperture of limited size [18,19]. Meanwhile, polymer-based liquid lenses allow the flexible selection of aperture size and are composed of an optical fluid and an elastomeric polymer membrane, usually polydimethylsiloxane (PDMS) [18–20]. These properties of ETL provide them with multiple unique advantages over other systems including greater compactness due to the use of less optomechanical components, a fast frequency response, less mechanic motion, and lower power consumption [17,18,21,22].

Furthermore, ETL are not only able to image the areas of interest but are also able to obtain different spatial depth perception (depth of field, DOF) [19,23–26]. Hu et al. [19] tackled the issue of shallow working volume caused by the limited DOF of conventional lenses by implementing an ETL to enlarge the 3D shape measurement range. With this approach, the authors were able to capture always in-focus phase-shifted fringe patterns by precisely synchronising the tunable lens attached to a camera with the image acquisition and the pattern projection. A prototype system was developed which was able to perform high-quality measurement for a depth range of approximately 1000 mm (400–1400 mm) and a measurement error of 0.05%. The authors concluded that such a system is suitable to be used for real-time 3D shape measurements by experimentally measuring moving objects.

A practical application of ETL was explored by Fahrbach et al. [23] who combined a light-sheet microscope with an ETL for the in vivo investigation of highly dynamic biological samples, for example, a beating zebrafish heart. The authors state that in order to obtain images of different planes using a conventional light-sheet microscope, it is necessary to move the sample relative to the fixed focal plane of the detection objective lens which limits speed, precision, and may be harmful to the sample. However, with the use of ETL, the authors were able to achieve flexible volume imaging at much higher speeds than previously reported and the DOF allowed them to image 17 planes within a beating zebrafish heart [19]. Similarly, a study conducted by Chen and Lin [26] demonstrated an electrically tunable endoscopic system by adopting a LC lens with a large spatial depth perception. As conventional endoscopic systems consist of several solid lenses and suffer from a fixed and limited DOF, the authors were able to image the object at different objective distances using the new system and the corresponding DOF helped in enlarging the total spatial depth perception in the scene [27].

However, although the usefulness of DOF and controlling the optical power in response to the electrical signal has been reported, we hypothesised these features can be exploited to investigate the parametric surface texture analysis of the materials. It is anticipated that the DOF in ETL will be able to provide accurate measurement of peaks and valleys present on the surface of materials thus allowing quantification of surface texture parameters. Hence, the aim of the present study was to develop and utilise a low-cost non-destructive ETL-based imaging system for 3D surface characterisation of materials. To ensure efficient and accurate analysis of the material surface, the system will firstly be calibrated using a widely adopted focus measure. The performance of the newly developed system will also be compared and comprehensively validated against a previously validated surface texture analysis technique—optical profilometry. Moreover, an image stitching feature [3] will also be integrated to demonstrate its ability to analyse the complete surface of the model material sample, a pharmaceutical tablet formulation. The surface properties of solid-state pharmaceutical formulations are of critical importance in determining the performance of the dosage form. The surface of dosage forms such as tablets, can be modified, either intentionally, for example by coating, or unintentionally through processing. A matrix tablet in its simplest form is a blend of a drug and a polymer that controls the rate of drug release and in current study this was the pharmaceutical system used as a model material.

2. Materials and Methods

Hydroxypropyl methylcellulose, HPMC, (K4M) and propranolol HCl were acquired from Colorcon, Dartford, U.K. and Sigma-Aldrich Chemical Co., St. Louis, MI, USA, respectively.

2.1. Electrically Tunable Lens (ETL)-Based Imaging Setup

The multiple components of the ETL-based imaging system are displayed in Figure 1. The system comprised a polymer-based liquid ETL (Phaseview, Verrières-le-Buisson, France) attached to a driver (Phaseview, Verrières-le-Buisson, France) which provided a voltage to the ETL and allowed tuning of this voltage (voltage range) enabling adjustment of the ETL focal power (focal power range). The ETL driver was connected to a computer used to control the voltage. Images were captured via a camera attached to the ETL system which relayed the live images to the computer for image processing to obtain 3D surface texture parameters. In the current imaging system, a charge-coupled device (CCD) camera with a light-emitting diode (light source) was employed. Furthermore, four objective lenses were available to be used with the system possessing magnification powers of 5×, 10×, 20×, and 50×. The specifications of the objective lenses are displayed in Table 1. As the focus of the present research was DOF which decreases, along with z-resolution, as numerical aperture increases, the objective lens utilised for the analyses in this study was 5× magnification. Moreover, to secure the sample under analysis, a high precision 0.25 μm automatic motorised, computer controlled, vertical translation stage was used.

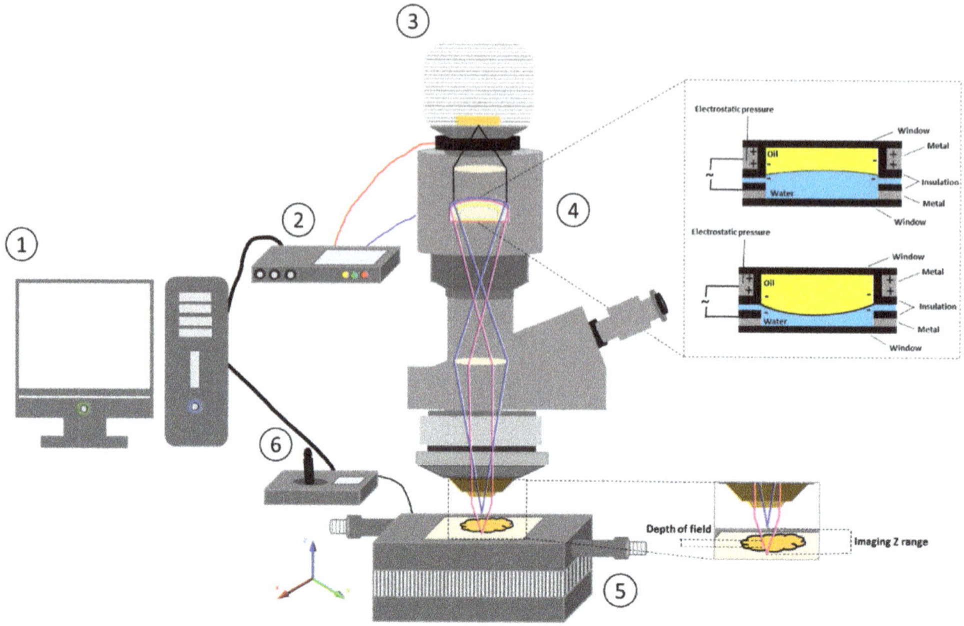

Figure 1. Schematic illustration of ETL-based imaging system. (1) computer; (2) ETL driver; (3) camera; (4) ETL; (5) automatic motorized vertical translation stage; and (6) stage controlling system.

Table 1. Specifications of electrically tunable lens integrated add-on camera system.

Objective Mag	Numerical Aperture	Z-Range (μm)	Z-Resolution (μm)	Z-Axis Accuracy (%)	Z-Axis Repeatability (%)	Max. Slope
5×	0.10	640	18.5	1	0.35	90°
10×	0.25	160	3			
20×	0.45	40	1			
50×	0.80	6.4	0.25			

Data obtained from the manufacturer.

2.2. System Calibration

Calibration of the ETL imaging system was conducted using a high precision 0.25 μm automatic motorised vertical translation stage. Briefly, this stage was set to different focus heights separated by regular intervals of 0.25 μm. For each focus height (μm), a series of images was captured while modifying the focal length setting by adjustment of the control current (mA) applied to the ETL and a focus measure was computed for each image using the Tenengrad operator (Equation (1)) [28,29]. Although multiple focus measure techniques are available, an ideal focus measure function produces its maximum value at the plane of best focus and values decrease in both directions as the defocus measure increases. This function produces a bell-shaped curve when focus measure values are plotted against image sequences with different focus distances. In this regard, the Tenengrad operator has been proven to be more accurate, robust, and less sensitive to noise than other focus measure functions [17–19].

$$\phi TEN(I(x,y)) = \sum_{M}\sum_{N}[S(x,y)]^2 \tag{1}$$

where,

$$S(x,y) = \sqrt{|G_x(x,y)|^2 + |G_y(x,y)|^2} \tag{2}$$

where $G_x(x,y)$ and $G_y(x,y)$ are the results of the Sobel operator which is used as an edge detection method, applied to the area of interest, *I*, along the *x* and *y* direction, respectively. The focus measure results were plotted against the corresponding control current (mA) applied which resulted in Gaussian-like curves with the peak representing the plane of best focus. To ensure the focus measure accurately detects the plane of best focus, the control current (mA) required to reach the greatest focus measure (a.u.) for each focus height (μm) (peak of each curve) was plotted against its corresponding focus height to establish a linear relationship.

2.3. Preparation of Pharmaceutical Tablet Formulation

A pharmaceutical tablet formulation (model sample material) was prepared as described by Nirwan et al., 2019 [3]. Briefly, a powder mixture of HPMC (30% *w/w*) and propranolol (70% *w/w*) was prepared using a tubular mixer (Artisan Technology Group®, Kansas City, MO, USA) at 50 rpm for 15 min, followed by content uniformity testing of the mixture using ultraviolet spectroscopy which confirmed uniform mixing. Tablets of 300 mg of the powder mixture were prepared using a Testometric™ M500–50 CT (Testometric™ Company Ltd., Rochdale, U.K.) apparatus directly connected to a computer with a 50 kN load cell equipped with a 13.00 mm Atlas Evacuable Pellet Die (Specac® Limited, Orpington, U.K.). Polished flat-faced punches were used, where the lower punch was held stationary while the upper punch moved at a speed of 3 mm min^{-1} during loading and 1 mm min^{-1} on unloading and an applied pressure of 150 MPa was used. Following ejection of the tablets, they were stored in an air-tight container over silica gel for 24 h to allow for elastic recovery before further investigation. All compaction work was conducted at a relative humidity and temperature in the range of 20% to 35% RH and 25 °C to 27 °C, respectively.

2.4. Imaging and Parametric Surface Texture Analysis of Matrix Tablet

Each tablet was fixed on an automatic motorised vertical translation stage using a carbon sticky dot. Following this, 35 nanoscale images were captured to allow analysis of the whole tablet surface using the ETL system in a controlled environment (temperature 18–25 °C and RH 25–33%) with an acquisition time of <1 min. The 35 individual images were stitched together to produce a single image revealing the overall tablet surface by implementing an image stitching algorithm using MATLAB® 2017 software (The MathWorks, Inc., Natick, MA, USA). The overall process consisted of two stages: (a) image stitching and (b) image blending. In the first stage, a relative position of the collected images was generated followed by the identification of the correlation point which was performed by sliding the adjacent edges of the image in both directions until a best match of edge features was achieved. This method involves the choice of an optimal search space as illustrated in Figure 2a, in which a detailed assessment was performed to identify the best correlation (which was a 20% overlap in the present study). The normalised cross-correlation can be expressed by the following equation, Equation (3):

$$\frac{\sum_{x=0}^{L-1}\sum_{y=0}^{K-1}(w(x,y)-w)(f(x+i,y+j)-f(i,j))}{\sqrt{\sum_{x=0}^{L-1}\sum_{y=0}^{K-1}(w(x,y)-w)^2}\sqrt{\sum_{x=0}^{L-1}\sum_{y=0}^{K-1}(f(x+i,y+j)-f(i,j))^2}} \tag{3}$$

where $w(x, y)$ represents a pixel value of the image, w is the mean value of the overall pixels included in the box area, $f(x + i, y + j)$ represents a pixel value of the composite image inside the box area, and $f(i, j)$ is the mean value of the composite image inside the box area. K and L represent the box dimensions. In the image blending stage, an advanced image blending algorithm was applied to improve the visual quality of the composite image. Figure 2b displays this process in which an overlap between a new image and the composite image is shown. For the collection of 3D quantitative surface texture parameters of all the stitched images, MATLAB® 2017 software (The MathWorks, Inc., USA) was used [30–33]. The mathematical equations (Equations (4)–(19)) and description of the surface texture parameters measured in this study are presented in Table 2. The surface texture parameters obtained using the ETL system were validated with a conventional optical interferometry system (Contour GT Elite K, Bruker Nano Surfaces, Bruker UK Ltd., Coventry, U.K.) previously validated by our group with singly acquired images (n = 5) [3].

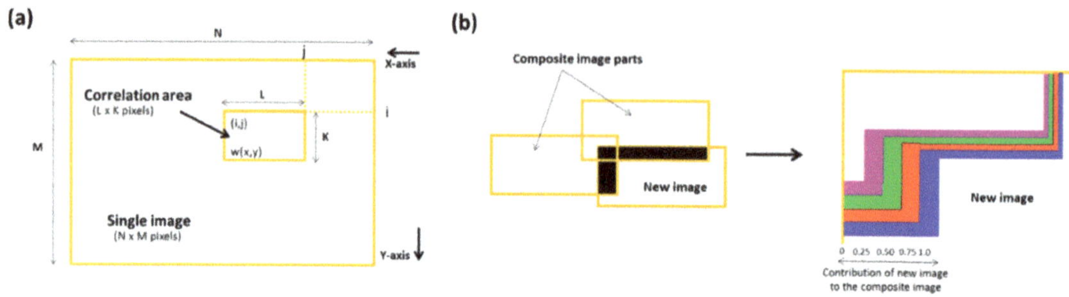

Figure 2. Schematic illustrations showing (**a**) image stitching and (**b**) blending of overlapping image intensities. This image is adopted from [3] with permission from the publisher, Royal Chemical Society.

Table 2. Summary of 3D surface texture parameters [34].

	Surface Topography Parameters	Description	Equation	
Amplitude	Root mean squared height (S_q, nm)	S_q presents an overall measure of the texture comprising the surface.	$S_q = \sqrt{\frac{1}{MN}\sum_{j=1}^{N}\sum_{i=1}^{M} n^2(x_i, y_j)}$	(4)
	Arithmetical average of surface (S_a, nm)	Provides an average of the overall roughness of the surface.	$S_a = \frac{1}{MN}\sum_{j=1}^{N}\sum_{i=1}^{M}\left\lvert n^2(x_i, y_j)\right\rvert$	(5)
	Skewness of height distribution (S_{sk})	Represents the degree of symmetry of the surface heights about the mean plane. The predominance of peaks comprising the surface is indicated by $S_{sk} > 0$, whereas the predominance of valley structures is indicated by $S_{sk} < 0$.	$S_{sk} = \frac{1}{MNS_q^3}\sum_{j=1}^{N}\sum_{i=1}^{M} n^3(x_i, y_j)$	(6)
	Kurtosis of height distribution (S_{ku})	Reveals the presence of inordinately high peaks/deep valleys ($S_{ku} > 3.00$) or lack thereof ($S_{ku} < 3.00$) comprising the surface.	$S_{ku} = \frac{1}{MNS_q^4}\sum_{j=1}^{N}\sum_{i=1}^{M} n^4(x_i, y_j)$	(7)
	Highest peak of the surface (S_p, nm)	The height of the highest point.	$S_p = \mathrm{MAX}\,(n_p)$	(8)
	Lowest valley of the surface (S_v, nm)	The depth of the lowest point expressed as a negative number.	$S_v = \mathrm{MIN}\,(n_v)$	(9)
	Deviation maximum height (S_z, nm)	The sum of S_p and S_v.	$S_z = (\lvert S_p\rvert + \lvert S_v\rvert)$	(10)

Table 2. *Cont.*

	Surface Topography Parameters	Description	Equation	
Spacing	Surface density summits (S_{ds}, $1/mm^2$)	Represents the number of summits per unit area comprising the surface. A summit is defined as at least 5% of S_z.	$S_{ds} = \frac{Number\ of\ summits}{(M-1)(N-1).\ \Delta x\ .\ \Delta y}$	(11)
	Surface aspect ratio (S_{tr})	Quantifies the spatial isotropy or directionality of the surface texture. The S_{tr} parameter will be closer to 0.00, whereas a spatially isotropic texture will produce a S_{tr} of 1.00.	$S_{tr} = \frac{\min\left(\sqrt{t_x^2+t_y^2}\right)}{\max\left(\sqrt{t_x^2+t_y^2}\right)}$	(12)
	Surface autocorrelation length (S_{al}, mm)	Represents the horizontal distance in the direction in which the autocorrelation function decays to a specified value (0.2 by default) the fastest. This provides a measure of the distance over the surface such that the new location will have minimal correlation with the original location.	$S_{al} = \min\left(\sqrt{t_x^2+t_y^2}\right)$	(13)
Hybrid	Surface root mean square (S_{dq})	Provides a measurement of the slopes which comprise the surface. The S_{dq} of a completely level surface is 0. When a surface has any slope, its S_{dq} value becomes larger.	$S_{dq} = \sqrt{\frac{1}{(M-1)(N-1)} \sum_{j=2}^{N} \sum_{i=2}^{M} p_{i,j}^2}$	(14)
	Interfacial area ratio (S_{dr}, %)	Represents a percentage the additional surface area contributed by the texture as compared with the projected surface. The S_{dr} of a completely level surface is 0 and increases with the spatial intricacy of the texture.	$S_{dr} = \frac{\sum_{j=1}^{N-1} \sum_{i=1}^{M-1} A_{i,j} - (M-1)(N-1)\Delta x\ .\ \Delta y}{(M-1)(N-1)\Delta x\ .\ \Delta y}\ .\ 100$	(15)
Volume	Material surface volume (V_m, $\mu m^3/mm^2$)	The material volume $V_m(p)$ at a given material ratio p is defined as the volume of the material per unit area calculated from the (inverse) areal material ratio function.	$V_m = K \int_{0\%}^{p} [S_{mc}(q) - S_{mc}(q)]\, dq$	(16)
	Core surface volume (V_{mc}, $\mu m^3/mm^2$)	The volume of material contained within the material ratio values of 10% and 80%.	$V_{mc} = \frac{V_m(h_{0.8}) - V_m(h_{0.10})}{(M-1)(N-1).\Delta x.\ \Delta y}$	(17)
	Void surface volume (V_{vc}, $\mu m^3/mm^2$)	The void volume $V_v(p)$ at a given material ratio p is defined as the volume of the voids per unit area calculated from the (inverse) areal material ratio function.	$V_v = K \int_{p}^{100\%} [S_{mc}(p) - S_{mc}(q)]\, dq$	(18)
	Core void surface volume (V_{vc}, $\mu m^3/mm^2$)	The volume of space contained within the material ratio values of 10% and 80%.	$V_{vc} = \frac{V_v(h_{0.10}) - V_m(h_{0.8})}{(M-1)(N-1).\Delta x.\ \Delta y}$	(19)
	Valley void surface volume (V_{vv}, nm^3/mm^2)	The volume of space of the surface from the height corresponding to a material ratio value (80% by default) to the lowest valley.	$V_{vv} = \frac{V_v(h_{0.8})}{(M-1)(N-1).\Delta x.\ \Delta y}$	(20)

2.5. Statistical Analysis

Analysis of variance (ANOVA) (confidence limit of $p < 0.05$) was used to investigate the statistical significance between the various surface texture parameters. Moreover, linear regression and root mean square error were also calculated for the 2D line scan data points.

3. Results and Discussion

To achieve high depth resolution, ensure accurate measurement of depth, and to account for and correct any influences from external factors including temperature on the ETL, calibration of the system is an essential process. The calibration process as described in Section 2.2 resulted in the Gaussian-like curves displayed in Figure 3a with each curve representing a different focus height as set on the automatic motorised vertical translation stage. The graph displays the focus measure (a.u.) achieved as calculated using the Tenengrad operator as the control current (mA) applied to the ETL which is adjusted to modify the focal length setting (Figure 3a). To ensure the focus measure accurately detects the plane of best focus, the control current (mA) required to reach the greatest focus measure (a.u.) for each focus height (μm) (peak of each curve) was plotted against its corresponding focus height (Figure 3b). This graph displayed a linear relationship between the control current applied to the ETL and the height at which the optical system focuses with a correlation coefficient of the linear regression of 0.999 indicating that the focus measure accurately detects the plane of best focus.

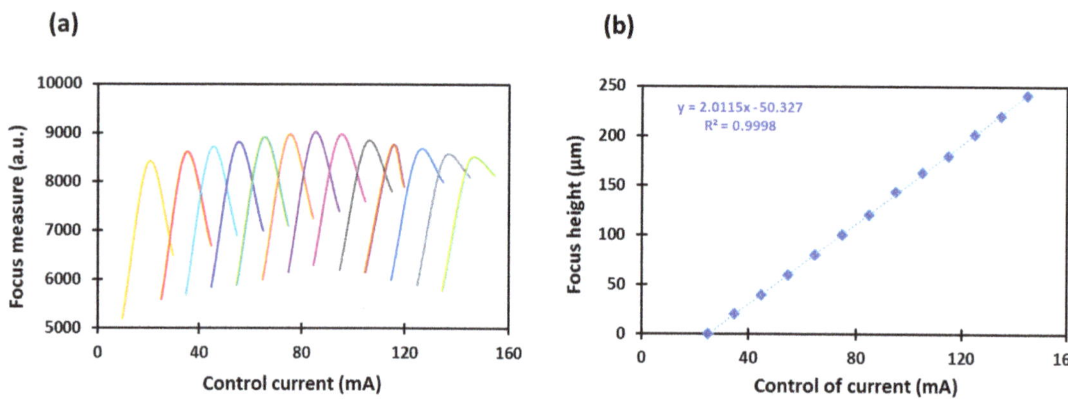

Figure 3. (**a**) Focus measure sweep as a function of tunable lens control current for different focus height and (**b**) focus height as a function of the tunable lens control current, calibration curve of the ETL imaging system.

Furthermore, the effectiveness of the Tenengrad operator as a focus measure is demonstrated in Figure 4 in which the Gaussian-like curve obtained for the 25 μm focus height is displayed along with the images captured at the peak and the extreme ends of the curve. These images reveal the most in-focus image at the peak of the curve and a decrease in focus on either side of the peak. This trend was especially evident upon visual analysis of the points of reference in each corresponding image which are highlighted with arrows in Figure 4 and display more in-focus features in the image captured at the peak of the curve compared with the images captured at either side.

Qualitative surface texture analysis was conducted using the ETL system as well as optical profilometry by producing 3D images of the surface of the tablet as presented in Figure 5a–d. The images acquired from both systems were successfully stitched, as shown in Figure 6a–d, using the method reported recently by our group [3]. The overall images have not showed any discrepancy. The comparability of both techniques for the analysis of the surface texture is shown in the rectangular boxes on the tablet surface which highlight the same area of the surface in the image captured using optical profilometry (Figure 5a) and the ETL system (Figure 5c). In these images, an almost identical surface texture can be seen within the respective rectangular boxes of both techniques with a similar location, size, and intensity of valleys as demonstrated by 2D line scans (Figure 7). The 2D line scan data points were further comparatively analysed and the findings showed a good agreement in data with RMSE (root mean square error) = 0.166 and R^2 (correlation coefficient) = 0.96,

Figure 7b. Overall, these images display a smooth tablet surface with minimal peaks and an equal distribution of valleys (Figure 5a–d). However, a higher degree of valleys can be seen in the image generated by the ETL-based system as displayed by the darker spots on the tablet surface (Figure 5c,d) and 2D line scan (Figure 7).

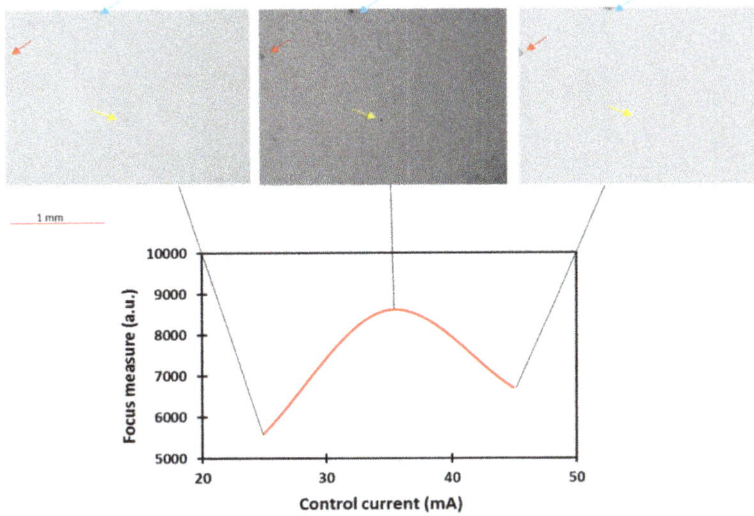

Figure 4. Gaussian-like curve illustrating the autofocusing ability of the system by presenting the images captured at the peak and the extreme ends of the curve.

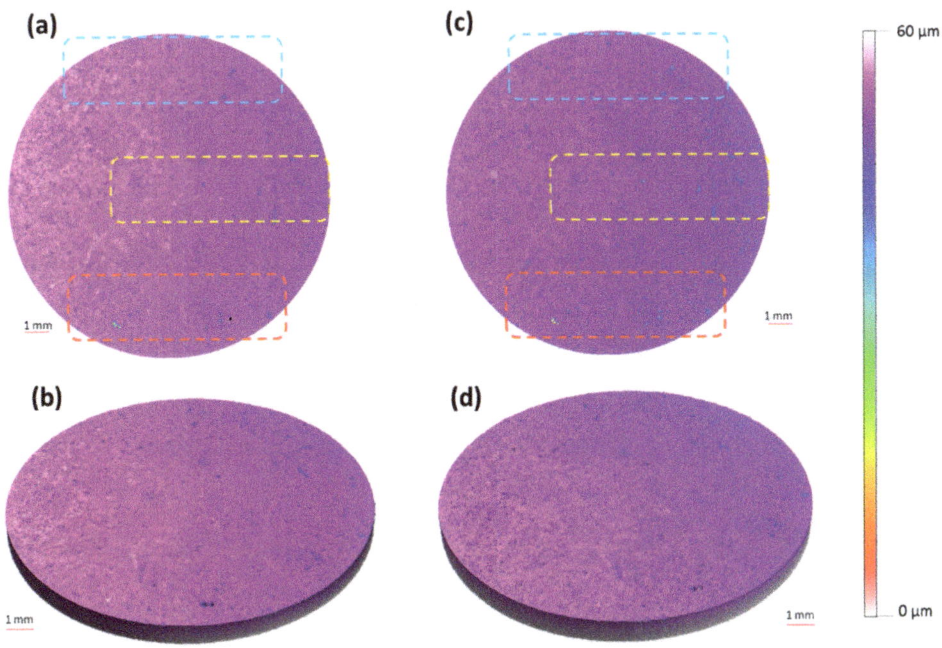

Figure 5. (**a**) 2D and (**b**) 3D surface texture images produced using optical profilometry. These images are adapted from [3] with permission from the publisher, Royal Society Chemistry. (**c**) 2D and (**d**) 3D surface texture images produced using the new developed ETL-based variable focus imaging system.

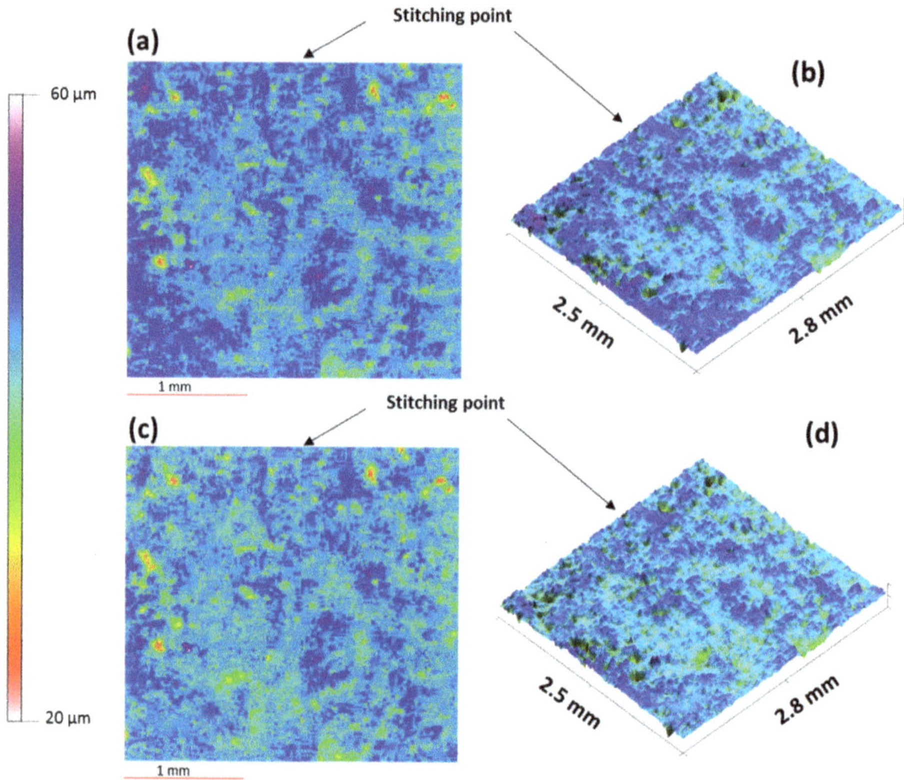

Figure 6. (**a**) 2D and (**b**) 3D surface texture images produced using optical profilometry, (**c**) 2D and, (**d**) 3D surface texture images produced using the newly developed ETL-based variable focus imaging system showing the zoomed area and the image-stitching points.

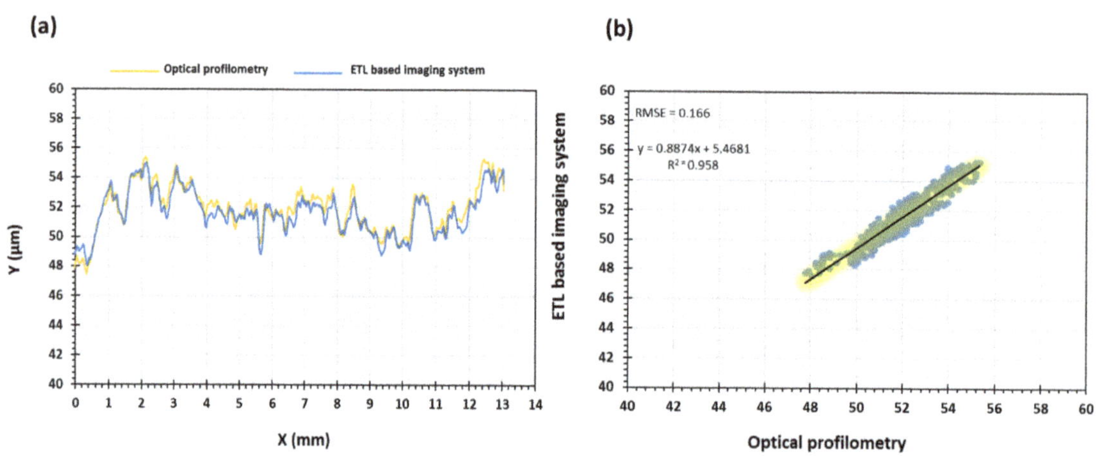

Figure 7. (**a**) 2D line scans of the tablets generated using a line passing through the exact centre point of the pharmaceutical tablet images generated using optical profilometry and newly developed ETL-based imaging system and (**b**) quantitative comparison by determining the root mean square error (RMSE) of the 2D line scan data points.

To further explore the comparability of both surface texture analysis techniques, quantitative analysis was also conducted by analysing surface texture parameters. The surface texture parameters explored in this study include amplitude, spacing, hybrid, and volume parameters. Amplitude parameters, which are also known as height parameters, provide analysis of vertical deviations of the roughness profile of the surface from the mean line, whereas spatial parameters focus on the direction of the plane. Hybrid parameters characterise both the height direction and the direction of the plane. Finally, volume parameters relate to the volume of material contained within specified material ratio values using the areal material ratio curve.

Surface texture analysis was conducted using the ETL system as well as optical profilometry and results are presented in the form of box and whisker plots to allow visual comparison of the surface texture parameters obtained using both techniques (Figures 8–10). The first of the amplitude parameters analysed were the arithmetical mean height (S_a) and the root mean square height (S_q). These measures assess the average standard deviation of the valleys and peaks in a surface profile and are typically used as a general evaluation of the surface roughness [35]. The mean S_a of the tablet surface obtained using the ETL system and optical profilometry was 3.35 μm and 3.32 μm, respectively, and the mean S_q was 4.28 μm and 4.23 μm, respectively. The maximum peak height (S_p), which is the height of the highest peak within the defined area, and the maximum pit height (S_v), which is the absolute value of the depth of the deepest pit within the defined area, were also analysed. The mean S_p and S_v of the surface obtained using the ETL system were 48.80 μm and 57.00 μm, respectively, and 51.86 μm and 54.48 μm, respectively, for optical profilometry. The sum of S_p and S_v were also calculated and defined as the maximum height (S_z). The mean S_z of the tablet surface obtained using the ETL system and optical profilometry was 110.20 μm and 106.34 μm, respectively. Furthermore, the skewness (S_{sk}), which represents the degree of symmetry of the surface heights about the mean plane, was also evaluated. The mean S_{sk} obtained using the ETL system and optical profilometry was −0.20 μm and −0.189 μm, respectively, indicating the predominance of valley structures and a height distribution skewed above the mean plane ($S_{sk} < 0$). The kurtosis (S_{ku}), which measures the sharpness of the roughness profile, indicated either a normal distribution ($S_{ku} = 3.00$) or slightly skewed towards a presence of inordinately high peaks/deep valleys ($S_{ku} > 3.00$) as determined by the mean S_{ku} obtained using the ETL system (3.00 μm) and optical profilometry (3.044 μm). Upon visual analysis of the box and whisker plots of the amplitude parameters (Figure 8), it may be observed that the surface texture parameters obtained using the ETL system generally resulted in a slightly larger interquartile range (Figure 8a–f) indicating greater variability in the results, however, not significant.

The spatial parameters explored in this study relate to the autocorrelation function which is a measurement of how regular or periodic the surface is. This is quantified by comparing a shifted image of the surface with the original image. An autocorrelation function of 1.00 indicates that the shifted surface is identical to the original surface, whereas an autocorrelation function of −1.00 suggests that all peaks align with corresponding valleys [30]. One application of the autocorrelation function in surface texture analysis is the measurement of the autocorrelation length (S_{al}) which represents the horizontal distance in the direction in which the autocorrelation function decays to a specified value (0.2 by default) the fastest. This provides a measure of the distance over the surface such that the new location will have minimal correlation with the original location. The mean S_{al} obtained used the ETL system and the optical profilometry system were 92.50 μm and 91.20 μm, respectively. Furthermore, the texture aspect ratio (S_{tr}) was also measured which quantifies the spatial isotropy or directionality of the surface texture. This value is calculated by dividing the value of S_{al} by the horizontal distance in the direction of the slowest decay of autocorrelation function to the value (0.2 by default). The mean values of S_{tr} were calculated as 0.901 and 0.880 for the ETL system and optical microscopy, respectively.

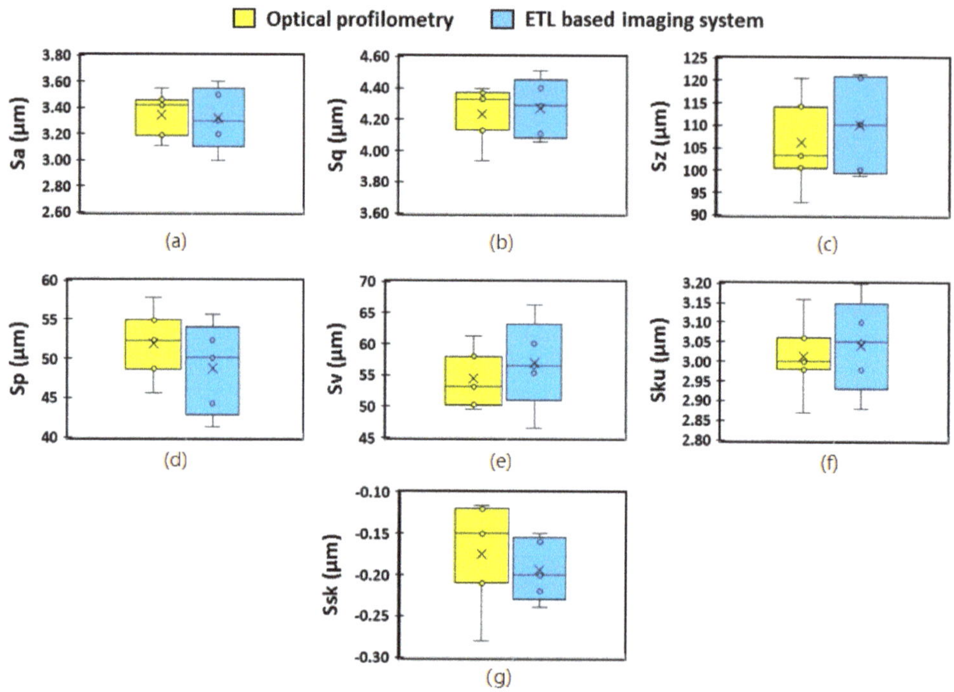

Figure 8. Amplitude and spacing parameters of 3D surface texture analysis. (**a**) S_a: arithmetical average of surface; (**b**) S_q: root mean squared height; (**c**) S_z: deviation maximum height; (**d**) S_p: highest peak of the surface; (**e**) S_v: lowest valley of the surface; (**f**) S_{ku}: kurtosis of height distribution; and (**g**) S_{sk}: skewness of height distribution.

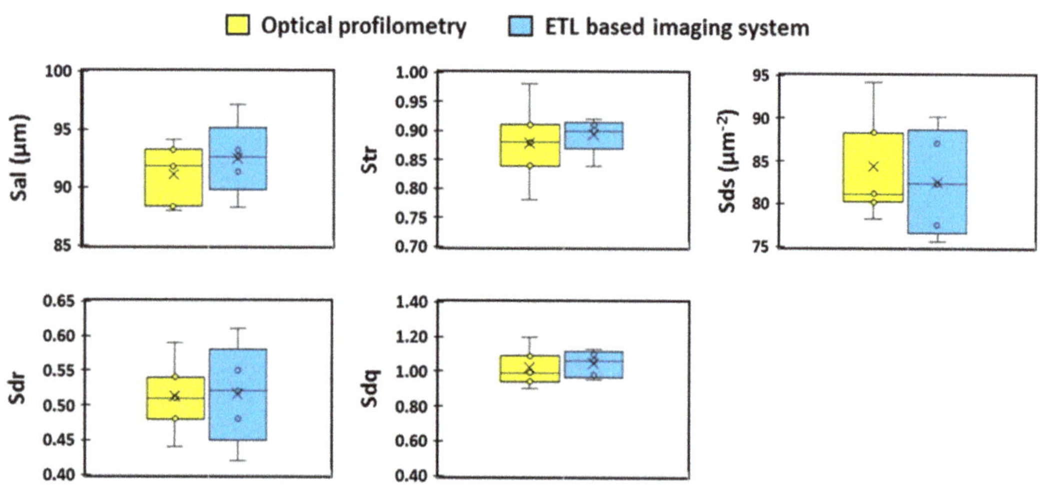

Figure 9. Spatial and hybrid parameters of 3D surface texture analysis. S_{al}: surface autocorrelation length; S_{tr}: surface aspect ratio; S_{ds}: surface density summits; S_{dr}: interfacial area ratio; and S_{dq}: surface root mean square.

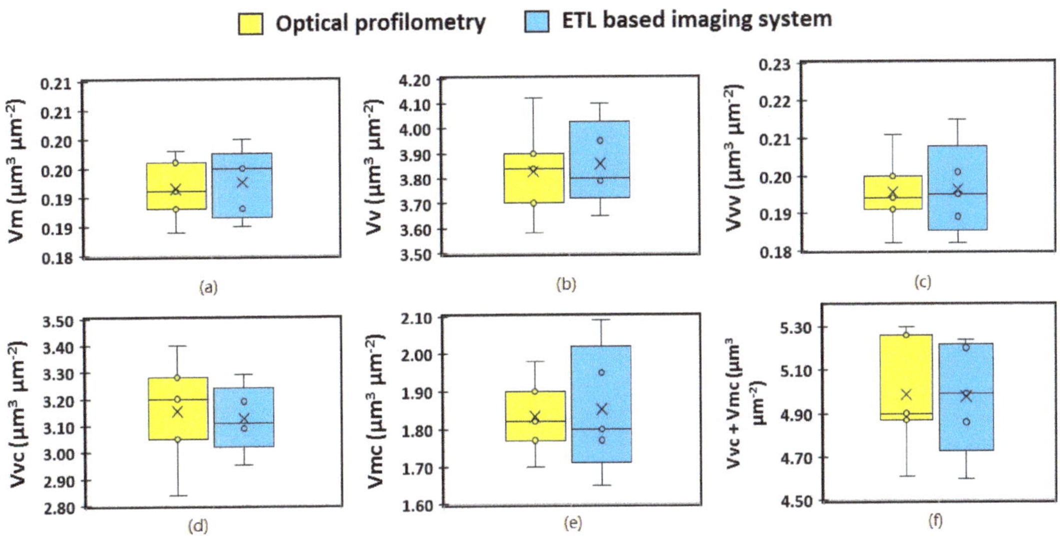

Figure 10. Spatial parameters of 3D surface texture analysis. (**a**) V_m: material surface volume; (**b**) V_v: surface void volume; (**c**) V_{vv}: valley void volume; (**d**) V_{vc}: core void volume; and (**e**) V_{mc}: core material surface volume; (**f**) V_{vc} + V_{mc}: core void plus core material surface.

The hybrid parameters characterised in the present study include the developed interfacial area ratio (S_{dr}), root mean square of the slope (S_{dq}), and summit density (S_{ds}). S_{dr} and S_{dq} are useful parameters for differentiating surfaces of similar amplitudes and average roughness and effectively quantify the spatial intricacy of the surface as well as the slope. The mean values of S_{dr} and S_{dq} computed using the ETL system and optical microscopy were 0.500 and 0.512, respectively, for S_{dr}, and 1.00 and 1.02, respectively, for S_{dq}. S_{ds}, on the other hand, is a measure of the number of summits comprising the surface per unit area using a definition of a summit as 5% of S_z. Use of the ETL system and optical profilometry produced mean S_{ds} values of 82.6 $1/\mu m^2$ and 84.2 $1/\mu m^2$, respectively. Both imaging techniques revealed no statistically significant difference between the methods (Figure 9).

The volume parameter defined as the material volume (V_m) provides a measure of the volume of material of the surface from the height corresponding to a material ratio value (which may be set to any value from 0% to 100%) to the highest peak of the surface. Similarly, V_{mc} (core material volume) is the volume of material contained within the material ratio values of 10% and 80%. The mean values calculated for these two parameters using the ETL system and optical microscopy were 0.190 $\mu m^3/\mu m^2$ and 0.192 $\mu m^3/\mu m^2$, respectively, for V_m, and 1.86 $\mu m^3/\mu m^2$ and 1.84 $\mu m^3/\mu m^2$, respectively, for V_{mc}. Other volume parameters explored in the present study include the void volume (V_v), dale void volume (V_{vv}), and core void volume (V_{vc}) which all provide a measure of the volume of material that would fill the surface between the chosen material ratio values [30]. The mean V_v, V_{vv}, and V_{vvc} values measured for the ETL system were 3.86 $\mu m^3/\mu m^2$, 0.198 $\mu m^3/\mu m^2$, and 3.12 $\mu m^3/\mu m^2$, respectively, and 3.82 $\mu m^3/\mu m^2$, 0.195 $\mu m^3/\mu m^2$, 3.15 $\mu m^3/\mu m^2$, respectively, for optical profilometry. Similar to the amplitude parameters, visual analysis of the box and whisker plots of the volume parameters (Figure 10) revealed slightly larger interquartile range for the parameters obtained using the ETL-based imaging system compared with optical microscopy (Figure 10a–c,e–f) indicating variability in the results, however, not statistically significant.

4. Conclusions

The current study successfully developed and employed an ETL variable focus imaging system for the quantitative surface texture analysis of a pharmaceutical solid dosage form (model material sample). The Tenengrad operator used to calibrate the system effectively identified the plane of best focus as demonstrated by the linear relationship between the control current applied to the ETL and the height at which the optical system focuses. The system was then successfully utilised to obtain surface texture parameters which related to the amplitude, spatial, hybrid, and volume characteristics of the surface. Moreover, validation of the ETL system against optical profilometry revealed comparable values of the parameters for both techniques with no statistically significant differences between the two. These results highlighted the potential application for ETL systems as an easily adaptable, low cost alternative non-destructive surface texture analysis method due to its enhanced DOF compared with conventional microscopy techniques including contact profilometers, AFM, coherence scanning interferometry, confocal microscopy, scanning laser microscope, and structure light scanning. It is anticipated that the benefits of this technique will have a wide-ranging impact in multiple industries as a powerful method for the surface analysis of materials including mechanical components, construction material, surgical implants, and pharmaceutical formulations.

Author Contributions: Conceptualization, M.U.G.; methodology, J.S.N. and M.U.G.; software, T.H., S.H., J.S.N. and M.U.G.; validation, J.S.N., S.L. and M.U.G.; formal analysis, J.S.N. and M.U.G.; investigation, J.S.N., B.R.C. and M.U.G.; data curation, J.S.N. and M.U.G.; writing—original draft preparation, J.S.N.; writing—review and editing, J.S.N., S.L., T.H., S.H., M.N., S.L., B.R.C. and M.U.G.; visualization, M.U.G.; supervision, M.U.G.; project administration, M.U.G. All authors have read and agreed to the published version of the manuscript.

Funding: This research received no external funding.

Data Availability Statement: Data will be made available on request.

Acknowledgments: The authors acknowledge the financial assistance provided by the University of Huddersfield.

Conflicts of Interest: The authors declare no conflict of interest.

References

1. Myshkin, N.K.; Grigoriev, A.Y.; Chizhik, S.A.; Choi, K.Y.; Petrokovets, M.I. Surface roughness and texture analysis in microscale. *Wear* **2003**, *254*, 1001–1009. [CrossRef]
2. Bhushan, B. *Modern Tribology Handbook, Two Volume Set*; CRC Press: Boca Raton, FL, USA, 2000; pp. 74–144.
3. Nirwan, J.S.; Conway, B.R.; Ghori, M.U. In situ 3D nanoscale advanced imaging algorithms with integrated chemical imaging for the characterisation of pharmaceuticals. *RSC Adv.* **2019**, *9*, 16119–16129. [CrossRef]
4. Reading, M.; Ghori, M.U.; Brown, D.R.; Fleming, L.T.; Antonijevic, M.D.; Grandy, D.B.; Hourston, D.J.; Song, M.; Conway, B.R. Chemical Imaging by Dissolution Analysis: Localized Kinetics of Dissolution Behavior to Provide Two-Dimensional Chemical Mapping and Tomographic Imaging on a Nanoscale. *Anal. Chem.* **2017**, *89*, 5882–5890. [CrossRef]
5. Paredes, J.I.; Villar-Rodil, S.; Tamargo-Martínez, K.; Martínez-Alonso, A.; Tascón, J.M.D. Real-Time Monitoring of Polymer Swelling on the Nanometer Scale by Atomic Force Microscopy. *Langmuir* **2006**, *22*, 4728–4733. [CrossRef] [PubMed]
6. Saurí, J.; Suñé-Negre, J.; Diaz, J.; Vilana, J.; Millán, D.; Ticó, J.; Miñarro, M.; Pérez-Lozano, P.; Garcia_Montoya, E. Relationships between surface free energy, surface texture parameters and controlled drug release in hydrophilic matrices. *Int. J. Pharm.* **2015**, *478*, 328–340. [CrossRef] [PubMed]
7. Bolean, M.; Borin, I.A.; Simão, A.M.; Bottini, M.; Bagatolli, L.A.; Hoylaerts, M.F.; Millán, J.L.; Ciancaglini, P. Topographic analysis by atomic force microscopy of proteoliposomes matrix vesicle mimetics harboring TNAP and AnxA5. *Biochim. Biophys. Acta (BBA)-Biomembr.* **2017**, *1859*, 1911–1920. [CrossRef] [PubMed]
8. Kemper, B.; Langehanenberg, P.; Von Bally, G. Digital Holographic Microscopy. *Opt. Photonik* **2007**, *2*, 41–44. [CrossRef]
9. Castañeda, R.; Garcia-Sucerquia, J. Single-shot 3D topography of reflective samples with digital holographic microscopy. *Appl. Opt.* **2018**, *57*, A12–A18. [CrossRef]
10. Trujillo, C.; Castañeda, R.; Piedrahita-Quintero, P.; Garcia-Sucerquia, J. Automatic full compensation of quantitative phase imaging in off-axis digital holographic microscopy. *Appl. Opt.* **2016**, *55*, 10299. [CrossRef]
11. McLeod, E.; Ozcan, A. Microscopy without lenses. *Phys. Today* **2017**, *70*, 50–56. [CrossRef]

12. Wu, Y.; Ozcan, A. Lensless digital holographic microscopy and its applications in biomedicine and environmental monitoring. *Methods* **2018**, *136*, 4–16. [CrossRef] [PubMed]
13. Lee, H.; Jeon, P.; Kim, D. 3D image distortion problem in digital in-line holographic microscopy and its effective solution. *Opt. Express* **2017**, *25*, 21969–21980. [CrossRef] [PubMed]
14. Latychevskaia, T.; Fink, H.-W. Solution to the Twin Image Problem in Holography. *Phys. Rev. Lett.* **2007**, *98*, 233901. [CrossRef]
15. Chen, B.K.; Chen, T.-Y.; Hung, S.G.; Huang, S.-L.; Lin, J.-Y. Twin image removal in digital in-line holography based on iterative inter-projections. *J. Opt.* **2016**, *18*, 065602. [CrossRef]
16. de Almeida, J.L.; Comunello, E.; Sobieranski, A.; da Rocha Fernandes, A.M.; Cardoso, G.S. Twin-image suppression in digital in-line holography based on wave-front filtering. *Pattern. Anal. Appl.* **2021**, *24*, 907–914. [CrossRef]
17. Algorri, J.F.; Morawiak, P.; Bennis, N.; Zografopoulos, D.; Urruchi, V.; Rodríguez-Cobo, L.; Jaroszewicz, L.R.; Sánchez-Pena, J.M.; López-Higuera, J.M. Positive-negative tunable liquid crystal lenses based on a microstructured transmission line. *Sci. Rep.* **2020**, *10*, 10153. [CrossRef] [PubMed]
18. Iwai, D.; Izawa, H.; Kashima, K.; Ueda, T.; Sato, K. Speeded-Up Focus Control of Electrically Tunable Lens by Sparse Optimization. *Sci. Rep.* **2019**, *9*, 12365. [CrossRef]
19. Hu, X.; Wang, G.; Zhang, Y.; Yang, H.; Zhang, S. Large depth-of-field 3D shape measurement using an electrically tunable lens. *Opt. Express* **2019**, *27*, 29697–29709. [CrossRef]
20. Ren, H.; Wu, S. *Introduction to Adaptive Lenses*; John Wiley & Sons: Hoboken, NJ, USA, 2012; Volume 75.
21. Dorronsoro, C.; Barcala, X.; Gambra, E.; Akondi, V.; Sawides, L.; Marrakchi, Y.; Rodriguez-Lopez, V.; Benedi-Garcia, C.; Vinas, M.; Lage, E.; et al. Tunable lenses: Dynamic characterization and fine-tuned control for high-speed applications. *Opt. Express* **2019**, *27*, 2085–2100. [CrossRef]
22. Beadie, G.; Sandrock, M.L.; Wiggins, M.J.; Lepkowicz, R.S.; Shirk, J.S.; Ponting, M.; Yang, Y.; Kazmierczak, T.; Hiltner, A.; Baer, E. Tunable polymer lens. *Opt. Express* **2008**, *16*, 11847–11857. [CrossRef]
23. Fahrbach, F.O.; Voigt, F.; Schmid, B.; Helmchen, F.; Huisken, J. Rapid 3D light-sheet microscopy with a tunable lens. *Opt. Express* **2013**, *21*, 21010–21026. [CrossRef] [PubMed]
24. Jiang, J.; Zhang, D.; Walker, S.; Gu, C.; Ke, Y.; Yung, W.H.; Chen, S.-C. Fast 3-D temporal focusing microscopy using an electrically tunable lens. *Opt. Express* **2015**, *23*, 24362–24368. [CrossRef] [PubMed]
25. Grewe, B.; Voigt, F.; Hoff, M.V.T.; Helmchen, F. Fast two-layer two-photon imaging of neuronal cell populations using an electrically tunable lens. *Biomed. Opt. Express* **2011**, *2*, 2035–2046. [CrossRef]
26. Lin, Y.-H.; Chen, H.-S. Electrically tunable-focusing and polarizer-free liquid crystal lenses for ophthalmic applications. *Opt. Express* **2013**, *21*, 9428–9436. [CrossRef]
27. Angot-Petit, L.J. Small scale surface profile recovery using a tunable lens based system. *Electron. Imaging* **2017**, *2017*, 39–48. [CrossRef]
28. Shah, M.; Mishra, S.; Sarkar, M.; Rout, C. Identification of robust focus measure functions for the automated capturing of focused images from Ziehl-Neelsen stained sputum smear microscopy slide. *Cytom. Part A* **2017**, *91*, 800–809. [CrossRef] [PubMed]
29. Mateos-Pérez, J.M.; Redondo, R.; Nava, R.; Valdiviezo, J.C.; Cristobal, G.; Escalante-Ramírez, B.; Ruiz-Serrano, M.J.; Pascau, J.; Desco, M. Comparative evaluation of autofocus algorithms for a real-time system for automatic detection of Mycobacterium tuberculosis. *Cytom. Part A* **2012**, *81*, 213–221. [CrossRef]
30. Hussain, A.; Mahmood, F.; Arshad, M.S.; Abbas, N.; Qamar, N.; Mudassir, J.; Ghori, M.U. Personalised 3D Printed Fast-Dissolving Tablets for Managing Hypertensive Crisis: In-Vitro/In-Vivo Studies. *Polymers* **2020**, *12*, 3057. [CrossRef]
31. Khizer, Z.; Nirwan, J.S.; Conway, B.R.; Ghori, M.U. Okra (Hibiscus esculentus) gum based hydrophilic matrices for controlled drug delivery applications: Estimation of percolation threshold. *Int. J. Biol. Macromol.* **2020**, *155*, 835–845. [CrossRef] [PubMed]
32. Ghori, M.U.; Nirwan, J.S.; Asim, T.; Chahid, Y.; Farhaj, S.; Khizer, Z.; Timmins, P.; Conway, B.R. MUCO-DIS: A New AFM-Based Nanoscale Dissolution Technique. *AAPS PharmSciTech* **2020**, *21*, 142. [CrossRef]
33. Khizer, Z.; Akram, M.R.; Sarfraz, R.M.; Nirwan, J.S.; Farhaj, S.; Yousaf, M.; Hussain, T.; Lou, S.; Timmins, P.; Conway, B.R.; et al. Plasticiser-Free 3D Printed Hydrophilic Matrices: Quantitative 3D Surface Texture, Mechanical, Swelling, Erosion, Drug Release and Pharmacokinetic Studies. *Polymer* **2019**, *11*, 1095. [CrossRef] [PubMed]
34. ISO 25178-2*Geometrical Product Specifications (GPS)—Surface Texture: Areal Part 2: Terms, Definitions and Surface Texture Parameters*, International Organisation for Standardization: Geneva, Switzerland, 1998.
35. Metrology, M. *Surface Texture Parameters Glossary*; Michigan Metrology LLC: Livonia, MI, USA, 2014.

micromachines

Article

Optical Fiber Sensor for Temperature and Strain Measurement Based on Multimode Interference and Square-Core Fiber

Kun Wang *, Xingchen Dong, Patrick Kienle, Maximilian Fink, Wolfgang Kurz, Michael H. Köhler, Martin Jakobi and Alexander W. Koch

Department of Electrical and Computer Engineering, Institute for Measurement Systems and Sensor Technology, Technical University of Munich, Theresienstraße 90, 80333 Munich, Germany; xingchen.dong@tum.de (X.D.); patrick.kienle@tum.de (P.K.); max.fink@tum.de (M.F.); wolfgang.kurz@tum.de (W.K.); michael.koehler@tum.de (M.H.K.); m.jakobi@tum.de (M.J.); a.w.koch@tum.de (A.W.K.)

* Correspondence: kun88.wang@tum.de; Tel.: +49-289-23356

Abstract: A variety of specialty fibers such as no-core fiber (NCF) have already been studied to reveal their sensing abilities. In this work, we investigate a specialty fiber, square-core fiber, for temperature and strain sensing. A simple single-mode–multimode–single-mode (SMS) fiber sensor was fabricated, consisting of a 30-cm-long square-core fiber. The experimental results indicate that the maximal wavelength-temperature and wavelength-strain sensitivities are −15.3 pm/°C and −1.5 pm/με, respectively, while the maximal power-temperature and power-strain sensitivities are 0.0896 dBm/°C and 0.0756 dBm/με. Analysis of the results suggests that the fiber sensor has the potential to be used as a high-sensitivity temperature sensor with a low strain sensitivity.

Keywords: multimode interference; square-core fiber; temperature measurement; strain measurement; optical fiber sensors

Citation: Wang, K.; Dong, X.; Kienle, P.; Fink, M.; Kurz, W.; Köhler, M.H.; Jakobi, M.; Koch, A.W. Optical Fiber Sensor for Temperature and Strain Measurement Based on Multimode Interference and Square-Core Fiber. *Micromachines* **2021**, *12*, 1239. https://doi.org/10.3390/mi12101239

Academic Editors: Luigi Sirleto and Giancarlo C. Righini

Received: 11 August 2021
Accepted: 8 October 2021
Published: 13 October 2021

1. Introduction

Optical fiber sensors have been a promising sensing device for decades due to their intrinsic characteristics, such as compact size, fast response, low cost, and insensitivity to electromagnetic fields [1–4]. A wide variety of configurations have been successfully reported according to the development of optical fiber fabrication techniques, for instance, based on fiber Bragg gratings (FBGs) [5], long-period gratings (LPGs) [6], Raman scattering [7], and Brillouin scattering [8]. The fabrication of these configurations usually requires sophisticated devices that result in relatively high costs. To address these factors, a simpler structure based on multimode interference was first reported by Mehta et al. in 2003 [9]. Since then, the multimode interference effect, which exists in multimode waveguides, has been comprehensively investigated. The basic optical fiber structure to fabricate a multimode interference device is a so-called single-mode–multimode–single-mode (SMS) fiber sensor, consisting of a short segment of multimode fiber (MMF) sandwiched between two single-mode fibers (SMFs).

Single-mode–multimode–single-mode (SMS) fiber structure has received significant attention in the last years because of its unique spectral features, high sensitivity, easy fabrication, and potential low costs [10,11]. It also shows great compatibility with other photonic devices and structures. For example, it can easily connect to other fiber structures with great flexibility to achieve better sensor performance such as higher sensitivity or multi-parameter measurement [12,13]. As such, it can perform flexibly as a variety of sensors in different scenarios, including temperature sensors [14], curvature sensor [15], refractive index sensors [16,17], humidity sensors [18], position sensors [19], strain sensors [20], and biomedical sensors [21].

Due to the high demand for sensing stability and measurement accuracy of fiber sensors, several effective post-processing technologies have been studied, such as tapering [22], etching [23], and polishing [24]. These approaches often require high-precision apparatus that drastically raise the manufacturing costs and difficulties, such as a high-accuracy fiber side-polishing system. An alternative approach, which can fulfill this demand with low costs and less complexity, is to consider specialty fibers by studying their feasibility and sensing performance. Some multimode-interference-based fiber sensors using specialty fibers have been reported to reveal the advantages of studying specialty fibers. Zhao et al. have fabricated a multimode-interference-based fiber sensor by splicing two segments of SMF and a 5-cm-long hollow core fiber (HCF) via two abrupt tapered joints. A temperature sensitivity of 0.0123 nm/°C and a curvature sensitivity of –5.05 dB/m^{-1} have been obtained [25]. Another multimode-interference-based fiber sensor consisting of cascaded MMF and photonic crystal fiber (PCF) fusion spliced with two sections of SMF was demonstrated by Yang et al., which has obtained a strain sensitivity of –14.89 pm/με and a temperature sensitivity of –243.8 pm/°C [26].

Here, we investigated the sensing performance of a simple multimode-interference-based fiber sensor containing a specialty fiber, the square-core fiber, for temperature and strain measurement. Square-core fiber produces an optical beam with uniform intensity over the core area because the shape of the core promotes mode mixing as light propagates through the fiber. It results in an even distribution of spatial modes in the output beam; therefore, it is ideal for applications such as laser machining. Further, square-core fiber also offers reduced focal ratio degradation, which means the square core shape is free from the input–output correlations, showing great potential for imaging and spectroscopy applications. These features of square-core fiber have been demonstrated by Velsink et al. by investigating the core shape, i.e., square core shape and circular core shape, for wavefront shaping application where a focus is formed at the output of the MMF [27]; however, the sensing ability of square-core fiber has not been experimentally studied. In this work, we demonstrated that the sensing performance of square-core fiber is promising. In addition, as the square core shape has not been experimentally proved, a comparison to other specialty fibers, which are mostly circular core shapes, is also presented. The experimental results also show that this fiber sensor can exhibit great potential applications in strain-insensitive temperature measurement.

2. Sensor Fabrication and Principle

A simple SMS fiber structure implementing the square-core fiber is proposed and studied. The general sensing principle of the SMS fiber sensor is based on the self-imaging effect in the multimode waveguide, i.e., multimode interference (MMI). The injected light is guided from the input SMF into the MMF and propagates along with the MMF core. The spot-size difference between the fundamental modes in the SMF and MMF excites a few lower modes in the MMF, which propagate with different propagation constants. The transmitted light in the MMF section generates multiple characteristic dips (or peaks) induced by multimode interference in its transmission spectrum [28]. At the second conjunction, these fields are coupled back to the fundamental mode of the output SMF. It needs to clarify that if the evanescent field of the SMS structure does not leak into the external media, it cannot sense the ambient changes. According to the detailed calculation [29,30], the sensing principle can be summarized as

$$P_{out} = | A_0^2 + A_1^2 e^{i(\beta_0 - \beta_1)L} + A_2^2 e^{i(\beta_0 - \beta_2)L} + \ldots |^2, \quad (1)$$

where P_{out} is the power in the output SMF, A_i is the field amplitude of the i-th mode at the first SMF/MMF boundary, β_i is the propagation constant of the i-th mode, and L is the length of the MMF. Equation (1) shows that the optical power in the output fiber is affected by physical changes caused by temperature and strain. A shift in the transmission spectrum occurs when temperature change and strain are applied to the MMF because they influence the inherent thermal expansion, photoelastic and thermo-optic effects of the fiber material, refractive indexes of the fiber, and fiber core diameter [31,32]; therefore, these changes can be quantitatively evaluated by measuring the shift in either power or spectral location of peaks (or dips).

Figure 1 shows the schematic diagram of the experimental setup for temperature and strain measurement. The square-core fiber (FP150QMT, Thorlabs) features a 150 ± 10 μm × 150 ± 10 μm square silica core, surrounded by a ϕ225 μm circular polymer cladding. The square-core fiber with a length of 30 cm is used as the sensing region, as shown in Figure 2. As the cladding material comprises hard polymer, it is connected to two silica SMFs (size: 9/125 μm) by butt-coupling, i.e., both ends of the square-core fiber are connected to the ends of SMFs via FC/PC mating adaptors [33]. One end of the SMF connects to a broadband light source (BLS, central wavelength 1550 nm) emitting the incident light, while the end of the other SMF connects to an optical spectrum analyzer (OSA), which simultaneously detects the changes in the light spectrum. To investigate the response of the proposed fiber sensor to axial strain variation, the sample is placed and clamped on two translation stages, which can be accurately controlled and hence apply varying axial tensile stress on the sample according to a standard strain calculation equation [31]:

$$\Delta\varepsilon = \frac{\Delta L}{L}, \tag{2}$$

where L is the initial distance between two translation stages and ΔL is the additional displacement when longitudinal stress is introduced to the proposed fiber sensor. The square-core fiber segment is also placed on a heating plate used as the heating device to control the temperature with 0.1 °C resolution.

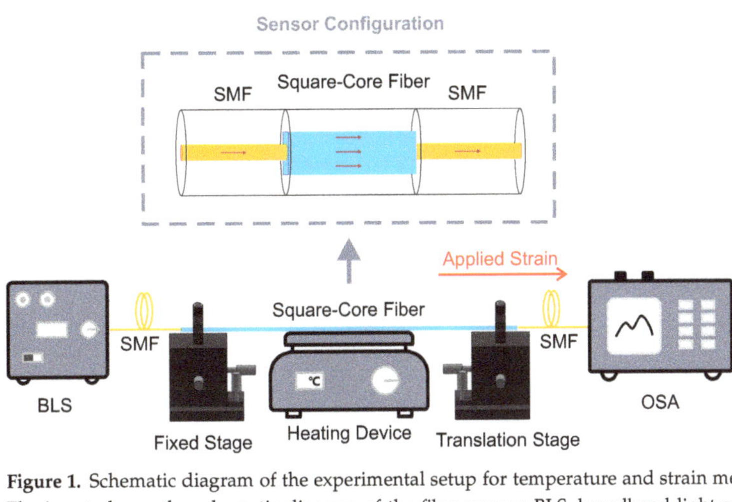

Figure 1. Schematic diagram of the experimental setup for temperature and strain measurement. The insert shows the schematic diagram of the fiber sensor; BLS, broadband light source; SMF, single-mode fiber; OSA, optical spectrum analyzer.

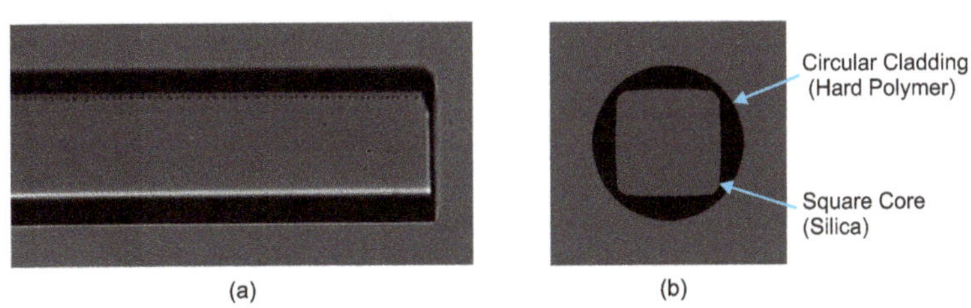

Figure 2. (**a**) Photo of square-core fiber from the top view. (**b**) Schematic diagram of the square core shape.

The transmission spectrum contains multiple dips caused by multimode interference within the spectrum range. In typical multimode interference structures, the interference fringes are being distorted due to multiple modes which are excited and participate in the interference. Although any of these dips (or peaks) can be used as a spectral indicator for sensing, the least distorted one is typically chosen to minimize the measurement error. In this work, we choose the one close to the central wavelength 1550 nm of the BLS. Note that the spectral distortion is significant in square-core fiber because of its imperfect end faces. In Figure 2a, it can be seen that the end face of square-core fiber is not perfect. This factor affects the coupling and the distortion in the transmission spectrum. Commonly, the specialty fibers can not be cleaved by standard cleaving devices, as the technology of cleaving specialty fibers has not been satisfactorily proved yet; however, if the imperfection is not in the fiber core region, the influence can be reduced or minimized. Further, using other methods, such as polishing, can improve the sensor performance. Despite these issues, this sensor structure is interesting because of its simple and inexpensive fabrication method and is therefore investigated in more detail.

3. Experiment and Discussion

The temperature dependence of the proposed fiber sensor was first measured without strain applied during the experiment. The square-core fiber section was heated in 10 °C steps to stabilize the sensing ability from 30 °C to 80 °C. As shown in Figure 3a, the wavelength and power of the chosen dip change as the temperature changes. When the temperature increased, the dip shifted to shorter wavelengths accompanied by the transmission power increasing. The resulting changes in dip wavelength and transmission power versus temperature are plotted and fitted in Figure 3b. The fitting results of wavelength-temperature (blue solid line) and power-temperature (green dash-dot line) both exhibit a high linear regression coefficient value (R^2) when subject to linear regression analysis. It was determined that the wavelength-temperature sensitivity is −15.3 pm/°C with an R^2 value of 0.954, and the power-temperature sensitivity is 0.0896 dBm/°C with an R^2 value of 0.970. The experiment was carried out for the second round in the same temperature range, and the spectral dip shifts on temperature and the fitting results are as shown in Figure 3c,d, respectively. The measured wavelength-temperature sensitivity is −14.8 pm/°C with an R^2 value of 0.930, and the power-temperature sensitivity is 0.0857 dBm/°C with an R^2 value of 0.978. The absolute value of wavelength-temperature sensitivity is comparable with other specialty fiber sensors, such as the fiber sensor based on cascaded single-mode–no-core–single-mode (SNS) and single-mode–multimode–single-mode (SMS) fiber structure, the absolute value of obtained wavelength-temperature sensitivity is around 1.7 times than that [34].

Furthermore, the experiment of strain measurement was conducted at room temperature (averagely 20 °C). The measured dependence of the spectral dip on strain is shown in Figure 4a. The spectral dip shifts to shorter wavelengths with increasing strain applied, whereas the transmission power increases. Figure 4b illustrates the dip wavelength and transmission power against strain, which also includes the linear fit of wavelength-strain (blue solid line) and power-strain (green dash-dot line), respectively. It can be seen that both fitting curves exhibit good linearity with high linear regression coefficient values (R^2). The wavelength-strain sensitivity is −1.2 pm/με with an R^2 value of 0.953 and the power-strain sensitivity is 0.0678 dBm/με with an R^2 value of 0.985. The experiment was carried out for the second round in the same strain range, and the spectral dip shifts on strain and the fitting results are as shown in Figure 4c,d, respectively. The measured wavelength-strain sensitivity is −1.5 pm/με with an R^2 value of 0.928, and the power-strain sensitivity is 0.0756 dBm/με with an R^2 value of 0.981. The absolute value of wavelength-strain sensitivity is comparable to the strain sensitivity of a thin core fiber implemented multimode interference fiber sensor with a core offset structure [35].

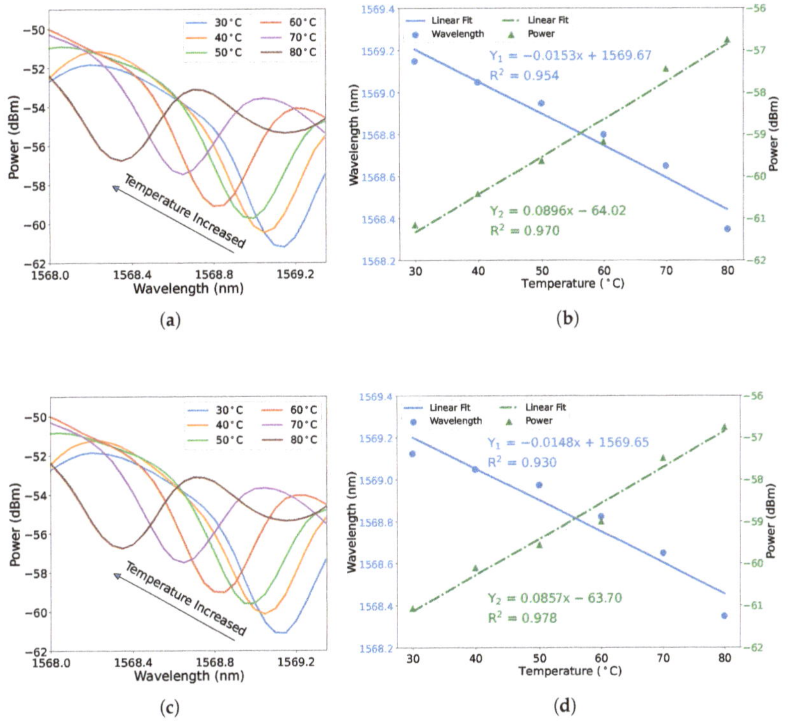

Figure 3. Measured dependence of spectral dip shifts on temperature in the range of 30 °C to 80 °C in steps of 10 °C, (**a**) first round and (**c**) second round. Dip shift as a function of temperature, (**b**) first round and (**d**) second round. The blue solid line represents dip wavelength shift against temperature, and the green dash-dot line represents dip power against temperature.

The two-round reproducibility of the temperature and strain measurement is presented. Figure 5a shows the average wavelength-temperature sensitivity and power-temperature sensitivity are −15.1 pm/°C and 0.0877 dBm/°C, respectively. The maximal standard deviation of wavelength shifts is 0.0177 nm and of transmission power is 0.2174 dBm. Figure 5b shows the average wavelength-strain sensitivity is −1.3 pm/$\mu\varepsilon$ and average power-strain sensitivity is 0.0717 dBm/$\mu\varepsilon$. The related maximal standard deviation of wavelength shifts and transmission power are 0.0176 nm and 0.5675 dBm, respectively.

Comparing Figures 3b and 4b, the absolute value of wavelength-temperature sensitivity shows high-temperature measurability, while the small absolute value of wavelength-strain indicates that the proposed fiber sensor is less sensitive to applied strain. Therefore, the multimode interference in the square-core fiber can be potentially exploited to develop high-sensitivity temperature sensors with reduced strain sensitivity. Table 1 presents a comparison of the temperature measurability for different specialty fibers, which shows that the temperature sensing ability of square-core fiber is promising with a simple structure.

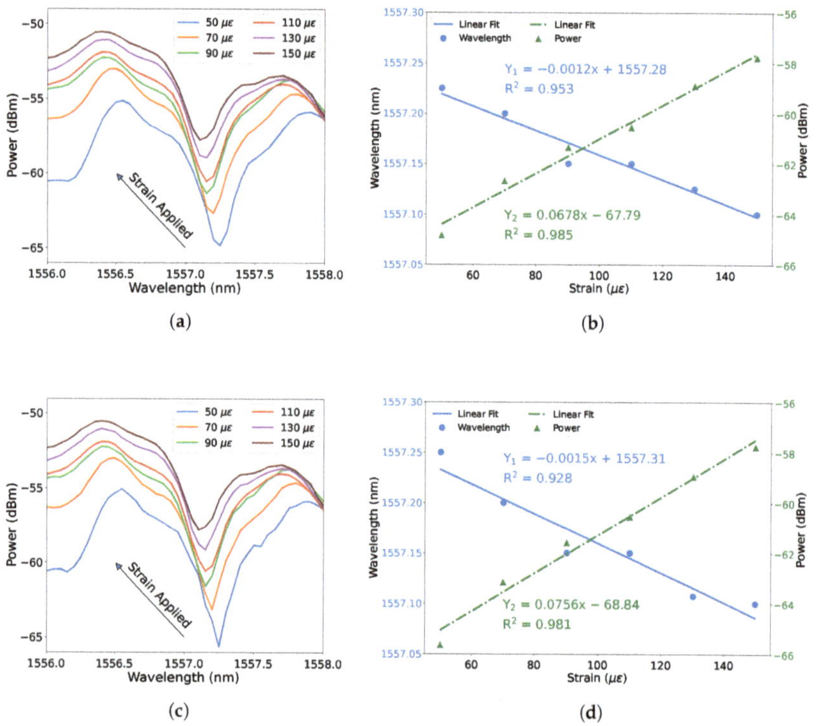

Figure 4. Measured dependence of spectral dip shifts on strain in the range of 50 με to 150 με in steps of 20 με, (**a**) first round and (**c**) second round. Dip shift as a function of strain, (**b**) first round and (**d**) second round. The blue solid line represents dip wavelength shift against strain, and the green dash-dot line represents dip power against strain.

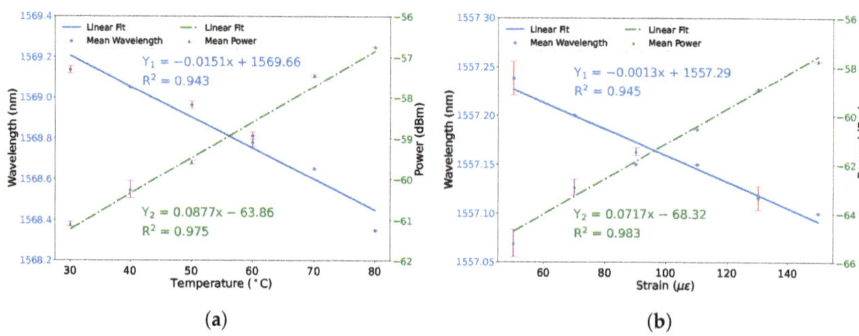

Figure 5. (**a**) Relation between average wavelength shifts and temperature, and the relation between average power changes and temperature. Error bar indicates the standard deviation. (**b**) Relation between average wavelength shifts and strain, and the relation between average power changes and strain. Error bar indicates the standard deviation.

Table 1. Comparison of temperature sensitivities with other multimode-interference-based optical fiber sensors using specialty fibers.

Optical Fiber Sensor Structure	Measurement Range	Sensitivity (max.)	Reference
SMF-HCF-SMF (with abrupt taper joints)	18–50 °C	12.3 pm/°C	[25]
SMF-No-Core Fiber (NCF)-SMF-MMF-SMF	30–90 °C	9.2 pm/°C	[34]
SMF-Tapered NCF-SMF	0–280 °C	16.56 pm/°C	[36]
SMF-NCF-SMF (with FBG)	0–50 °C	12.8 pm/°C	[37]
Tapered SMF-Micro MMF-Tapered SMF	35–60 °C	0.028 dB/°C	[38]
SMF-Square-Core Fiber-SMF	30–80 °C	−15.3 pm/°C 0.0896 dBm/°C	This work

4. Conclusions

In conclusion, the sensing characteristics of square-core fiber were experimentally investigated in a simple SMS structure for temperature and strain measurement. The obtained maximal wavelength-temperature and wavelength-strain sensitivities are −15.3 pm/°C and −1.5 pm/με, respectively, with the maximal power sensitivities of 0.0896 dBm/°C and 0.0756 dBm/με for temperature and strain. The measurement results proved that square-core fiber is promising for sensing. Further, it also indicates that this sensor can exhibit great potential applications in strain-insensitive temperature measurement.

Author Contributions: Conceptualization, K.W.; methodology, K.W.; validation, K.W.; formal analysis, K.W.; investigation, K.W.; data curation, K.W.; writing—original draft preparation, K.W.; writing—review and editing, X.D., P.K., M.F., W.K., M.H.K., and M.J.; supervision, A.W.K. All authors have read and agreed to the published version of the manuscript.

Funding: This research received no external funding.

Institutional Review Board Statement: Not applicable.

Informed Consent Statement: Not applicable.

Data Availability Statement: Not applicable.

Acknowledgments: This work was supported by China Scholarship Council (CSC) (Grant Nos. 201808340074 (Kun Wang) and 201706050026 (Xingchen Dong)) and the Technical University of Munich (TUM) in the framework of the Open Access Publishing Program.

Conflicts of Interest: The authors declare no conflict of interest.

References

1. Wang, X.-D.; Wolfbeis, O.S. Fiber-Optic Chemical Sensors and Biosensors (2015–2019). *Anal. Chem.* **2020**, *92*, 397–430. [CrossRef]
2. Wang, X.; Tian, K.; Yuan, L.; Lewis, E.; Farrell, G.; Wang, P. A High-Temperature Humidity Sensor Based on a Singlemode-Side Polished Multimode-Singlemode Fiber Structure. *J. Light. Technol.* **2018**, *36*, 2730–2736. [CrossRef]
3. Mizuno, Y.; Numata, G.; Kawa, T.; Lee, H.; Hayashi, N.; Nakamura, K. Multimodal Interference in Perfluorinated Polymer Optical Fibers: Application to Ultrasensitive Strain and Temperature Sensing. *IEICE Trans. Electron.* **2018**, *101-C*, 602–610. [CrossRef]
4. Wang, K.; Dong, X.; Köhler, M.H.; Kienle, P.; Bian, Q.; Jakobi, M.; Koch, A.W. Advances in Optical Fiber Sensors Based on Multimode Interference (MMI): A Review. *IEEE Sens. J.* **2021**, *21*, 132–142. [CrossRef]
5. Hill, K.; Meltz, G. Fiber Bragg grating technology fundamentals and overview. *J. Light. Technol.* **1997**, *15*, 1263–1276. [CrossRef]
6. Wang, Y. Review of long period fiber gratings written by CO_2 laser. *J. Appl. Phys.* **2010**, *108*, 081101. [CrossRef]
7. Alahbabi, M.N.; Cho, Y.T.; Newson, T.P. Simultaneous temperature and strain measurement with combined spontaneous Raman and Brillouin scattering. *Opt. Lett.* **2005**, *30*, 1276–1278. [CrossRef] [PubMed]
8. Mizuno, Y.; Hayashi, N.; Fukuda, H.; Song, K.Y.; Nakamura, K. Ultrahigh-speed distributed Brillouin reflectometry. *Light. Sci. Appl.* **2016**, *5*, e16184. [CrossRef] [PubMed]

9. Mehta, A.; Mohammed, W.; Johnson, E. Multimode interference-based fiber-optic displacement sensor. *IEEE Photonics Technol. Lett.* **2003**, *15*, 1129–1131. [CrossRef]
10. Zhao, Y.; Cai, L.; Li, X.G. High Sensitive Modal Interferometer for Temperature and Refractive Index Measurement. *IEEE Photonics Technol. Lett.* **2015**, *27*, 1341–1344. [CrossRef]
11. Olivero, M.; Vallan, A.; Orta, R.; Perrone, G. Single-Mode–Multimode–Single-Mode Optical Fiber Sensing Structure With Quasi-Two-Mode Fibers. *IEEE Trans. Instrum. Meas.* **2018**, *67*, 1223–1229. [CrossRef]
12. Zhao, Y.; Cai, L.; Li, X.G.; Meng, F.C. Liquid concentration measurement based on SMS fiber sensor with temperature compensation using an FBG. *Sens. Actuators B Chem.* **2014**, *196*, 518–524. doi:10.1016/j.snb.2014.01.075. [CrossRef]
13. Chai, Q.; Liu, Y.; Zhang, J.; Yang, J.; Chen, Y.; Yuan, L.; Peng, G.D. Asymmetric transmission and reflection spectra of FBG in single-multi-single mode fiber structure. *Opt. Express* **2015**, *23*, 11665–11673. [CrossRef]
14. Lu, C.; Su, J.; Dong, X.; Sun, T.; Grattan, K.T.V. Simultaneous Measurement of Strain and Temperature with a Few-Mode Fiber-Based Sensor. *J. Light. Technol.* **2018**, *36*, 2796–2802. [CrossRef]
15. Gong, Y.; Zhao, T.; Rao, Y.J.; Wu, Y. All-Fiber Curvature Sensor Based on Multimode Interference. *IEEE Photonics Technol. Lett.* **2011**, *23*, 679–681. [CrossRef]
16. Zhang, F.; Li, S.; Yan, X.; Zhang, X.; Wang, F.; Suzuki, T.; Ohishi, Y.; Cheng, T. A Refractive Index Sensitive Liquid Level Monitoring Sensor Based on Multimode Interference. *Photonics* **2020**, *7*, 89. [CrossRef]
17. Tong, Z.; Su, J.; Cao, Y.; Zhang, W. Simultaneous Measurement Based on Composite Interference Structure. *IEEE Photonics Technol. Lett.* **2014**, *26*, 1310–1313. [CrossRef]
18. Wang, X.; Farrell, G.; Lewis, E.; Tian, K.; Yuan, L.; Wang, P. A Humidity Sensor Based on a Singlemode-Side Polished Multimode–Singlemode Optical Fibre Structure Coated with Gelatin. *J. Light. Technol.* **2017**, *35*, 4087–4094. [CrossRef]
19. Sun, K.; Ding, Z.; Zhang, Z. Fiber directional position sensor based on multimode interference imaging and machine learning. *Appl. Opt.* **2020**, *59*, 5745–5751. [CrossRef]
20. Tian, K.; Farrell, G.; Wang, X.; Yang, W.; Xin, Y.; Liang, H.; Lewis, E.; Wang, P. Strain sensor based on gourd-shaped single-mode-multimode-single-mode hybrid optical fibre structure. *Opt. Express* **2017**, *25*, 18885–18896. [CrossRef] [PubMed]
21. Socorro, A.B.; Santamaría, E.; Fernández-Irigoyen, J.; Villar, I.D.; Corres, J.M.; Arregui, F.J.; Matias, I.R. Fiber-Optic Immunosensor Based on an Etched SMS Structure. *IEEE J. Sel. Top. Quantum Electron.* **2017**, *23*, 314–321. [CrossRef]
22. Wang, P.; Brambilla, G.; Ding, M.; Semenova, Y.; Wu, Q.; Farrell, G. High-sensitivity, evanescent field refractometric sensor based on a tapered, multimode fiber interference. *Opt. Lett.* **2011**, *36*, 2233–2235. [CrossRef] [PubMed]
23. Cardona-Maya, Y.; Del Villar, I.; Socorro, A.B.; Corres, J.M.; Matias, I.R.; Botero-Cadavid, J.F. Wavelength and Phase Detection Based SMS Fiber Sensors Optimized With Etching and Nanodeposition. *J. Light. Technol.* **2017**, *35*, 3743–3749. [CrossRef]
24. Tang, J.; Zhou, J.; Guan, J.; Long, S.; Yu, J.; Guan, H.; Lu, H.; Luo, Y.; Zhang, J.; Chen, Z. Fabrication of Side-Polished Single Mode-Multimode-Single Mode Fiber and Its Characteristics of Refractive Index Sensing. *IEEE J. Sel. Top. Quantum Electron.* **2017**, *23*, 238–245. [CrossRef]
25. Zhao, Y.; Cai, L.; Li, X.g. In-fiber modal interferometer for simultaneous measurement of curvature and temperature based on hollow core fiber. *Opt. Laser Technol.* **2017**, *92*, 138–141. [CrossRef]
26. Yang, F.; Wang, Z.; Wang, D. A highly sensitive optical fiber strain sensor based on cascaded multimode fiber and photonic crystal fiber. *Opt. Fiber Technol.* **2019**, *47*, 102–106. [CrossRef]
27. Velsink, M.C.; Lyu, Z.; Pinkse, P.W.H.; Amitonova, L.V. Comparison of round- and square-core fibers for sensing, imaging, and spectroscopy. *Opt. Express* **2021**, *29*, 6523–6531. [CrossRef]
28. Matias, I.R.; Ikezawa, S.; Corres, J. *Fiber Optic Sensors: Current Status and Future Possibilities*; Springer: Berlin/Heidelberg, Germany, 2016; Volume 21.
29. Kumar, A.; Varshney, R.K.; Siny Antony, C.; Sharma, P. Transmission characteristics of SMS fiber optic sensor structures. *Opt. Commun.* **2003**, *219*, 215–219. [CrossRef]
30. Tripathi, S.M.; Kumar, A.; Varshney, R.K.; Kumar, Y.B.P.; Marin, E.; Meunier, J.P. Strain and temperature sensing characteristics of single-mode–multimode–single-mode structures. *J. Light. Technol.* **2009**, *27*, 2348–2356. [CrossRef]
31. Li, E. Temperature compensation of multimode-interference-based fiber devices. *Opt. Lett.* **2007**, *32*, 2064–2066. [CrossRef]
32. Zhang, Y.; Xue, L.; Wang, T.; Yang, L.; Zhu, B.; Zhang, Q. High Performance Temperature Sensing of Single Mode-Multimode-Single Mode Fiber With Thermo-Optic Polymer as Cladding of Multimode Fiber Segment. *IEEE Sens. J.* **2014**, *14*, 1143–1147. [CrossRef]
33. Wang, K.; Dong, X.; Köhler, M.H.; Kienle, P.; Bian, Q.; Fink, M.; Jakobi, M.; Koch, A.W. Optical fiber sensors based on multimode interference using square-core fiber for temperature measurement. In *Photonic Instrumentation Engineering VIII*; Soskind, Y., Busse, L.E., Eds.; International Society for Optics and Photonics, SPIE: Bellingham, WA, USA, 2021; Volume 11693, pp. 120–126.
34. Chen, Y.; Wang, Y.; Chen, R.; Yang, W.; Liu, H.; Liu, T.; Han, Q. A hybrid multimode interference structure-based refractive index and temperature fiber sensor. *IEEE Sens. J.* **2015**, *16*, 331–335. [CrossRef]
35. Zhang, X.; Liu, C.; Liu, J.; Yang, J. Single modal interference-based fiber-optic sensor for simultaneous measurement of curvature and strain with dual-differential temperature compensation. *IEEE Sens. J.* **2018**, *18*, 8375–8380. [CrossRef]
36. Andre, R.M.; Biazoli, C.R.; Silva, S.O.; Marques, M.B.; Cordeiro, C.M.B.; Frazao, O. Strain-Temperature Discrimination Using Multimode Interference in Tapered Fiber. *IEEE Photonics Technol. Lett.* **2013**, *25*, 155–158. [CrossRef]

37. Li, C.; Ning, T.; Wen, X.; Li, J.; Zheng, J.; You, H.; Chen, H.; Zhang, C.; Jian, W. Strain and temperature discrimination using a fiber Bragg grating and multimode interference effects. *Opt. Commun.* **2015**, *343*, 6–9. [CrossRef]
38. Sun, Q.; Luo, H.; Luo, H.; Lai, M.; Liu, D.; Zhang, L. Multimode microfiber interferometer for dual-parameters sensing assisted by Fresnel reflection. *Opt. Express* **2015**, *23*, 12777–12783. [CrossRef] [PubMed]

micromachines

MDPI

Article

Development of LCEs with 100% Azobenzene Moieties: Thermo-Mechanical Phenomena and Behaviors

Domenico Sagnelli *,†, Massimo Rippa †, Amalia D'Avino, Ambra Vestri, Valentina Marchesano and Lucia Petti

Institute of Applied Sciences and Intelligent Systems of CNR, 80072 Pozzuoli, Italy
* Correspondence: domenico.sagnelli@isasi.cnr.it; Tel.: +39-0818675127
† These authors have contributed equally to this work.

Abstract: Azobenzene is one of the most investigated photo-responsive liquid crystalline molecules. It can isomerize between two different isoforms, trans (E) and cis (Z) configurations, when stimulated by light. It is used as a molecular engine in photo-mobile materials (PMPs). The use of liquid crystals (LCs) as building blocks enhances the mechanical properties of the PMPs. It is not easy to obtain PMPs with monodomain configurations when the LCs are 100% azobenzene. In this work, we studied three LC mixtures, describing the thermo/mechanical phenomena that regulate the actuation of such materials. The nematic temperature of the LC elastomers was measured and the PMPs carefully characterized for their bending and speed capability. Our finding suggests that the ratio between linear and cross-linker monomer greatly influences the nematic temperature of the mixture. Furthermore, 100% azobenzene materials polymerized using dicumyl peroxide can be useful to design polarization-selective switches.

Keywords: azobenzene; photo-mobile polymers; liquid crystals; thermal properties

Citation: Sagnelli, D.; Rippa, M.; D'Avino, A.; Vestri, A.; Marchesano, V.; Petti, L. Development of LCEs with 100% Azobenzene Moieties: Thermo-Mechanical Phenomena and Behaviors. *Micromachines* **2022**, *13*, 1665. https://doi.org/10.3390/mi13101665

Academic Editor: Yi Zhang

Received: 5 August 2022
Accepted: 30 September 2022
Published: 3 October 2022

1. Introduction

Conversion of light into mechanical work has been a hot topic for over 50 years. This approach is of great interest because actuators can be temporally, spatially, and remotely controlled by light, which can be beneficial in various applications [1–7].

One of the most investigated photo-responsive liquid crystalline molecules is the azobenzene [5,8–11] that isomerizes between trans (E) and cis (Z) configurations [12,13] when stimulated by polarized light. This isomerization can be exploited to use azobenzene as an engine to impart phototropic properties to materials. Azobenzene has been included in materials or as a dopant dye or as moiety in the polymeric network. In 1967, Lovrien [14] was the first to dope a polymer with photosensitive chromophores (such as azobenzene) generating a photomechanical outcome. However, phase separation of the chromophore and the matrix might occur, which is the major drawback. Initially in the early works [15], the synthesis of polymer networks with embedded azobenzene had a limited contractile strain (less than 1%) and the materials used to record optical gratings [16,17]. Later, with the use of liquid crystal (LCs)-based acrylate monomers, it was possible to improve the strength and strain of such materials [18,19]. This approach is of interest, since the characteristic anisotropy of LCs increases the mechanical properties of the polymer. The use of azobenzene molecules embedded as moieties in liquid crystal elastomers (LCEs) yields photo-mobile materials that efficiently convert light into mechanical output [5,19]. Subsequently, the focus in the literature was to decrease their concentration to allow the use of photo-initiators, which are greener and easier to use compared to traditional initiators. Photo-initiators cannot absorb enough energy to start polymerization in the presence of an azobenzene concentration higher than 20%. Furthermore, it is difficult to obtain PMPs with monodomain configurations with 100% azobenzene, since the nematic temperature range of azo-based LCs is quite narrow. In this work, three LC mixtures comprised of 100% of azobenzene moieties were developed and characterized. In particular, the thermo/mechanical phenomena that regulate the actuation of such materials are described. Also, as compared to traditional methods [5,6], the polymerization was achieved using dicumyl peroxide. Our finding suggests that such material can be useful for designing polarization-selective switches. The polarization effect on the activity of the switch seems to be independent from thermotropic components.

2. Materials and Methods

2.1. *Preparation of Cell Reactor*

For preparing reactor cells, we used glass slides and plastic spacers. Each glass slide was cut at needed measurements using a diamond tip. Subsequently, they were washed using, in sequence, first, glassware liquid cleaner (10 min sonic bath), then ultra-pure water (10 min sonic bath), acetone (20 min sonic bath), and finally isopropanol (5 min sonic bath). Once dry, the glass slides were spin-coated (3 s 4000 rpm, 30 s 4000 rpm, 3 s 0 rpm) using Elvamide (1%, 3%, 6%, *w*/*w* concentration in methanol). Finally, the slides were rubbed automatically using an automatized home-made machine. The rubbed sides of two glasses were faced anti-parallelly (respect to the rubbed direction) and spaced with a 50 μm thick Kapton layer to obtain a cell reactor.

2.2. *Themo-Optical Characterization of Monomer Mixtures*

To evaluate the orientational order of the mixture of LC monomers, we estimated the order parameter using linear dichroism measurements by transmission of light through the sample, with a set-up composed of: polarized laser (wavelength 633 or 457 nm), a heater chamber with an optical window (Mettler Toledo FP82HT), and a power detector (Coherent Model M-2). To evaluate the order parameter from polarized transmitted light, we used the dichroic ratio as reported in Equation (1) [10], with a transition moment parallel to the principal axis of the molecules:

$$S = \frac{(\mathrm{A} \parallel - \mathrm{A}\perp)}{(\mathrm{A} \parallel + 2\mathrm{A}\perp)} \quad (1)$$

where A⊥ and AII are the absorbances of the cell containing the molten monomers mix, which have a director (defined as the mean value of the directions of the molecular long axes) *n* perpendicular or parallel, respectively, to the direction of the laser polarization. The absorbance was calculated as: $\log_{10} \frac{1}{T}$, where T represents the cell transmittance.

The birefringence of the molten mix was investigated by introducing a polarizer before the power detector, and measuring the transmittance when the cell was perpendicular, parallel, or tilted 45 degrees as compared to the direction of the laser polarization.

2.3. *Synthesis of the Photo-Mobile Polymer*

The PMP films were prepared using three mixtures of LC monomers. We used three different molar ratios between A6zA6 (cross-linker) and A6z2 (linear) (BEAM co. www.beamco.com–structures in Supplementary Materials Figure S3) of 1:8, 1:1, and 8:1. The initiator used was the dicumyl peroxide, and this was 6.6 mol% of the total. The mixture of LCs used in this work are achiral and nematic molecules.

The mixtures were solubilized in DCM to help the homogenization at room temperature. The solvent was left to evaporate overnight under the fume hood.

Once dry, the mixture was injected by capillarity in the cells and the initiator activated. Subsequently, the cells were heated to the nematic temperature using the heating stage of Mettler Toledo FP82HT.

2.4. *Spectrophotometry Measurement*

Spectrophotometer JASCO V-650 (accuracy 0.5 nm, range 190–850 nm, Oklahoma, OK, USA) was used to investigate the optical properties of the PMP films. Both total transmittance T (%) and total reflectance R (%) were measured using the integrating sphere JASCO ISN-722 (inside diameter 60 mm, range 200–870 nm).

2.5. Photo-Actuation Characterization of PMPs

The PMPs actuators were cut as cantilevers (5 mm × 1 mm) and irradiated at 405, 457, 532, and 785 nm with 100:1 polarized laser. The set-up was composed of a neutral density filter, a retarder waveplate (λ/2), a focusing lens, and a sample holder mounted on a 3D translator.

The movements of the cantilevers were recorded at 60 fps. The movies were unpacked using virtualDub (v1.10.4). The bending angle was measured from the frames collected, considering both the initial and the final position of the cantilever respective to a graph paper placed behind the sample. The angle was converted first in radiant then in arclength (mm), and the speed calculated in m/s. The seconds were obtained by considering the number of frames in which the cantilever reaches its maximum bending.

2.6. Thermographic Measurements

Thermal response of the PMP film under laser irradiation was investigated by thermographic measurements. Analysis was performed using a LWIR AVIO TVS500 camera with an uncooled microbolometric detector (spectral range 8–14 µm, FPA 320 × 240 pixels and NETD ~60 mK at 25 °C) mounted with a 22 mm focal lens with iFOV 1.68 mrad. Thermal movies of the PMP films with a frame rate of 10 Hz were recorded while they were irradiated with a laser wavelength of 457 nm and power in the range 10–80 mW. The commercial software, IRT Analyzer (GRAYESS Inc.), which was supplied with the camera, was used for monitoring the temperature in real-time and for basic operations.

2.7. Polarized Optical Microscopy

The experiments were performed with a polarized upright microscope (Olympus BX51).

3. Results and Discussion

3.1. Fabrication and Synthesis Optimization

In the first step of this work, the possibility of preparing the reactor cells with Elvamide-coated glasses was investigated. Three different concentrated solutions of Elvamide were studied for this purpose (1%, 3%, or 6% *w*/*w*). After the spin-coating, the Elvamide films were characterised with different thicknesses to understand their influence on the rubbing. In particular, with 1% solution, the thickness of the spin-coated layer is 25 nm, and such a low level of thickness does not allow the formation of proper rubbing paths; in fact, the PMPs prepared do not bend properly (Figure S1 in Supplementary Materials). When the concentration of Elvamide is increased up to 3% and 6%, the thickness is between 75 and 200 nm (Table 1). The best results in terms of alignment are obtained with a 6% Elvamide coating. In fact, the PMPs produced at 6% are optimal in term of bending (Figure S1 in Supplementary Materials).

Table 1. Thickness of Elvamide coating layer before rubbing.

Name	Thickness
1% Elvamide	25 nm
3% Elvamide	75 nm
6% Elvamide	200 nm

Afterwards, the polymerization temperatures were investigated to obtain optimal PMPs using A6zA6 and A6z2 mixtures (A6zA6:A6z2 = 8:1 or 1:1 or 1:8 molar ratio).

The first approach used to understand the nematic range of the mixtures in use was DSC (Figure 1). Unfortunately, such a technique does not show enough information to pinpoint a precise nematic temperature (Figure 1). In fact, either the nematic region is too short, meaning it is covered by the isotropic melting signal, or too large to select a single temperature. For these reasons, another approach was followed, as described in Figure 2.

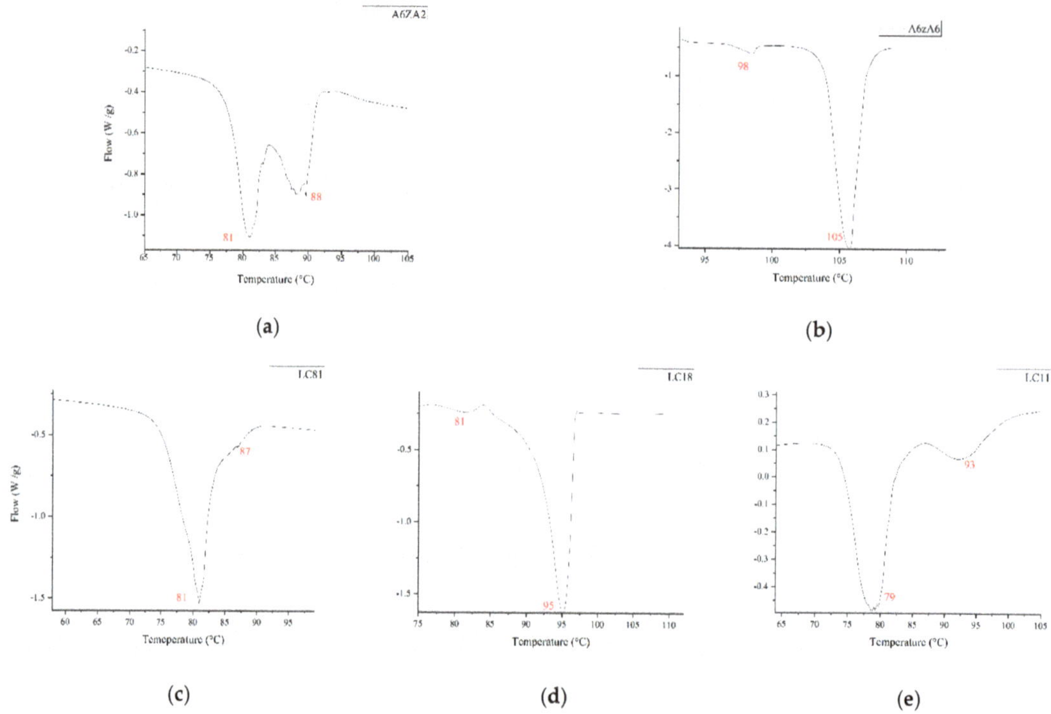

Figure 1. Melting profiles of LCs employed: (**a**,**b**) show the melting profile of A6zA2 and A6zA6, respectively, these are the monomer LCs used for the mixtures. (**c**) Melting profile of LC81 that does not show any nematic range, possibly because it is covered by the isotropic temperature. (**d**) Melting profile of LC18: in this case, again the isotropic temperature covers the nematic range. (**e**) Melting profile of LC11: in this case, it is possible to see the nematic phase between the two melting temperatures, indicating a range centered around 84 °C.

This information was extrapolated from data collected using the set-ups described in Figure 2.

For the nematic phase, the order state can be defined as the preferred direction and a degree of order of the liquid crystals [20]. Such parameters can be used as a quantitative tool. Due to the opacity of our PMPs, we use them not as a quantitative method, but as a qualitative method, to narrow down the range of temperatures. For this reason, the S was used together with the transmittance of the LCs to pinpoint precisely the nematic temperature of the mixture.

In particular, the state of order of the mixtures was calculated as a function of the temperature. When the transition moment (defined as the electric dipole moment associated with the transition between the two isomers) is oriented parallel to the molecular long axis, the order parameter has the relationship (1) noted in *Materials and Methods*.

The transition moment is oriented parallel to the molecular long axis thanks to the rubbing of the cells, and subsequently parallel or perpendicular to the polarized light.

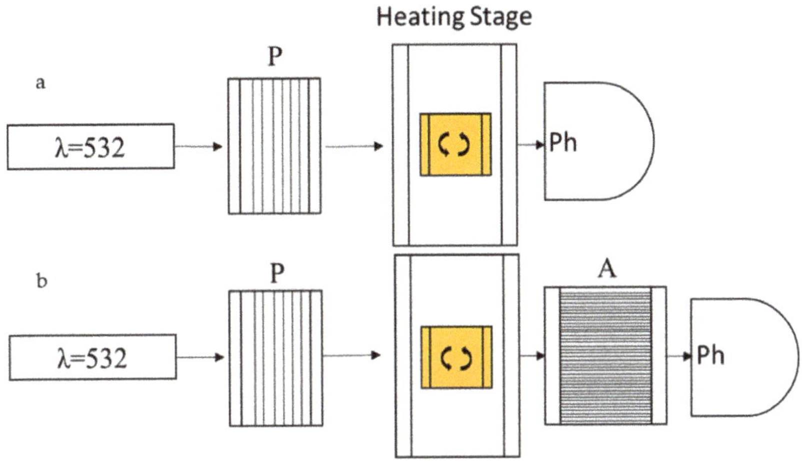

Figure 2. Schematic configuration of the transmittance measurement experiment. (**a**) Configuration to measure the state of order adjusting the position of the reactor cell parallel or orthogonal to the table. (**b**) Crossed polarized configuration to measure the transmittance when the rubbed cell is tilted 45 degrees as compared to the table. P and A are polarizer and analyzer, respectively. Ph is a photodiode used to measure the transmittance.

In particular, the cell reactors are infiltrated with the monomer mixtures in isotropic conditions (98–110 °C), in the absence of the initiator. Subsequently, the transmittance of the mixtures is measured between 75 and 100 °C with a rate of 0.5 °C/measure, using a 100:1 polarized laser with a wavelength of 532 nm (Figure 2a). The laser wavelength is chosen to avoid, as much as possible, the trans–cis isomerization of the azobenzene, and to have, at the same time, an appreciable absorbance. The transmitted light was measured with a power meter and converted in percentage, considering, as reference, the transmittance of an empty cell. Additionally, to evaluate the order state of the melted monomer mix, the measurements were performed with the director both parallel and orthogonal to the table. To precisely determine the polymerization temperatures, the maximum light-transmittance was measured with the cell between two crossed polarizers. Using this approach when the LC cell is rotated of 45 degrees in respect to the polarized light, and the LCs are nematic, the polarized light is rotated, and rotated again by the second polarizer, then having a maximum peak of transmission. In case the cell is in an isotropic state, or parallel to one of the two polarizers, the light is not rotated and cancelled by the second polarizer.

These measurements were performed for all formulations to investigate the effect of the ratio between linear (L) and cross-linker (C) monomer on the LCs organization. In Figure 3a–c, we report the transmittance measurements for the three mixtures. First, the mix with a ratio of 8:1 (LC81) was measured. Interestingly, the transmission of light polarized parallelly and perpendicularly shows an opposite trend between 87.5 °C and 80 °C, indicating that in that range of temperatures, the mixtures of liquid crystals arrange from an isotropic to a nematic to a crystal phase. During the cooling, the transmittance signal decreases around 85 °C (Figure 3a) when the director is parallel to the laser polarization, showing that the molecules are recrystallizing from an isotropic state towards the nematic phase. For this reason, order state calculation was performed to pinpoint the current nematic temperature range. Consequently, the order state (S) calculations increase for the LCs in the range between 81.5 and 84 °C (Figure S2a in Supplementary Materials). To narrow this range down, an analyser was set parallel to the optical table between the heating stage and the power meter (Figure 2b). When the polarization of the laser and the axis of the analyser are perpendicular, the power meter shows a minimum in transmission, as long the nematic director of the cell is parallel to one of the two directions. When the cell is rotated 45 degrees instead at the isotropic–nematic transition, the maximum temperature in transmission is recorded. This happens because the liquid crystals induce a rotation to the polarization axis of the light. In fact, we found that only at 83, 82, and 81.5 °C does the mixture have a higher order state (Figure S2b in Supplementary Materials).

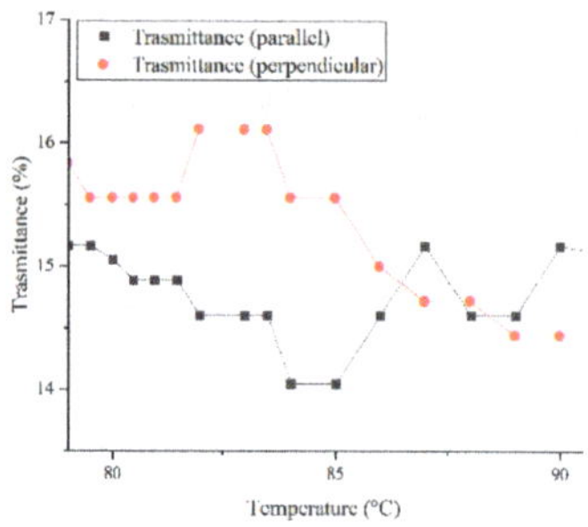

(a)

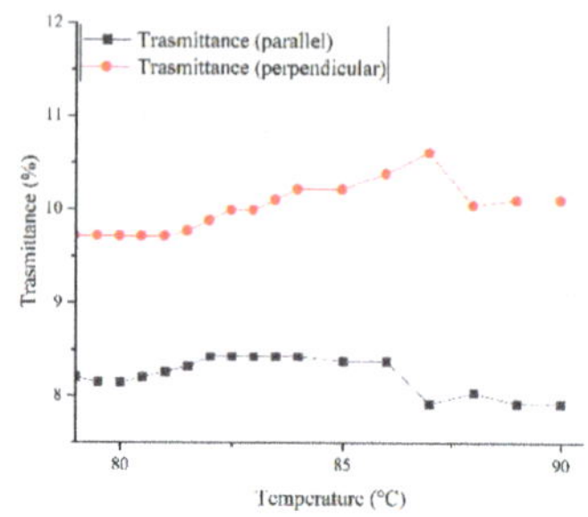

(b)

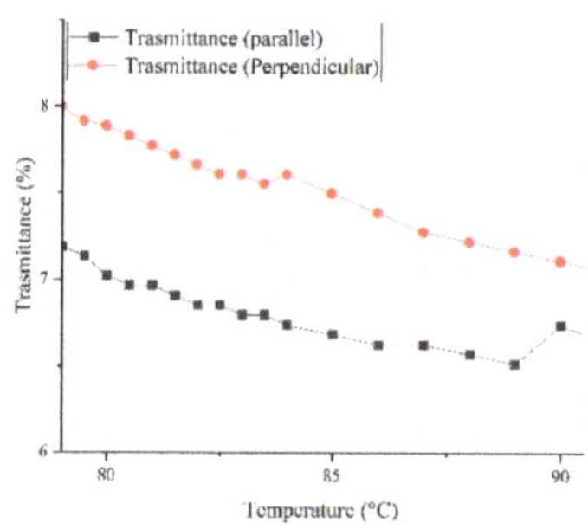

(c)

Figure 3. (**a**) Transmittance measurement of LC81. (**b**) Transmittance measurement of LC18. (**c**) Transmittance measurement of LC11.

The same measurements were performed for the other mixtures. The mixture LC18 displays a flex in transmittance around 88 °C, possibly indicating that the passage from the isotropic to nematic phase starts there (Figure 3b). After this, we used the order parameter to narrow the temperature range (Figure 4a). In Figure 4a, it is possible to see when the LCs transition from isotropic to nematic. In fact, we clearly see this between 88 and 86 °C, when the signal rises from the baseline, indicating an increment in order. Again, to clarify the data obtained via S calculation, the transmittance of the cell was measured between two crossed polarizers. Using this approach, the nematic range is narrowed down to only two temperatures, 87 and 88 °C (Figure 4b). The last mixture (LC11) does not show any clear change in the transmittance graph (Figure 3c). When the order parameter is measured, a wide range of temperatures are found (Supplementary Figure S2c). Unfortunately, not even measuring the transmission between the two polarizers helps narrowing down the range (Figure S2d). At a molar ratio of 1:1, the two LCs possibly compete for the mix's stabilization, never reaching an equilibrium. For these reasons, the temperature chosen for LC11 is based on the observation of the direct transmittance (Figure 3c), showing a slight peak at 84 °C.

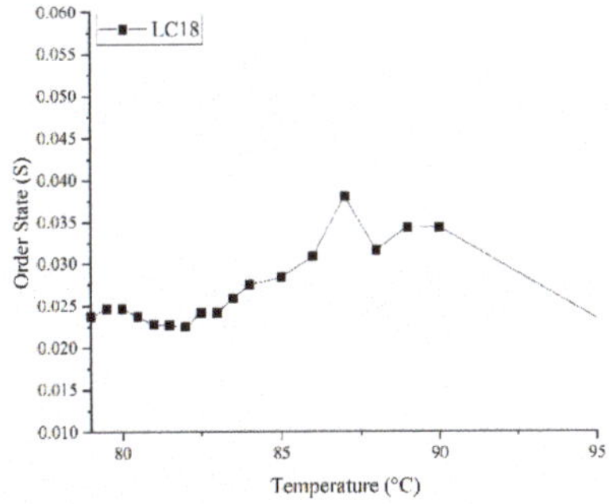

(**a**)

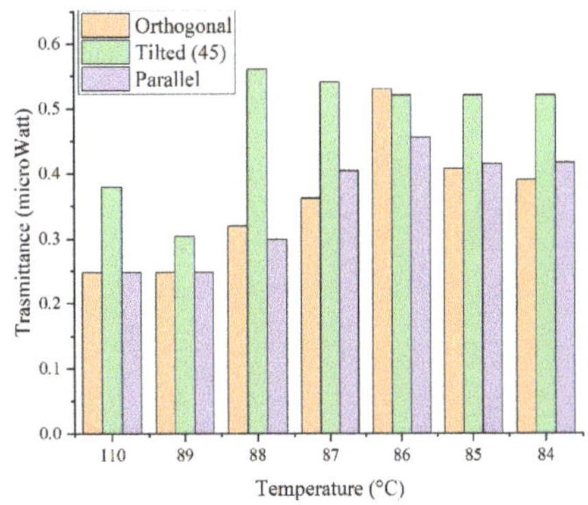

(**b**)

Figure 4. (**a**) Example of order state graph of LC18 and (**b**) transmittance of the cell between two crossed polarizers.

3.2. Polymerization and Alignment Confirmation

The polymerization of monomer mixtures (A6Z2 and A6ZA6, Beamco; monomers were used as received) is achieved by injecting an initiator together with the mixture (Dicumyl peroxide-DPX-6.6 mol%) in the rubbed cells. The mixtures are infiltrated in an isotropic state (temperature $\geq$ 100 °C), avoiding the activation of the initiator (temperature > 110°). Traditionally, these type of monomers mixtures are polymerized with an initiator such azobisisobutyronitrile (AIBN), and infiltrated in a vacuum oven to remove bubbles [5]. DPX avoids the use of a vacuum oven; in fact, in the conditions used, gas bubbles do not form in the PMP. This approach greatly simplifies the synthesis of PMPs with 100% azobenzene moieties.

To obtain an isotropic configuration, the initiator was activated at 130 °C for 3 h and the sample kept at a temperature greater than 100 °C. To achieve a nematic organisation, the mixtures were kept at 160 °C for 20 s to initiate polymerization. Afterward, the PMP was placed on a stabilized heating plate or on a heating stage at the temperature measured previously.

After the polymerization occurred, a polarized microscope was used to check the birefringence properties of the PMPs. Interestingly, the samples LC11 and LC18 are semi-transparent, and it is possible to demonstrate their ability to rotate the polarized light (Figure 5a). In contrast, the PMP LC81 is opaque and does not have considerable transmittance to measure its birefringence.

For this reason, the alignment of the azobenzene moieties in LC81 was studied performing a transmittance experiment, as previously explained for the monomer mixtures.

In brief, the transmittivity of the film was measured between 25 and 100 °C using a polarized He/Neon laser with a wavelength of 633 nm, for which the absorbance of the azobenzene it is close to zero.

Interestingly, the nematic PMP (Figure 5b 8:1 ratio) shows a highly ordered state (S = 0.9) between 20 and 30 °C. After increasing the temperature, the order state parameter decreases in all samples measured, indicating a correlation between temperature and LCs order state. This experiment shows that when the PMPs are polymerized at a lower temperature (nematic temperature), the S of the azobenzene moieties is higher than when polymerized at higher temperatures (isotropic states). In fact, when the factor S of the films are compared, a clear difference between the two PMP is observed (Figure 4a). The PMPs polymerized at the LC nematic temperature (A6zA2:A6zA6 = 8:1) show a higher order state up to 70 °C. Instead, the PMP polymerized in isotropic state show a low order state starting from room temperature.

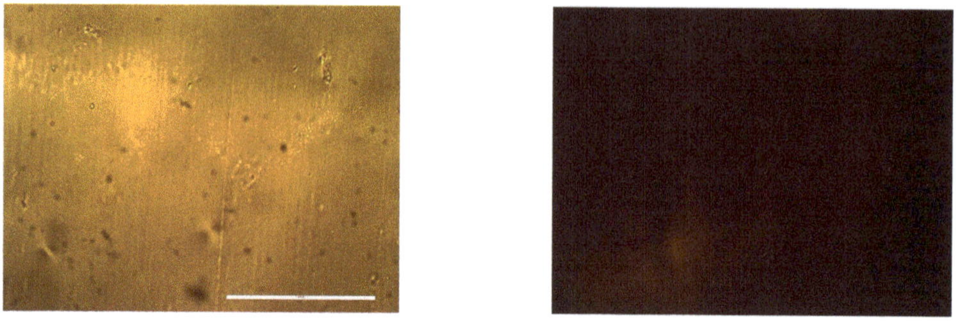

(a)

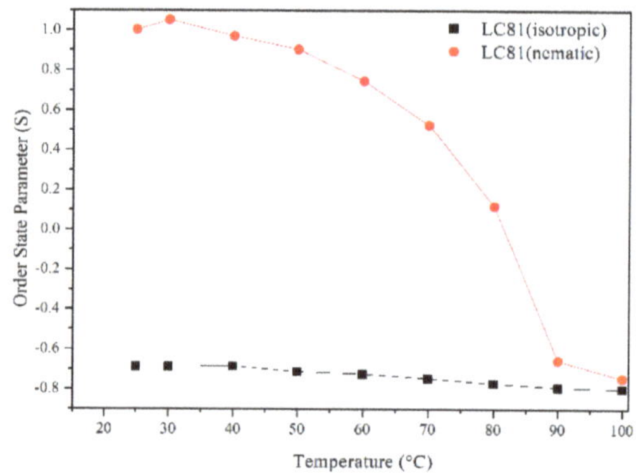

(b)

Figure 5. (**a**) Polarized microscopy images of LC18, showing the birefringence properties of the materials. The bar represents 200 um. (**b**) Order state parameter of LC81 measured at various temperatures (probe wavelength λ = 633 nm).

3.3. PMP Absorbance and Thermo-Mechanical Properties

The optical properties of PMPs were also investigated by spectrophotometric analysis. In Figure 6, the transmittance (blue line) and the reflectance (black line) measured for a PMP film with a thickness of 50 μm are reported. The absorbance of the film is successively calculated using the well-known relation A(%) = 100 − T(%) − R(%), and represented in the same figure with a red line. The absorbance estimated is higher than 90% in the wavelength range UV/VIS 200–500 nm, then the values decrease quickly to a few percentages in the near-infrared region (Figure 6). These results show how a laser source with wavelength of 457 nm can be an optimal choice to investigate the behavior of the film bending, because of the absorption of optical energy. Interestingly, the PMP shows reflectivity up to 30% after 550 nm. This effect is probably due to various factors. Polymeric material glossiness is related to its smoothness [21]. Probably, because of our preparation/polymerization procedure, the material does not show much reflectivity, since the rubbing causes the formation of a very rough surface. Another factor is correlated to the presence of azobenzene, which in our analysis condition, shows a light orange color. Azobenzene, even in different conditions than ours, was previously used to switch optical properties of liquid crystals mixtures and elastomers [22].

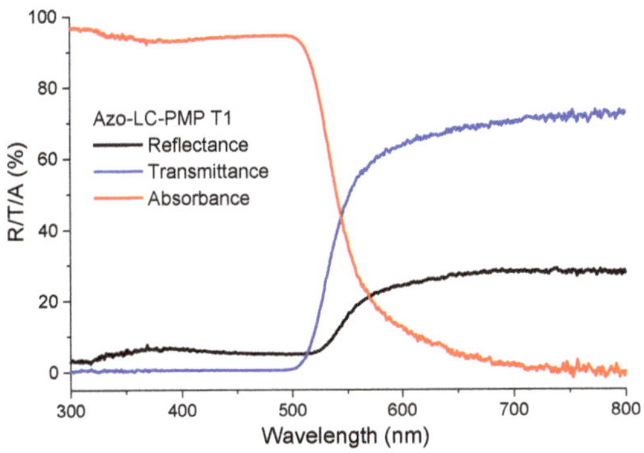

Figure 6. Spectral characterization of LC81: reflectance (black line), transmittance (blue line), and absorbance (red line).

Subsequently, the three PMPs were characterized for their mechanical response to polarized light (457 nm). The samples were cut as cantilevers (5 × 1 mm) and positioned with the rubbing direction perpendicular to the optical table.

First, the results show that the PMP LC81 is better at converting light into mechanical energy (Figure 7a). Second, the performances of the film is better when the laser polarization orientation is parallel to the rubbing direction compared to the perpendicular orientation.

For the first observation, it is obvious that the ability of the PMP LC81 to convert light into mechanical work is due to the degree of cross-linking. In fact, LC11 and LC18 have, respectively, four and eight times more the content of cross-linker. This results in more brittle and rigid materials.

The second observation is more difficult to justify, because it is deeply correlated to azobenzene properties. In fact, azobenzene behavior is correlated to the way it interacts with the polarized light. When the molecule long axis is parallel to the polarization of the laser, it absorbs the energy and uses it to isomerize to a more energetic isoform (trans>cis). Through its relaxation time (laser switched off) or heat, the molecule would back-isomerize to trans. This cycle happens a certain number of times (trans>cis>trans), until the molecular long axis of the azobenzene turns perpendicular to the polarization [23]. At that point, the molecule does not interact anymore with the laser [23]. That is clearly shown in Figure 7b, where the performance of the material is better when the orientation of the polarized light is parallel to the rubbing direction. This is more proof that our synthesis approach allows a good alignment and organization of LCs in the PMP.

LC81 is also characterized for its bending speed (up to 0.02 m/s), showing a fast and dose-dependent response to polarized light (Figure 7c). Interestingly, when the laser is turned off, the material returns to its starting position as fast as when bending with the light on.

This material is notably polarization-dependent, as already explained, but in this configuration, and at this specific polymerization temperature, we observe a new behavior (Figure 8). In fact, the PMP can be programmed to act as a switch using the polarization of the laser. If the PMP is hit by a laser polarized parallel to the rubbing director, it bends toward the laser and then goes back to its initial position when the laser is turned off (Figure 8a,b). When the PMP is firstly hit with a laser polarized perpendicularly and turned off and then hit with parallelly polarized light, the PMP changes bending direction (Figure 8c). Furthermore, this behavior can be reverted to the original by continuously irradiating the PMP first with parallel polarized laser, and then changing the polarization to perpendicular. In this case, the PMP reverts back to its initial configuration. This behavior is mainly observed with LC81, meaning that the degree of cross-linking plays a fundamental role in such behavior. Less cross-linking means a looser and more elastic network. Furthermore, the observed behavior, is probably correlated with the intrinsic nature of the materials. In fact, the PMPs studied are LCEs that notably retain the properties of their building blocks: liquid crystals. In this case, the sole liquid crystal used has azobenzene moieties that rotate its molecular long axis in function of the light polarization. For these reasons, we argue that the combination of azobenzene properties and the degree of cross-linking is the main cause of such behavior. Such behavior could possibly be finely modulated in depending on azobenzene concentration and degree of cross-linking.

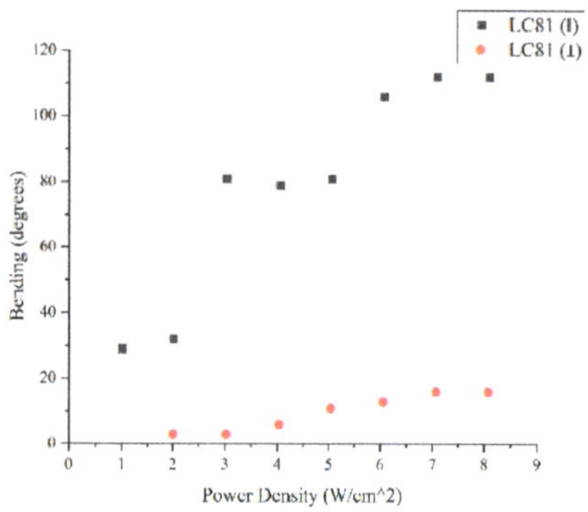

(**a**)

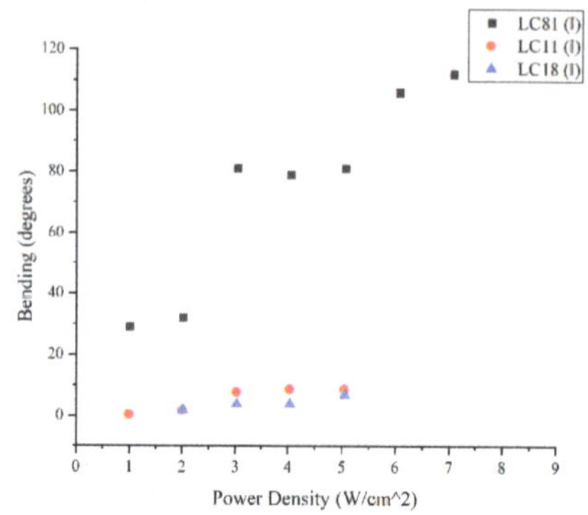

(**b**)

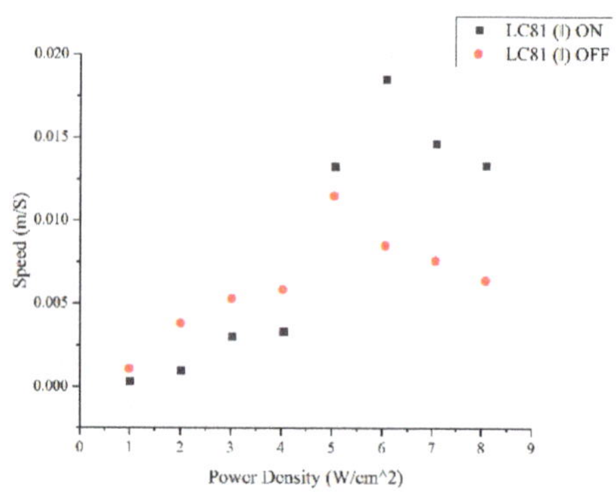

(**c**)

Figure 7. (**a**) Bending of LC81 when irradiated with polarized light parallel and perpendicular to the rubbing direction, (**b**) bending of LC81, LC18, and LC11 irradiated by polarized light parallel to the rubbing direction, and (**c**) bending speed of LC81 when the laser is ON and OFF.

Finally, we investigated the thermal response of the LC81 PMPs by thermographic analysis when irradiated with a laser wavelength of 457 nm. In Figure 9a, the maximum temperature increases (ΔT) that the film reaches in the range 0–80 mW are shown, considering both parallel and perpendicular laser polarization. As visible from the graph, the temperature trends achieved for the two different polarizations are similar, and can be considered quite linear with respect to the laser power. The temperature values with parallel polarization are only slightly higher than those measured with the perpendicular polarization. These results relate to the thermal behavior of the sample, and do not justify the differences in mechanical response (Figure 7b) associated with the two laser polarizations, indicating how the thermal effects do not play a predominant role in the film's movement. In Figure 9b, an example of a thermal image acquired during the measurements is shown.

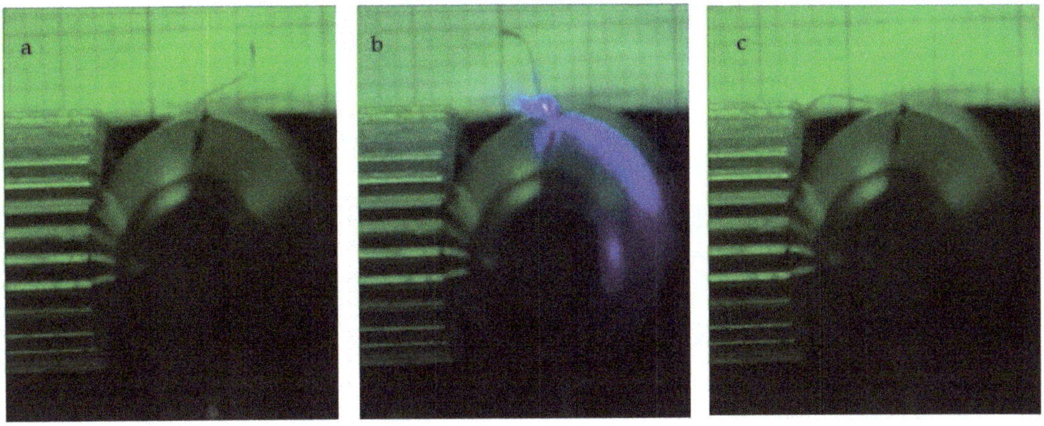

Figure 8. Movement modulation of the PMP as a function of the polarization. If the PMP is hit by a laser polarized parallel to the rubbing director, it bends toward the laser and then goes back to its initial position when the laser is turned off (**a**,**b**). When the PMP is firstly hit with a laser polarized perpendicularly and turned off and then hit with parallelly polarized light, the PMP changes bending direction (**c**).

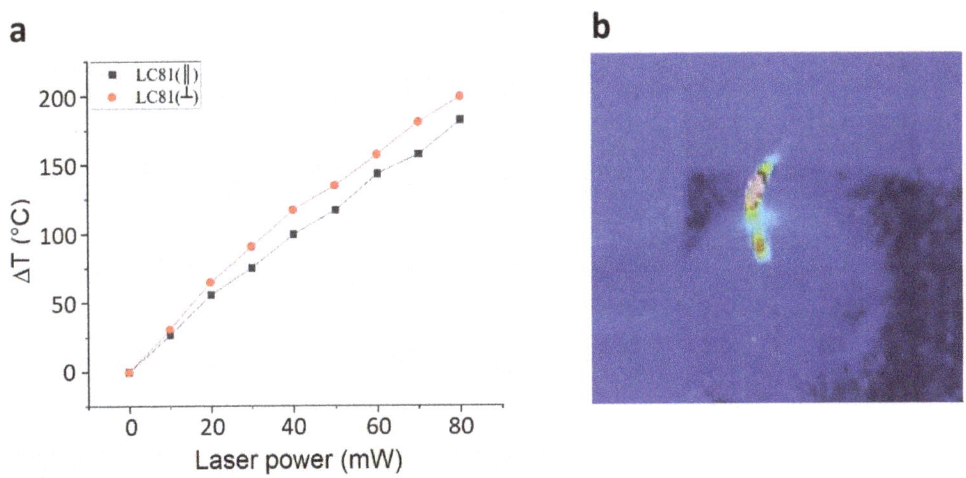

Figure 9. Thermographic analysis: (**a**) maximum temperature increases (ΔT) that the film reaches when irradiated with a laser wavelength of 457 nm with power in the range 0–80 mW, (**b**) example of thermal image acquired during the measurements.

4. Conclusions

We developed and characterized three azobenzene-based photo-mobile polymer mixtures to understand how the ratio between the linear (L) and cross-linker (C) monomer influences their final properties. We found that the nematic temperature of the LCs mixture is greatly influenced by the ratio of L and C. In fact, increasing the molar ratio in favor of the cross-linker (LC18), considerably increases the nematic temperature (up to 88 °C).

Furthermore, we characterised their light-to-mechanical properties, and show that when the ratio is in favor of the linear monomer, the PMPs bend more and the process is faster compared to the other mixtures. Finally, we show that the thermal effects do not play a predominant role in the movement of the film.

Supplementary Materials: The following supporting information can be downloaded at: https://www.mdpi.com/article/10.3390/mi13101665/s1, Figure S1: (Left) PMP polymerized in cells with 1% Elvamide. (Right) PMP polymerized in cells with 6% Elvamide. Figure S2: (**a**) order state of LC81, (**b**) transmittance of LC81 between crossed polarizers, (**c**) order state of LC11, (**d**) transmittance of LC11 between crossed polarizers. Figure S3: Here are depicted the LC monomers used in this work. The monomers are mixed in two different molar ratios (8:1 1:1 1:8). (Up) monomer A6zA6 (down) monomer A6z2.

Author Contributions: D.S.: conceptualization, methodology, formal analysis, investigation, data curation, writing—original draft preparation, and writing—review and editing; M.R.: investigation, data curation, and writing—review and editing; A.D.: formal analysis, investigation, and data curation, L.P.: supervision (PI), resources, and writing—review and editing; A.V. and V.M.: writing—review and editing. All authors have read and agreed to the published version of the manuscript.

Funding: This project received funding from the European Union's Horizon 2020 research and innovation programme under H2020-FETOPEN-2018-2019-2020-01Call.

Institutional Review Board Statement: Not applicable.

Informed Consent Statement: Not applicable.

Data Availability Statement: Not applicable.

Acknowledgments: The authors gratefully acknowledge support for this work from the Project PULSE-COM (Grant agreement No. 863227). Furthermore, we would like to thank the researcher Mariacristina Cocca (CNR–IPCB), who helped us in performing the DSC measurements.

Conflicts of Interest: The authors declare no conflict of interest.

References

1. Ube, T.; Ikeda, T. Cross-Linked Liquid-Crystalline Polymers as Photomobile Materials. In *Mechanically Responsive Materials for Soft Robotics*; Wiley: Hoboken, NJ, USA, 2019; pp. 209–232.
2. Fukuhara, K.; Nagano, S.; Hara, M.; Seki, T. Free-Surface Molecular Command Systems for Photoalignment of Liquid Crystalline Materials. *Nat. Commun.* **2014**, *5*, 3320. [CrossRef] [PubMed]
3. Yu, Y.; Nakano, M.; Ikeda, T. Photoinduced Bending and Unbending Behavior of Liquid-Crystalline Gels and Elastomers. *Pure Appl. Chem.* **2004**, *76*, 1467–1477. [CrossRef]
4. Harris, K.D.; Cuypers, R.; Scheibe, P.; Van Oosten, C.L.; Bastiaansen, C.W.M.; Lub, J.; Broer, D.J. Large Amplitude Light-Induced Motion in High Elastic Modulus Polymer Actuators. *J. Mater. Chem.* **2005**, *15*, 5043–5048. [CrossRef]
5. White, T.J.; Tabiryan, N.V.; Serak, S.V.; Hrozhyk, U.A.; Tondiglia, V.P.; Koerner, H.; Vaia, R.A.; Bunning, T.J. A High Frequency Photodriven Polymer Oscillator. *Soft Matter* **2008**, *4*, 1796–1798. [CrossRef]
6. Serak, S.; Tabiryan, N.; Vergara, R.; White, T.J.; Vaia, R.A.; Bunning, T.J. Liquid Crystalline Polymer Cantilever Oscillators Fueled by Light. *Soft Matter* **2010**, *6*, 779–783. [CrossRef]
7. Jenkins, D.J.A.; Wolever, T.M.S.; Jenkins, A.L.; Thorne, M.J.; Lee, R.; Kalmusky, J.; Reichert, R.; Wong, G.S. The Glycaemic Index of Foods Tested in Diabetic Patients: A New Basis for Carbohydrate Exchange Favouring the Use of Legumes. *Diabetologia* **1983**, *24*, 257–264. [CrossRef]
8. Mamiya, J.I.; Kuriyama, A.; Yokota, N.; Yamada, M.; Ikeda, T. Photomobile Polymer Materials: Photoresponsive Behavior of Cross-Linked Liquid-Crystalline Polymers with Mesomorphic Diarylethenes. *Chem. A Eur. J.* **2015**, *21*, 3174–3177. [CrossRef]
9. Ube, T.; Ikeda, T. Photomobile Polymer Materials with Crosslinked Liquid-Crystalline Structures: Molecular Design, Fabrication, and Functions. *Angew. Chem. Int. Ed.* **2014**, *53*, 10290–10299. [CrossRef]
10. Sagnelli, D.; Calabrese, M.; Kaczmarczyk, O.; Rippa, M.; Vestri, A.; Marchesano, V.; Kortsen, K.; Crucitti, V.C.; Villani, F.; Loffredo, F.; et al. Photo-Responsivity Improvement of Photo-Mobile Polymers Actuators Based on a Novel Lcs/Azobenzene Copolymer and Zno Nanoparticles Network. *Nanomaterials* **2021**, *11*, 3320. [CrossRef]
11. Todorov, T.; Tomova, N.; Nikolova, L. High-Sensitivity Material with Reversible Photo-Induced Anisotropy. *Opt. Commun.* **1983**, *47*, 123–126. [CrossRef]
12. Kondo, M. Photomechanical Materials Driven by Photoisomerization or Photodimerization. *Polym. J.* **2020**, *52*, 1027–1034. [CrossRef]
13. Chang, V.Y.; Fedele, C.; Priimagi, A.; Shishido, A.; Barrett, C.J. Photoreversible Soft Azo Dye Materials: Toward Optical Control of Bio-Interfaces. *Adv. Opt. Mater.* **2019**, *7*, 1–25. [CrossRef]
14. Lovrien, R. The Photoviscosity Effect. *Proc. Natl. Acad. Sci. USA* **1967**, *57*, 236–242. [CrossRef]
15. Eich, M.; Reck, B.; Ringsdorf, H.; Wendorff, J.H. Reversible Digital And Holographic Optical Storafe In Polymeric Liquid Crystals (PLC). *Mol. Polym. Optoelectron. Mater.* **1987**, *0682*, 93. [CrossRef]

16. Rochon, P.; Batalla, E.; Natansohn, A. Optically Induced Surface Gratings on Azoaromatic Polymer Films. *Appl. Phys. Lett.* **1995**, *66*, 136–138. [CrossRef]
17. Rochon, P.; Bissonnette, D.; Natansohn, A.; Xie, S. Azo Polymers for Reversible Optical Storage III Effect of Film Thickness on Net Phase Retardation and Writing Speed. *Appl. Opt.* **1993**, *32*, 7277. [CrossRef]
18. Lee, K.M.; Koerner, H.; Vaia, R.A.; Bunning, T.J.; White, T.J. Relationship between the Photomechanical Response and the Thermomechanical Properties of Azobenzene Liquid Crystalline Polymer Networks. *Macromolecules* **2010**, *43*, 8185–8190. [CrossRef]
19. White, T.J.; Serak, S.V.; Tabiryan, N.V.; Vaia, R.A.; Bunning, T.J. Polarization-Controlled, Photodriven Bending in Monodomain Liquid Crystal Elastomer Cantilevers. *J. Mater. Chem.* **2009**, *19*, 1080–1085. [CrossRef]
20. Saupe, A. Recent Results in the Field of Liquid Crystals. *Angew. Chem. Int. Ed. Engl.* **1968**, *7*, 97–112. [CrossRef]
21. Yamada, H.; Kukino, M.; Wang, Z.A.; Miyabara, R.; Fujimoto, N.; Kuwabara, J.; Matsuishi, K.; Kanbara, T. Preparation and Characterization of Green Reflective Films of Polyaniline Analogs Containing Azobenzene Units. *J. Appl. Polym. Sci.* **2015**, *132*, 41275. [CrossRef]
22. Hrozhyk, U.A.; Serak, S.V.; Tabiryan, N.V.; Bunning, T.J. Optical Tuning of the Reflection of Cholesterics Doped with Azobenzene Liquid Crystals. *Adv. Funct. Mater.* **2007**, *17*, 1735–1742. [CrossRef]
23. Wu, Y.; Demachi, Y.; Tsutsumi, O.; Kanazawa, A.; Shiono, T.; Ikeda, T. Photoinduced Alignment of Polymer Liquid Crystals Containing Azobenzene Moieties in the Side Chain. 3. Effect of Structure of Photochromic Moieties on Alignment Behavior. *Macromolecules* **1998**, *31*, 4457–4463. [CrossRef]

micromachines

MDPI

Article

Signal-to-Noise Ratio Analysis for the Voltage-Mode Read-Out of Quartz Tuning Forks in QEPAS Applications

Michele Di Gioia [1,2], Luigi Lombardi [2], Cristoforo Marzocca [2,*], Gianvito Matarrese [2], Giansergio Menduni [1], Pietro Patimisco [1] and Vincenzo Spagnolo [1,2]

1 PolySense Lab, Dipartimento Interateneo di Fisica, University and Politecnico of Bari, Via Amendola 173, 70126 Bari, Italy
2 Dipartimento di Ingegneria Elettrica e Dell'Informazione, Politecnico of Bari, Via Edoardo Orabona 4, 70126 Bari, Italy
* Correspondence: cristoforo.marzocca@poliba.it; Tel.: +39-080-5963313

Abstract: Quartz tuning forks (QTFs) are employed as sensitive elements for gas sensing applications implementing quartz-enhanced photoacoustic spectroscopy. Therefore, proper design of the QTF read-out electronics is required to optimize the signal-to-noise ratio (SNR), and in turn, the minimum detection limit of the gas concentration. In this work, we present a theoretical study of the SNR trend in a voltage-mode read-out of QTFs, mainly focusing on the effects of (i) the noise contributions of both the QTF-equivalent resistor and the input bias resistor R_L of the preamplifier, (ii) the operating frequency, and (iii) the bandwidth (BW) of the lock-in amplifier low-pass filter. A MATLAB model for the main noise contributions was retrieved and then validated by means of SPICE simulations. When the bandwidth of the lock-in filter is sufficiently narrow (BW = 0.5 Hz), the SNR values do not strongly depend on both the operating frequency and R_L values. On the other hand, when a wider low-pass filter bandwidth is employed (BW = 5 Hz), a sharp SNR peak close to the QTF parallel-resonant frequency is found for large values of R_L ($R_L > 2$ MΩ), whereas for small values of R_L ($R_L < 2$ MΩ), the SNR exhibits a peak around the QTF series-resonant frequency.

Keywords: quartz-enhanced photoacoustic spectroscopy; quartz tuning fork; voltage-mode read-out; front-end electronics; signal-to-noise ratio; gas sensing

Citation: Di Gioia, M.; Lombardi, L.; Marzocca, C.; Matarrese, G.; Menduni, G.; Patimisco, P.; Spagnolo, V. Signal-to-Noise Ratio Analysis for the Voltage-Mode Read-Out of Quartz Tuning Forks in QEPAS Applications. *Micromachines* **2023**, *14*, 619. https://doi.org/10.3390/mi14030619

Academic Editors: Luigi Sirleto and Giancarlo C. Righini

Received: 7 February 2023
Revised: 3 March 2023
Accepted: 4 March 2023
Published: 8 March 2023

1. Introduction

Quartz-enhanced photoacoustic spectroscopy (QEPAS) is a well-known technique used for the detection of specific trace gases in complex mixtures [1–7]. The high performances in terms of selectivity and sensitivity allow for the exploitation of this technique in a wide range of applications, such as environmental monitoring [8–10], chemical analysis [11,12], and advanced biomedical diagnostics [13,14]. In QEPAS, acoustic waves are generated between the prongs of a quartz tuning fork (QTF) by the absorption of modulated light from the gas molecules, via non-radiative relaxation processes [2,15]. QTFs are employed as piezoelectric sensitive elements to transduce pressure waves in an electric signal [2,6]. This technique was firstly introduced in 2002 and exploited standard QTFs resonating at 32 kHz [16]. These quartz resonators are characterized by a good immunity to environmental acoustic noise because of their high quality factors (Q) and compact dimensions. Due to the sharp resonance, external noise sources outside of the resonator's small bandwidth (~4 Hz at atmospheric pressure) do not influence the QTF signal [17].

Suitable front-end electronics must be designed to read-out the signal generated by the QTF. The common read-out architecture employed in QEPAS sensors is the transimpedance amplifier (TIA) [8–14], schematically depicted in Figure 1.

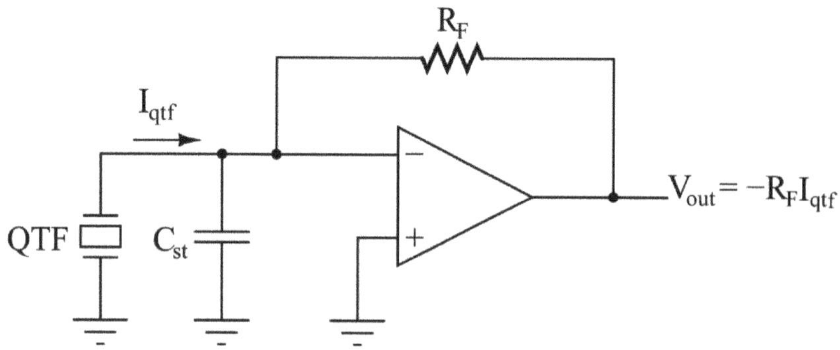

Figure 1. QTF read-out by means of a transimpedance preamplifier.

In this configuration, the current generated by the QTF (I_{qtf}) due to the charge displacement caused by piezoelectric effect when pressure waves put prongs in vibration is forced to flow entirely in the feedback resistor R_F. Due to the virtual ground established on the inverting node of the operational amplifier (OPAMP), the measurement is insensitive to the stray capacitance C_{st} and, in turn, the output voltage (V_{out}) is proportional to I_{qtf}.

Recently, it has been demonstrated that a voltage-mode approach for the read-out electronics of QTFs can be advantageous in terms of signal-to-noise ratio (SNR) [18–21]. In this case, the piezoelectric transducer is coupled to a voltage amplifier, realized by means of a classic OPAMP in a non-inverting configuration, as illustrated in Figure 2, and characterized by very high input impedance. Since the QTF is an open circuit at DC, a load resistor R_L must necessarily be connected to the non-inverting input of the amplifier to establish its DC voltage to ground, when the input bias current of the OPAMP can be neglected.

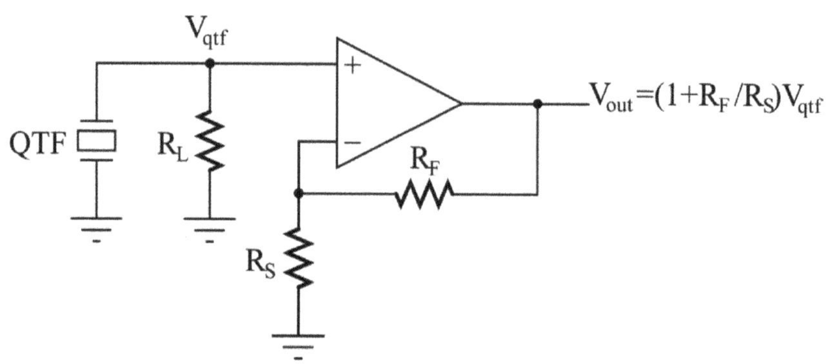

Figure 2. Voltage-mode read-out of the QTF (OPAMP in non-inverting configuration).

The thermal noise of the resistor R_L contributes to the overall noise spectral density of the circuit output. Furthermore, the QTF experiences the loading effect due to R_L; then, the input voltage V_{qtf} of the amplifier results from the product of the current generated by the QTF under this load and the value of R_L itself. As a consequence, R_L has a direct impact on both the level of the useful signal V_{out} and the noise at the output of the preamplifier.

QEPAS requires synchronous detection techniques based on lock-in amplifiers (LIAs) to efficiently extract the useful signal component from the noise floor [2,22–26]. In LIAs, the amplifier signal is first multiplied with both a sinewave and a 90° phase-shifted copy of that sinewave at a selected operating frequency (f_{op}); then, a low-pass filter (LPF) is used to retrieve the signal component at f_{op} or its multiples [22,26]. The phase noise in high-Q mechanical oscillators, such as QTFs, could be detrimental when LIAs are employed, since the amplitude of the output signal, and, in turn, the SNR, will be affected [27].

Both QTF signal and noise sources undergo the signal conditioning chain composed of the front-end voltage amplifier, the multiplication with a sine wave at frequency f_{op}, and the low-pass filter. Therefore, a study on the SNR as a function of (i) the parameters of the amplifier components, (ii) the operating frequency, and (iii) the low-pass filter bandwidth must be carefully conducted to fully exploit the resonance properties of the QTF and maximize the performances of a QEPAS sensor.

In this work, we reported the effect of the R_L value on the frequency where the SNR exhibits its peak at the output of the lock-in amplifier, i.e., the optimum operating frequency for the QEPAS system. Starting from the well-known Butterworth–Van Dyke model for the QTF [17,28–30], we derived an analytical expression for the amplitude of the signal generated by the QTF as a function of frequency at different R_L values. For our purposes, only the main electronic noise contributions were considered, whereas phase noise was not taken into account. Furthermore, a mathematical modelling was developed with MATLAB for all the relevant contributions to the total noise spectral density at the output of the preamplifier, making possible a comparison among their relative weights. Finally, the behavior of the SNR as a function of the lock-in demodulation frequency was studied at different R_L values and LPF bandwidths, namely, the acquisition time. All the proposed analytical expressions were validated by comparing the developed MATLAB model with SPICE simulations, carried out considering a realistic model for the OPAMP used in the preamplifier. As a result of the study, general guidelines for the choice of the resistor R_L and the most suitable operating frequency for the QEPAS system implementing the voltage-mode read-out of QTFs can be derived. Furthermore, the proposed analysis allows for the study of the noise contributions at different bandwidths to optimize the acquisition time of QEPAS measurements.

2. Signal Response of a Quartz Tuning Fork Read Out in Voltage Mode

As mentioned above, the Butterworth–Van Dyke circuit and an equivalent Thevenin source were employed to model the QTF when excited by an acoustic wave [17] to find an analytical expression of the output signal V_{out} as a function of frequency for the circuit in Figure 2. For this investigation, we considered (i) an OPAMP with sufficiently high gain-bandwidth product so that the gain of the non-inverting configuration is $A_v = V_{out}/V_{qtf} = 1 + R_F/R_S$, as shown in Figure 2; and (ii) the capacitance C_{in} between the non-inverting input and ground. Thus, the circuit to be studied is sketched in Figure 3.

Here, the parasitic shunt capacitance C_p between the terminals of the QTF is in parallel with the input capacitance of the voltage amplifier C_{in}, so that the total parasitic capacitance which loads the QTF is $C_{pt} = C_p + C_{in}$. The voltage source V_{in} represents the internal electric signal generated by the piezoelectric effect, when the QTF is excited by an acoustic wave. Straightforward calculations provide the expression of the transfer function $H_v(j\omega)$ between the internal voltage $V_{in}(j\omega)$ and the output voltage $V_{out}(j\omega)$:

$$\begin{aligned} H_v(j\omega) &= \frac{V_{out}(j\omega)}{V_{in}(j\omega)} = \\ &= A_v \frac{j\omega C_s R_L}{1-\omega^2\left(LC_s+R_pR_LC_{pt}C_s\right)+j\omega\left[\left(R_p+R_L\right)C_s+R_LC_{pt}-\omega^2 LC_{pt}C_sR_L\right]}. \end{aligned} \quad (1)$$

The squared modulus of this transfer function, which describes how the squared amplitude of the circuit response depends on the frequency, can be written as follows:

$$|H_v(j\omega)|^2 = \frac{A_v^2}{\frac{\left(1-\frac{\omega^2}{\omega_R^2}\right)^2}{\omega^2 R_L^2 C_s^2} + \left(1+\frac{C_{pt}}{C_s}\right)^2\left[\left(1-\frac{\omega^2}{\omega_P^2}\right)+\frac{R_P}{R_L}\frac{C_s}{C_{pt}+C_s}\right]^2}, \quad (2)$$

where $\omega_R = \frac{1}{\sqrt{LC_s + R_p R_L C_{pt} C_s}}$ is the angular frequency in which the transfer function $H_v(j\omega)$ exhibits a real value and $\omega_P = \frac{1}{\sqrt{L\frac{C_s C_{pt}}{C_s + C_{pt}}}} = \frac{1}{\sqrt{LC_{eq}}}$ is the parallel-resonant angular frequency of the QTF as loaded in the circuit shown in Figure 3.

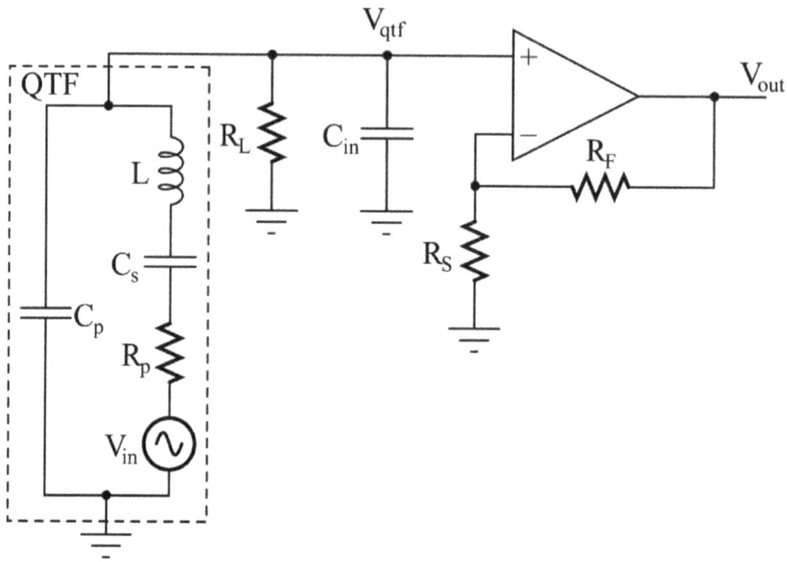

Figure 3. Butterworth–Van Dyke model for the QTF in the voltage-mode read-out circuit.

AC SPICE simulations have been carried out to confirm the validity of the expression in Equation (2). The set of typical parameters listed in Table 1 has been used for the QTF in both the analytical model and SPICE simulations. Moreover, in SPICE simulations, C_{in} is included in the model of the AD8067, provided by Analog Devices. AD8067 is a low-noise, high speed, FET input operational amplifier. Thanks to its very low input bias current, it is suitable for precision and high gain applications [31].

Table 1. Parameter values used to compare the results of expression in Equation (2) to SPICE simulations.

Parameter	Value
C_p	5 pF
C_s	5.2424 fF
L	4.5 kH
R_p	92.7 kΩ
C_{in}	1.5 pF
R_F	47 kΩ
R_S	1 kΩ

The series-resonant frequency f_S of the QTF is:

$$f_S = \frac{\omega_S}{2\pi} = \frac{1}{2\pi\sqrt{LC_s}} = 32768 \text{ Hz},$$

whereas its quality factor Q is

$$Q = \frac{1}{\omega_s R_p C_s} = 10^4,$$

which are typical values for a standard QTF used in QEPAS sensors [2,16,17,26]. Moreover, the gain of the non-inverting configuration is $A_V = 48$ and the parallel-resonant frequency of the QTF is $f_P = \omega_P/2\pi = 32{,}781$ Hz. The value of $C_P \cong C_P + C_S$ could be found by measuring the equivalent capacitance of the QTF at low frequencies using the capacitance-voltage profiling technique. The ratio C_S/C_P is extracted by the ratio f_P/f_S after measuring the parallel and series-resonant frequencies of the QTF, where the sensor exhibits the minimum and maximum admittance, respectively. Last, L can be found from f_S, knowing C_S, and R_P is extracted from the quality factor Q of the QTF [21]. Figure 4 shows the comparison between the results provided by the SPICE simulations and the analytical model for three different values of R_L (100 kΩ, 0.5 MΩ, and 2.5 MΩ).

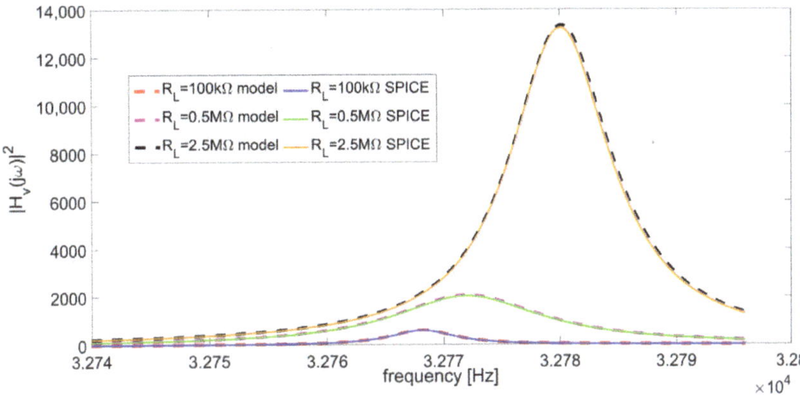

Figure 4. Comparison between SPICE simulations and analytical model in Equation (2) of the frequency response of the circuit in Figure 3 for R_L = 100 kΩ, 0.5 MΩ and 2.5 MΩ.

The perfect matching between two modellings demonstrates that Equation (2) can be used to accurately represent the behavior of the circuit in Figure 3. The maximum difference between the peak frequencies of corresponding curves is in the order of a few hundredths of Hz and the mean absolute percentage error between corresponding curves is about 1.8%.

2.1. *Peak Frequency of the QTF Signal Response as a Function of R_L*

As shown in Figure 4, the output signal amplitude and the peak frequency f_{peak} strongly depend on the value of the resistor R_L. This is particularly relevant for an optimal choice of the operating frequency in the QEPAS technique, aimed at exploiting as much as possible the resonance properties of the QTF.

Figure 5 shows the f_{peak} trend as a function of R_L. This curve was retrieved computing $|H_v(j\omega)|^2$ for different values of R_L and then applying MATLAB "max" function to yield f_{peak} values.

For R_L values lower than 100 kΩ, f_{peak} tends to the series-resonant frequency f_S = 32768 Hz; whereas, for values of R_L higher than 2 MΩ, f_{peak} tends to assume the values of the parallel-resonant frequency f_P, in our case equal to 32781 Hz.

The behavior of the peak position as a function of the value of R_L can be explained considering the two terms which compose the denominator of the function $|H_v(j\omega)|^2$ in Equation (2), reported here, below, for ease of reading, as Den1 and Den2:

$$\mathrm{Den1}(\omega) = \frac{\left(1-\frac{\omega^2}{\omega_R^2}\right)^2}{\omega^2 R_L^2 C_s^2},\ \mathrm{Den2}(\omega) = \left(1+\frac{C_{pt}}{C_s}\right)^2\left[\left(1-\frac{\omega^2}{\omega_P^2}\right)+\frac{R_P}{R_L}\frac{C_s}{C_{pt}+C_s}\right]^2 \quad (3)$$

The behavior of Den1 and Den2 as a function of frequency is reported in Figure 6a,b for R_L = 100 kΩ and R_L = 10 MΩ, respectively.

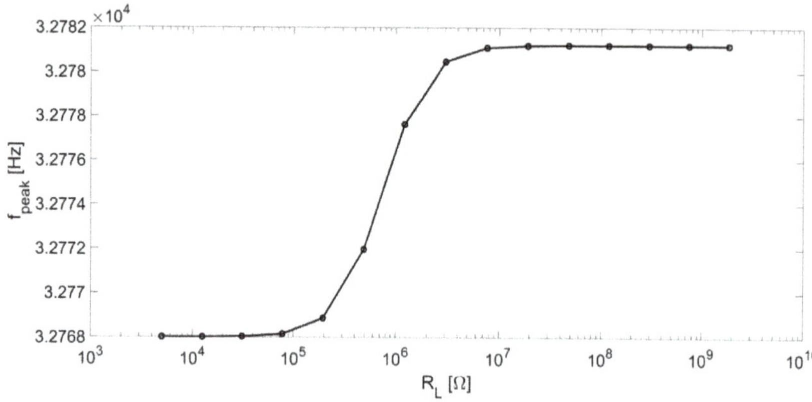

Figure 5. Peak frequency f_{peak} of $|H_v|^2$ as a function of R_L.

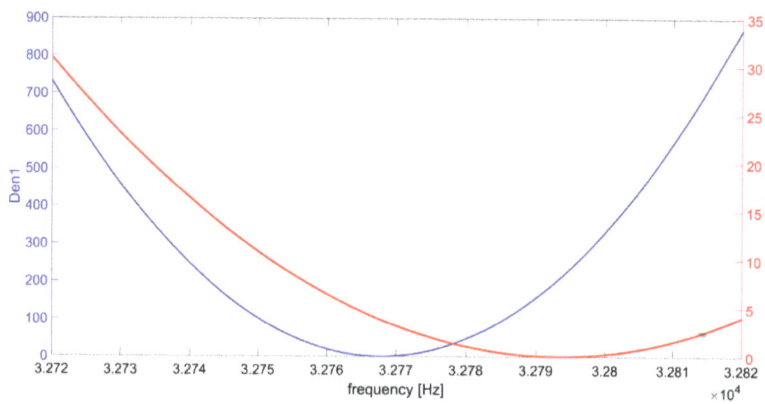

(**a**)

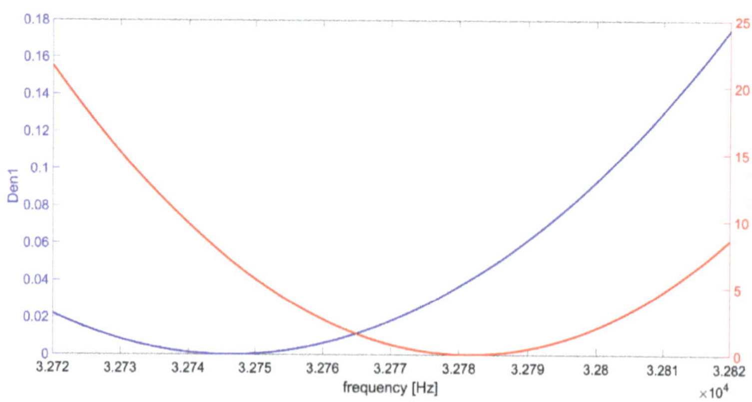

(**b**)

Figure 6. Behavior of Den1(f) and Den2(f) in Equation (3) as a function of frequency for (**a**) R_L = 100 kΩ and (**b**) R_L = 10 MΩ.

Results show that Den1(ω) is strongly dependent on the value of R_L. In the investigated frequency range: (i) the maximum value of Den1 varies from 868.3 in Figure 6a to 0.17 in Figure 6b; (ii) the blue curve at 100 kΩ varies more rapidly than the Den1 curve at 10 MΩ in the frequency range close to the minimum; and (iii) the frequency where the Den1 minimum occurs decreases by more than 20 Hz, from 32,768 Hz in Figure 6a to 32,746 Hz in Figure 6b. Conversely, the dependence of Den2(ω) on R_L is only due to the term $\frac{R_P}{R_L}\frac{C_s}{C_{pt}+C_s}$, which is negligible when $R_L \geq 1$ MΩ, since $C_s << C_{pt}$ and $R_P << R_L$. Indeed: (i) the red curve maximum value varies from 30.8 in Figure 6a to 21.5 in Figure 6b; (ii) Den2 value variations in the frequency range close to the minimum value are slightly different at 100 kΩ and at 10 MΩ; and (iii) the difference frequency where the Den2 minimum occurs is about 13 Hz, from 32,794 Hz in Figure 6a to 32,781 Hz in Figure 6b. As a result, for $R_L < 100$ kΩ, the contribution of Den1(ω) becomes dominant, so that when R_L decreases, the minimum value of the denominator of $|H_v(j\omega)|^2$ tends to the zero of Den1(ω) function, located at $\omega = \omega_R$. Moreover, in this range of R_L values, ω_R could be approximated to ω_S:

$$\omega_R = \frac{1}{\sqrt{LC_s + R_p R_L C_{pt} C_s}} \cong \frac{1}{\sqrt{LC_s}} = \omega_S,$$

and the peak of $|H_v(j\omega)|^2$ tends to the series-resonant frequency of the QTF. Instead, for $R_L > 2$ MΩ, Den1(ω) becomes less relevant and Den2(ω) tends to be dominant in the sum of the two terms. As a consequence, the minimum of the sum tends to the zero of Den2(ω). Finally, in this range of R_L, this zero is very close to $\omega = \omega_P$, thus, the peak of $|H_v(j\omega)|^2$ is almost coincident with the parallel-resonant frequency of the QTF.

2.2. Peak of the QTF Signal Response as a Function of R_L

The trend of the peak value of $|H_v(j\omega)|^2$ as a function of R_L value is shown in Figure 7. The same method applied for Figure 5 was used to obtain this curve.

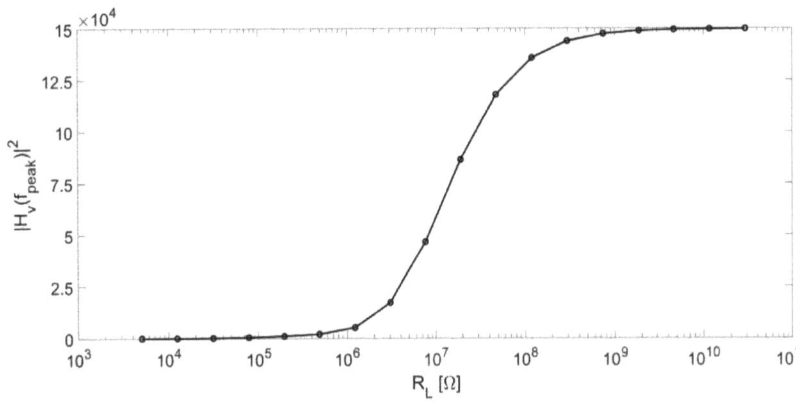

Figure 7. Peak value of $|H_v|^2$ as a function of R_L.

The peak value of $|H_v(j\omega)|^2$ is an increasing function of R_L up to a saturation value at $R_L > 1$ GΩ. As discussed in the previous section, for large values of R_L, $\omega_{peak} = 2\pi f_{peak}$ is very close to the parallel-resonant angular frequency ω_P and the peak of $|H_v(j\omega)|^2$ tends to the following value:

$$\left|H_v\left(j\omega_{peak}\right)\right|^2 \cong |H_v(j\omega_P)|^2 \cong A_v^2 \frac{\omega_P^2 R_L^2 C_s^2}{\left(1 - \frac{\omega_P^2}{\omega_R^2}\right)^2} \quad (4)$$

The denominator of Equation (4) can be rewritten as:

$$1-\frac{\omega_P^2}{\omega_R^2}=1-\frac{LC_s+R_pR_LC_{pt}C_s}{LC_{eq}}=\frac{L(C_{eq}-C_s)-R_pR_LC_{pt}C_s}{LC_{eq}}\cong-\frac{R_pR_LC_{pt}}{L}$$

Since $C_{Equation}\cong C_s$ and R_L is very large, it results

$$\left|H_v\left(j\omega_{peak}\right)\right|^2\cong A_v^2\frac{\frac{1}{LC_{eq}}R_L^2C_s^2}{\frac{R_p^2R_L^2C_{pt}^2}{L^2}}\cong A_v^2\frac{LC_s}{R_p^2C_{pt}^2},$$

which is independent on R_L.

However, the performance of the QEPAS technique will depend on the SNR obtained at the output of the preamplifier, not only on the amplitude of the signal. Therefore, it is mandatory to carry out a detailed study of the electronic noise contributions that are involved in the circuit, with the purpose of finding out the optimal operating frequency maximizing the SNR.

3. Contributions to the Output Noise Spectral Density

The most relevant contributions to the total electronic noise at the output of the voltage-mode preamplifier are shown in Figure 8.

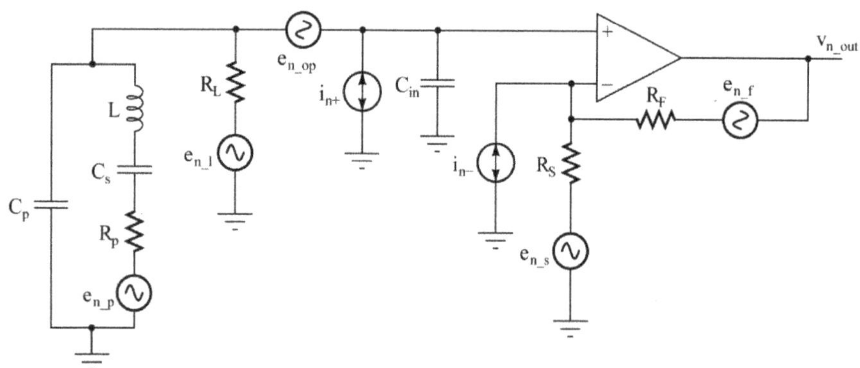

Figure 8. Noise contributions in the circuit of Figure 3.

In our calculation, the phase noise was neglected [27] and only the main electronic noise contributions were considered. Each resistor R_i of the circuit has been associated to its thermal noise voltage source, $e_{n\_i}^2 = 4kTR_i$. The OPAMP noise has been characterized by means of the classic equivalent input noise voltage (e_{n_op}) and current (i_{n+} and i_{n-}) sources. To simplify the study without losing accuracy, it is possible to neglect the noise contributions of R_F and R_S, composing the feedback network, due to the small values of these resistors. For the same reasons, the equivalent noise current i_{n-} associated to the inverting input of the OPAMP likewise does not give any relevant contribution. Moreover, all the sources in Figure 8 can be considered independent, so that the total output noise power spectral density $S_{ntot}(\omega)$ can be evaluated as follows:

$$S_{ntot}(\omega)\cong S_{np}(\omega)+S_{nL}(\omega)+S_{nop}(\omega)+S_{nin+}(\omega), \quad (5)$$

where the terms of the right-hand side come from R_p, R_L, e_{n_op}, and i_{n+}, respectively. In Equation (5), the single terms are the product of the spectral density of each noise source multiplied by the squared modulus of the transfer function between the noise source and the output of the circuit, determined using the superposition principle [32].

Concerning the noise associated to the resistor R_p, the transfer function $S_{np}(\omega)$ between the source e_{n_p} and the output of the preamplifier is $H_v(j\omega)$, thus:

$$S_{np}(\omega) = 4kTR_p|H_v(j\omega)|^2.$$

As a consequence, the behavior of $S_{np}(\omega)$ as a function of both the frequency and R_L is the same discussed in Section 2.

Let us now consider the noise contribution from the resistor R_L. The transfer function between the source e_{n_l} and the output of the front-end is:

$$\begin{aligned} H_L(j\omega) &= \frac{V_{n\_out}(j\omega)}{e_{n\_l}(j\omega)} \\ &= A_v \frac{1-\omega^2 LC_s + j\omega R_p C_s}{1-\frac{\omega^2}{\omega_R^2} + j\omega\left[(R_L+R_p)C_s + R_L C_{pt} - \omega^2 LC_{pt}C_s R_L\right]}. \end{aligned} \quad (6)$$

The denominators of $H_v(j\omega)$ and $H_L(j\omega)$ are the same and the two transfer functions differ only in their numerator. The contribution of the thermal noise of the resistor R_L to the total output noise spectral density is:

$$S_{nL}(\omega) = 4kTR_L|H_L(j\omega)|^2. \quad (7)$$

Figure 9 shows the behavior of S_{nL} as a function of the frequency, for four different values of R_L (100 kΩ, 500 kΩ, 6 MΩ, and 20 MΩ).

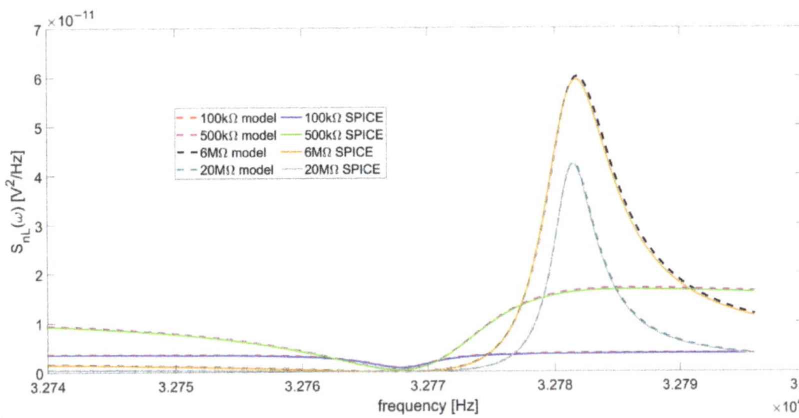

Figure 9. Comparison between SPICE simulations and analytical model described in Equation (7) of the output noise spectral density contribution from the thermal noise of R_L, for four different values of the resistor: 100 kΩ, 500 kΩ, 6 MΩ, and 20 MΩ.

The curves calculated with Equation (7) and those obtained by SPICE simulation are in excellent agreement. The function $|H_L(j\omega)|$ has a minimum located at $\omega = \omega_S$, since, at the series-resonant frequency, the Butterworth–Van Dyke impedance model of the QTF is reduced to R_p, which is its minimum value. Accordingly, also $S_{nL}(\omega)$ exhibits a minimum at the same frequency. Furthermore, a peak appears around the parallel-resonant frequency for increasing values of R_L. Figure 10 reports the behavior of the peak value of $S_{nL}(\omega)$ as a function of R_L, which varies from 500 kΩ to 20 MΩ.

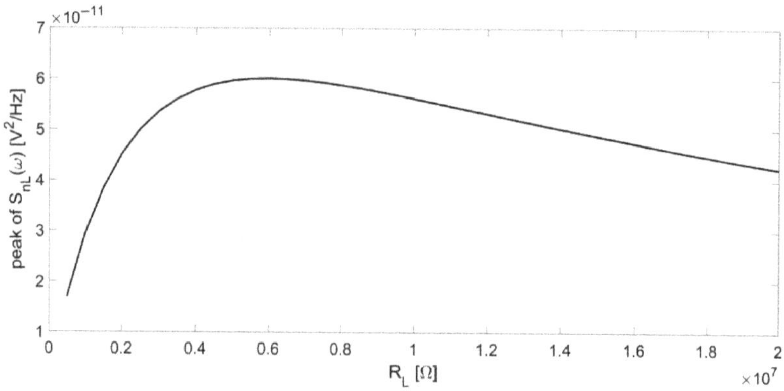

Figure 10. Peak of the output noise spectral density contribution due to R_L as a function of R_L itself.

Using the set of parameters listed in Table 1, starting from low values of R_L, this peak appears when $R_L \cong 500\ k\Omega$ (dashed magenta and green curves in Figure 9), then increases up to $R_L \cong 6\ M\Omega$ (Figure 10). Beyond this value, the peak decreases for increasing values of R_L and the function $S_{nL}(\omega)$ assumes lower values in the frequency range under investigation.

Figures 7 and 10 suggest that, at least for what concerns the noise contributions considered up to now, it is convenient to work with large values of the resistor R_L because the amplitude of the output signal increases with R_L (see Figure 7) and the noise contribution due to this resistor decreases (see Figure 10), whereas the contribution from the thermal noise of R_p has exactly the same frequency behavior of the signal.

Let us now consider the contribution of the input equivalent voltage noise of the OPAMP to the total output noise:

$$S_{nop}(\omega) = e_{n\_op}^2 |H_{en}(j\omega)|^2. \tag{8}$$

Since the flicker noise has been considered negligible in the narrow bandwidth of interest, namely, around the QTF resonance frequency, the input equivalent noise voltage has a constant power spectral density, i.e., it can be considered as white noise. The value of e_{n_op} has been set at 6.6 nV/$\sqrt{\text{Hz}}$, as reported in the data sheet of the AD8067 [31]. The transfer function $H_{en}(j\omega)$ is the following:

$$\begin{aligned} H_{en}(j\omega) &= \frac{V_{n\_out}(j\omega)}{e_{n\_op}(j\omega)} \\ &= A_v \frac{1-\omega^2(LC_s+R_pR_LC_pC_s)+j\omega\left[(R_p+R_L)C_s+R_LC_p-\omega^2LC_pC_sR_L\right]}{1-\omega^2(LC_s+R_pR_LC_{pt}C_s)+j\omega\left[(R_p+R_L)C_s+R_LC_{pt}-\omega^2LC_{pt}C_sR_L\right]}, \end{aligned} \tag{9}$$

in which the numerator differs from the denominator only for the capacitance $C_{pt} = C_p + C_{in}$ replaced by C_p. This contribution can be considered as constant in the frequency range investigated for QEPAS applications.

The contribution of the input equivalent current noise i_{n+} to the overall output noise spectral density $S_{ntot}(\omega)$ is

$$S_{nin+}(\omega) = i_{n+}^2(\omega) R_L^2 |H_L(j\omega)|^2. \tag{10}$$

Using Equations (10) and (7), the contributions to $S_{ntot}(\omega)$, respectively due to i_{n+} and R_L, can be compared:

$$\frac{S_{nin+}(\omega)}{S_{nL}(\omega)} = \frac{i_{n+}^2(\omega) R_L}{4kT}.$$

The contribution of $S_{in+}(\omega)$ can be neglected with respect to $S_{nL}(\omega)$ if

$$i_{n+}^2(\omega) \ll \frac{4kT}{R_L}.$$

If we consider a large R_L value, i.e., 100 MΩ, $S_{in+}(\omega)$ will be negligible at room temperature if i_{n+} << 13 fA/$\sqrt{\text{Hz}}$. Since the AD8067 has FET inputs, $i_{n+}(\omega)$ cannot be considered white, but is a linear function of the frequency, in the range where the flicker noise can be neglected [33]. In our model, the value of $i_{n+}(\omega)$ at 10 kHz has been set to 1 fA/$\sqrt{\text{Hz}}$, a slightly higher value than the one reported in the datasheet of the OPAMP, which is 0.6 fA/$\sqrt{\text{Hz}}$ [31], and the slope of the linear function has been set to +20 dB/dec [33]. Figure 11 shows the excellent correspondence between the analytical model used for $i_{n+}(\omega)$ and the input equivalent noise current resulting from a simulation carried out with the SPICE model of the AD8067.

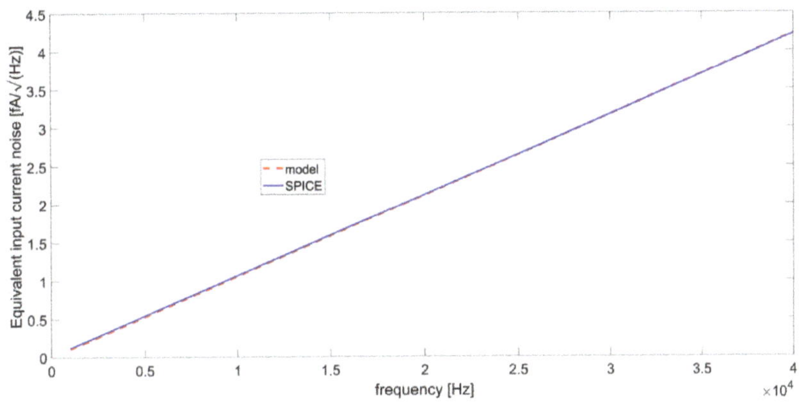

Figure 11. Fitting between the analytical model and SPICE simulations of the input equivalent noise current of the AD8067.

It is worth noticing that around the resonance frequencies of the QTF, the level of $i_{n+}(\omega)$ remains lower than the limit of 13 fA/$\sqrt{\text{Hz}}$ determined above. Thus, we can conclude that the contribution $S_{in+}(\omega)$ can be neglected with respect to $S_{nL}(\omega)$ in the noise analysis of our circuit without losing accuracy.

The analysis carried out so far allows for a comparison of the terms in Equation (5) to understand which ones are dominant for the output noise spectral density of the preamplifier. Figure 12 compares the spectral contributions of the OPAMP and the resistors R_p and R_L to the overall $S_{ntot}(\omega)$ at the output of the circuit. The terms due to e_{n_op} and i_{n+}, namely, $S_{nop}(\omega)$ and $S_{nin+}(\omega)$, respectively, have been summed, resulting in the overall noise contribution of the OPAMP ($S_{opamp}(\omega)$) and allowing for a comparison of the analytical model with the results of SPICE simulations, in which the two terms are not distinguishable. The same R_L values of Figure 9 have been considered (100 kΩ, 500 kΩ, 6 MΩ, and 20 MΩ).

For all the considered cases, the results provided by the analytical model and the SPICE simulations exhibit a very good agreement. In addition, the contribution of the OPAMP to the total output noise is always negligible compared to the sum of the contributions from R_L and R_p. $S_{nL}(\omega)$ tends to prevail over $S_{np}(\omega)$ for R_L < 6 MΩ (Figure 12a,b), whereas the opposite happens when the value of R_L is larger than 6 MΩ (Figure 12c,d). This confirms the previous conclusion about the advantage of working with large values of the resistor R_L in order to achieve good performance in terms of SNR.

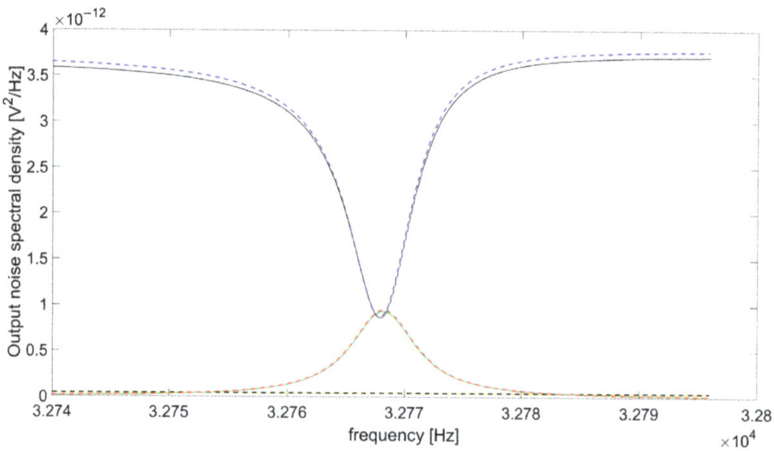

(**a**)

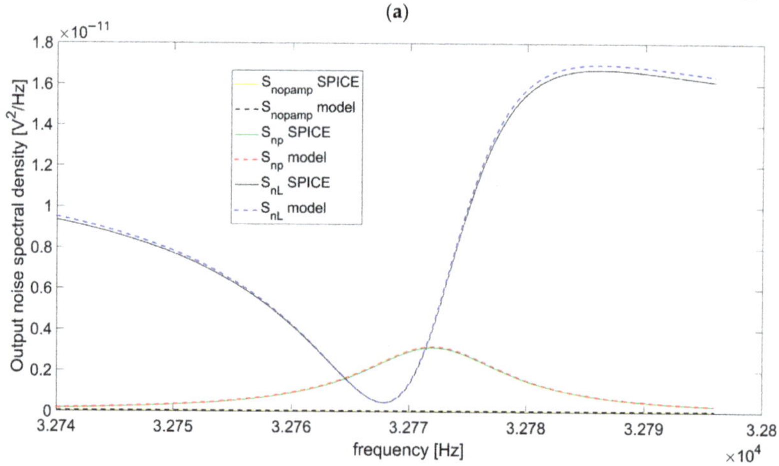

(**b**)

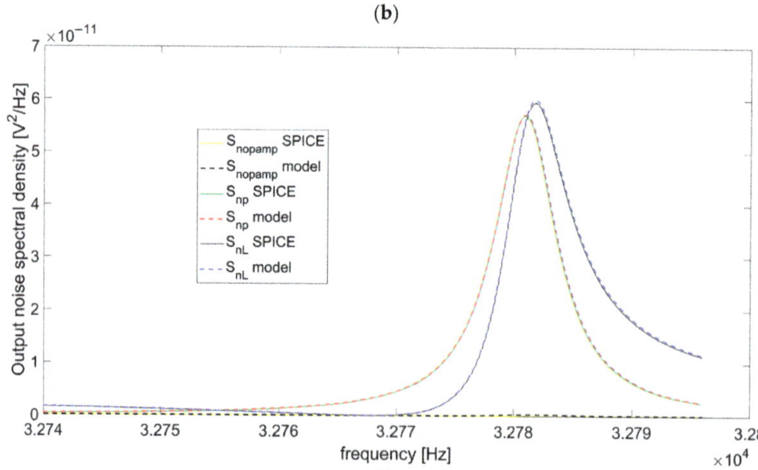

(**c**)

Figure 12. *Cont.*

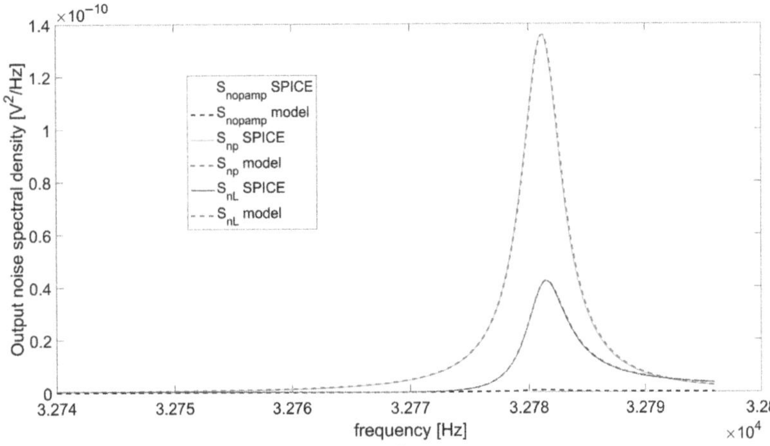

(d)

Figure 12. Contributions to the overall output noise spectral density due to the OPAMP and resistors R_p and R_L, for (**a**) R_L = 100 kΩ, (**b**) R_L = 500 kΩ, (**c**) R_L = 6 MΩ, (**d**) R_L = 20 MΩ: comparison between the results obtained with the analytical model and SPICE simulations.

4. Signal-to-Noise Ratio at the Output of the Lock-in Amplifier

Synchronous detection techniques based on laser modulation and lock-in amplifier are always used in QEPAS sensors to increase the SNR of the measurements. In the previous sections, the amplitude of the useful signal and the noise spectral density at the output of the voltage-mode preamplifier are expressed as a function of frequency. These expressions can be conveniently exploited to evaluate the SNR at the output of the LIA. Thanks to synchronous demodulation and narrow-band low-pass filtering, the LIA output signal is DC-level proportional to the amplitude of the preamplifier response at the operating frequency. When the preamplifier response is acquired at the QTF resonance frequency, the trend of the SNR can be investigated in a small range around f_s. Thus, if the preamplifier response is acquired at a certain frequency f_{op} close to f_s, the output signal is proportional to $|H_v(f_{op})|$. For what concerns noise, as a result of demodulation, the LIA transfers around DC the noise spectrum at the output of the preamplifier, centered around the LIA reference frequency. This noise is then filtered by means of the LIA narrow low-pass filter. Thus, the LIA behaves in practice like a narrow band-pass filter centered around its reference frequency. Hence, LIA was modelled as a biquadratic band-pass filter with a transfer function:

$$H_{LIA}(j\omega) = \frac{j\frac{\omega_{op}}{Q_{filt}}\omega}{\omega_{op}^2 - \omega^2 + j\frac{\omega_{op}}{Q_{filt}}\omega}, \tag{11}$$

characterized by unity gain at center frequency and −3dB bandwidth BW = ω_{op}/Q_{filt} [34].

Therefore, we can describe the behavior of the SNR at the output of the LIA as a function of the chosen operating frequency $f_{op} = \omega_{op}/2\pi$ by means of the following function:

$$\begin{aligned} SNR_n(\omega_{op}) &= \frac{|H_v(\omega_{op})|}{\sqrt{\int_{-\infty}^{\infty} S_{ntot}(\omega)|H_{LIA}(j\omega)|^2 d\omega}} = \\ &\cong \frac{|H_v(\omega_{op})|}{\sqrt{\int_{-\infty}^{\infty} S_{np}(\omega)|H_{LIA}(j\omega)|^2 df + \int_{-\infty}^{\infty} S_{nL}(\omega)|H_{LIA}(j\omega)|^2 d\omega}} \end{aligned} \tag{12}$$

where the amplitude of the input signal V_{in} (see Figure 3) has been normalized to 1 V.

By separating the noise contributions from R_p and R_L, we can define two functions:

$$SNR_{np}^2(\omega_{op}) = \frac{|H_v(\omega_{op})|^2}{\int_{-\infty}^{\infty} S_{np}(\omega)|H_{LIA}(j\omega)|^2 d\omega}$$

and

$$SNR_{nL}^2(\omega_{op}) = \frac{|H_v(\omega_{op})|^2}{\int_{-\infty}^{\infty} S_{nL}(\omega)|H_{LIA}(j\omega)|^2 d\omega}$$

Thus, the total normalized squared-SNR can be rewritten as follows:

$$SNR_n^2 = \frac{1}{\frac{1}{SNR_{np}^2} + \frac{1}{SNR_{nL}^2}}. \tag{13}$$

First, let us consider a very narrow-bandwidth BW = 0.5 Hz, corresponding to a settling time of about 1.3 s for an equivalent first-order LIA low-pass filter. In this case, the behavior of the integrated noise as a function of f_{op} is expected to be very similar to the output noise spectral density. Thus, the SNR_{np} contribution is expected to be almost spectrally flat since the noise from R_P and the signal have the same transfer function $H_v(f)$. Nonetheless, considering a representative value for R_L = 100 kΩ, a peak of SNR_{np} emerges around the peak frequency of $|H_v(f)|$, because of the effect of integration, as reported in Figure 13. As shown in Figure 13, for a bandpass filter, the effect of the integration bandwidth on the input noise spectral density can be more intuitively represented by means of a simple brick-wall filter.

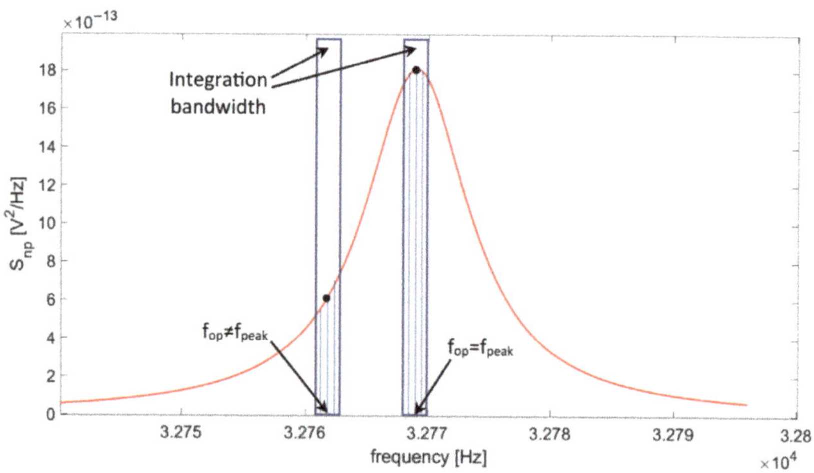

Figure 13. Integration of the contribution S_{np} (red line), due to R_P, to the total output spectral noise density at a low-pass filter bandwidth of 0.5 Hz around the operating frequency f_{op}, for $f_{op} = f_{peak}$ and $f_{op} \neq f_{peak}$ (R_L = 100 kΩ).

Indeed, in the narrow integration bandwidth, the function $S_{np}(f)$ takes monotonically decreasing values on both sides of $f_{op} = f_{peak}$; whereas, for any other frequency value, $S_{np}(f)$ has decreasing values in one direction, but increasing values in the other, as illustrated in Figure 13. As a consequence, the value of the integrated noise normalized to the value of the signal reaches a minimum value at $f_{op} = f_{peak}$, generating a peak value in the function $SNR_{np}(f_{op})$.

The presence of a peak value for SNR_{np} is also visible in Figure 14, which reports SNR_{np} as a function of the frequency f_{op}, for different values of R_L. Moreover, for increasing values of R_L, the peak moves from f_S to f_P; for large values of R_L, it becomes slightly more pronounced, even though the overall behavior always remains rather flat.

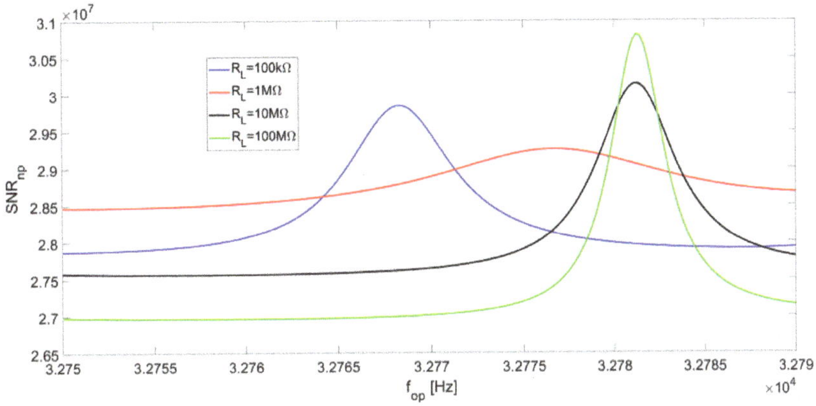

Figure 14. R_P noise contribution to the SNR_n at the LIA output as a function of the operating frequency, for different values of R_L and at a low-pass filter bandwidth of BW = 0.5 Hz.

As for the term SNR_{nL}, its behavior as a function of f_{op} is strongly affected by the minimum of the noise spectral density S_{nL} at f_S. This effect prevails over the peak value of the signal, which moves towards f_P for increasing values of R_L, resulting in a local maximum point on SNR_{nL} which is always located at the series-resonant frequency f_S for any value of R_L. Figure 15 shows the behavior of SNR_{nL} as a function of the operating frequency for different R_L values.

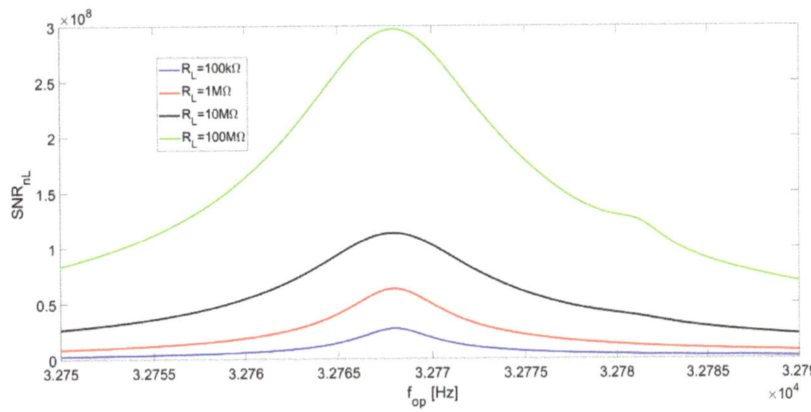

Figure 15. R_L noise contribution to the SNR_n at the LIA output as a function of the operating frequency, for different values of R_L and at a low-pass filter bandwidth of BW = 0.5 Hz.

Higher values of SNR_{nL} are obtained for increasing values of R_L values. Therefore, as reported in Equation (13), SNR_{np} becomes dominant in the calculation of the overall SNR for large values of R_L.

We can now consider the overall SNR at the LIA output given by Equation (13) as a function of f_{op}. Figure 16 reports the SNR_n as a function of f_{op} for different values of R_L with BW = 0.5 Hz. Here, a comparison between the results given by the analytical model and the corresponding SPICE simulation is also proposed, highlighting an excellent correspondence.

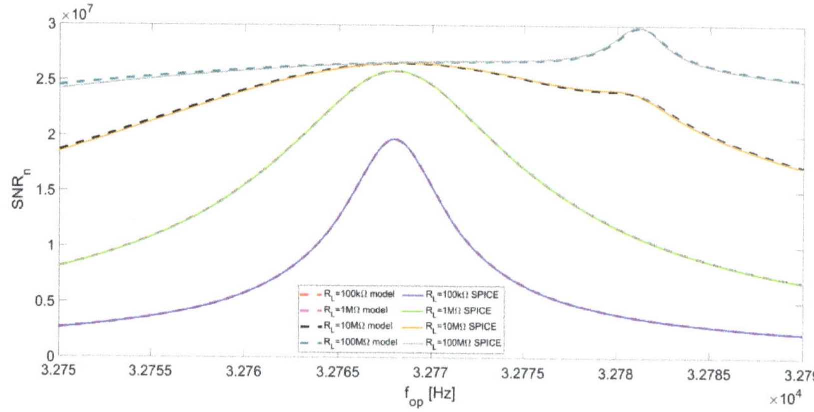

Figure 16. Total normalized signal-to-noise ratio SNR_n at the LIA output as a function of the operating frequency, for different values of R_L and at a low-pass filter bandwidth of BW = 0.5 Hz. The corresponding results of SPICE simulations are also reported.

For low values of R_L, SNR_{np} is almost flat (see Figure 14) and the peak of the total SNR_n coincides with the peak of SNR_{nL}, so the best operating frequency for QEPAS is the series-resonant frequency f_S of the QTF. When R_L is increased, as already pointed out, the contribution of SNR_{nL} becomes less relevant, so a local peak in the SNR_n emerges, corresponding to the peak of SNR_{np}, which occurs at the parallel-resonant frequency f_P.

From Figure 16, it is evident that the SNR_n peak feature at f_S becomes less sharp as R_L increases. For R_L = 100 MΩ, the SNR_n becomes quite flat around f_S and a sharp peak appears at f_P. In general, slightly better results in terms of SNR are obtained with large values of R_L.

The study above has been carried out for a LIA with a very narrow-band filter, which is useful when the requirements in terms of noise rejection are very harsh and the long time needed for a single measurement can be tolerated. With fast measurements [35,36], the bandwidth of the LIA filter must necessarily be increased, and the results of the previous analysis must be revised. Moreover, the QTF response time represents a further issue to deal with when fast measurements are needed. The response time is given by:

$$\tau = \frac{Q}{\pi f_s} \cong 100\ \text{ms}.$$

This implies that, in conventional QEPAS applications, a long integration time (300 to 400 ms) is required to acquire the LIA output signal. Nevertheless, specific QEPAS techniques such as Beat Frequency (BF) QEPAS exploit the fast transient response of an acoustically excited QTF to retrieve the gas concentration, the resonance frequency, and the quality factor of the QTF [7]. This approach overcomes the limitations imposed by the time response of the QTF, allowing shorter acquisition times and faster measurements.

As concerns the contribution to the total SNR at the LIA output due to the resistor R_p, since the integration bandwidth is larger, the effect of the integration around f_{peak}, described above (see Figure 13), is more relevant. Consequently, the peak of SNR_{np} as a function of the operating frequency, always placed at f_{peak}, becomes sharper and more pronounced. Figure 17 shows the behavior of SNR_{np} as a function of f_{op} when BW = 5 Hz, corresponding to a settling time of about 130 ms for an equivalent first order LIA low-pass filter for the same values of R_L considered in Figure 14.

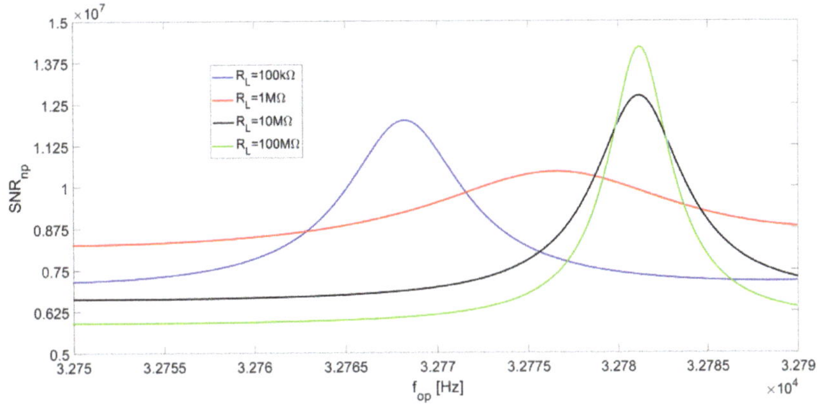

Figure 17. R_P noise contribution to the SNR_n at the LIA output as a function of the operating frequency, for different values of R_L and at a low-pass filter bandwidth of BW = 5 Hz.

We now analyze the R_L noise contribution to the SNR for increasing values of the LIA bandwidth. The effect of the minimum of S_{nL}, placed at f_S, tends to be less relevant, because of the increase of the integration bandwidth. Therefore, the function SNR_{nL} becomes flatter around the series-resonant frequency. Nonetheless, for small values of R_L, the peak of the signal is located at f_S and, consequently, the peak of SNR_{nL} is still placed at f_S. When R_L is increased, the peak of the signal moves towards the parallel-resonant frequency f_P and the effect of the minimum of S_{nL} becomes less relevant, so SNR_{nL} tends to be flat. For large values of R_L, S_{nL} values decrease (see Figure 12c,d), whereas the peak of the signal placed at f_P increases; as a result, SNR_{nL} exhibits a peak at f_P, which becomes sharper as R_L increases. The behavior of SNR_{nL} as a function of the operating frequency f_{op} is shown in Figure 18 for BW = 5 Hz.

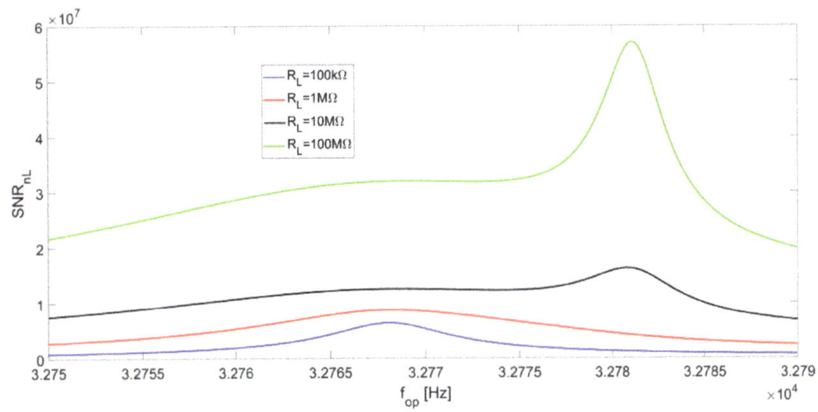

Figure 18. R_L noise contribution to the SNR_n at the LIA output as a function of the operating frequency, for different values of R_L and at a low-pass filter bandwidth of BW = 5 Hz.

Finally, from Equation (13), Figure 19 shows that the peak value of SNR_n results at f_S, where the peak of both SNR_{np} and SNR_{nL} is located, for low values of R_L.

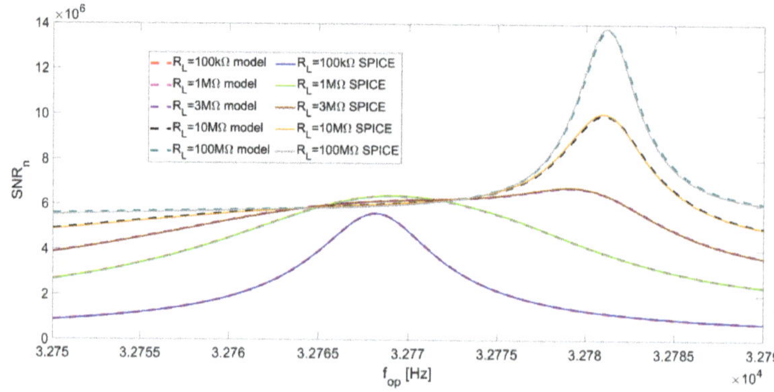

Figure 19. Total normalized signal-to-noise ratio SNR_n at the LIA output as a function of the operating frequency, for different values of R_L and at a low-pass filter bandwidth of BW = 5 Hz. The corresponding results of SPICE simulations are also reported.

For increasing values of R_L, the SNR_n peak feature starts to flatten out until, for values of R_L larger than about 2 MΩ, a peak emerges close to f_P, where both SNR_{np} and SNR_{nL} have their maximum value. This peak becomes as sharper as R_L increases, as shown in Figure 19, which represents the behavior of SNR as a function of f_{op} for BW = 5 Hz. Additionally in this case, SPICE simulations are in very good agreement with the results provided by the analytical model. The SNR_n peak at 100 MΩ is more than two times higher than the values close to f_s.

Finally, the presented study can be applied to compute the Normalized Noise Equivalent Absorption (NNEA), an important parameter to compare QEPAS sensors [2,4,5,7,10]. NNEA is defined as follows:

$$\mathrm{NNEA} = \frac{P \cdot \alpha}{\mathrm{SNR} \cdot \sqrt{\Delta f}}$$

where P is the laser optical power, α is the absorption coefficient of the gas under investigation, SNR is the signal-to-noise ratio, and Δf is the integration bandwidth.

In [10], an NNEA of 5.0×10^{-9} $W \cdot cm^{-1}/\sqrt{Hz}$ for the detection of CO_2 in an N_2 mixture, employing a transimpedance amplifier with a 10-MΩ feedback resistor and a narrow-bandwidth LIA filter, was demonstrated. Assuming the same value of NNEA for a voltage preamplifier with a 10-MΩ bias resistor and an integration bandwidth of 0.5 Hz, a 100-MΩ bias resistor and the same integration bandwidth would lead to an NNEA of 4.4×10^{-9} $W \cdot cm^{-1}/\sqrt{Hz}$, thus an improvement of a factor of 1.1, as suggested from Figure 16.

Considering a 5 Hz LIA filter bandwidth, an NNEA of 5.8×10^{-9} $W \cdot cm^{-1}/\sqrt{Hz}$ can be calculated for a bias resistor of 10 MΩ. An improvement of a 1.4 factor can be calculated when a bias resistor of 100 MΩ is employed, leading to a NNEA of 4.1×10^{-9} $W \cdot cm^{-1}/\sqrt{Hz}$. Table 2 summarizes the NNEA values obtained for different values of R_L and Δf: it can be observed that increasing R_L at a fixed filter bandwidth always results in an improvement of the NNEA.

Table 2. NNEA for different values of R_L and Δf, calculated starting from the value reported in [10] for the detection of CO_2 in an N_2 mixture.

R_L [MΩ]	Δf [Hz]	NNEA [$W \cdot cm^{-1}/\sqrt{Hz}$]
10	0.5	5.0×10^{-9} [10]
100	0.5	4.4×10^{-9}
10	5	5.8×10^{-9}
100	5	4.1×10^{-9}

5. Discussion and Conclusions

In this work, we investigated the SNR trend as a function of the operating frequency for a voltage-mode read-out of QTFs, comparing the results obtained by the developed model with SPICE simulations. The effects of the R_L values, the operating frequency, and the bandwidth of the LIA low-pass filter on the overall SNR were investigated. We firstly demonstrated that the contribution of the OPAMP to the output noise of the voltage preamplifier can be always neglected and the output noise power density is dominated by the contributions of the QTF-equivalent resistor R_p and R_L. Moreover, the dominant contribution to the total noise changes from R_L, when $R_L < 10$ MΩ, to R_p, for $R_L > 10$ MΩ.

According to our model, the peak value of the SNR tends to increase along with the bias resistor R_L.

When a narrowband low-pass filter of the LIA is employed (e.g., bandwidth of 0.5 Hz), the curves shown in Figure 16 suggest that only a limited increase of the SNR can be obtained by either increasing the value of the resistor R_L or varying the operating frequency. For $R_L < 10$ MΩ, the best operating frequency for the QEPAS technique is the series-resonant frequency of the QTF f_S, where SNR function exhibits a peak value. This peak becomes less pronounced when R_L is increased. A 10-MΩ bias resistor ensures a 1.3 times higher SNR at the series-resonant frequency f_S, with respect to a 100 kΩ-bias resistor. For large values of R_L, SNR tends to be flat around f_S and a small peak emerges at the parallel-resonant frequency f_P. For instance, increasing the value of R_L up to 100 MΩ leads to a further increase of the SNR peak by a factor of 1.4. Thus, for these large values of R_L, the choice of the optimal operating frequency for QEPAS is not very critical, in case of a narrow-band LIA filter.

When a wider LPF bandwidth of the LIA is employed (e.g., bandwidth of 5 Hz), the SNR peak as a function of the operating frequency is still placed at f_S for small values of R_L, as in the case of narrow-band filter (see Figure 19). For large values of R_L, the SNR peak at f_P tends to be sharper as compared to the case of narrow-band LIA filter, thus the operating frequency for the QEPAS technique must be chosen as close as possible to f_P to maximize SNR. As an example, for R_L = 100 MΩ, the SNR peak is 1.4 times higher with respect to R_L =10 MΩ.

In addition, the parallel-resonant frequency f_P of the QTF depends on the input capacitance of the OPAMP, so it is not an intrinsic property of the sensitive element. Thus, suitable techniques must be used to measure f_P in presence of C_{in} to exploit large values of R_L, increase SNR, and optimize the performance of the QEPAS sensor, especially when short acquisition times are needed. Instead, for long acquisition times, large values of R_L are not very effective for the increase of the overall SNR, and an accurate setting of the operating frequency close to f_s is not critical because of the flatness of SNR as a function of f_{op}. Of course, the value of R_L cannot be made too large, since the input bias current of the OPAMP would cause unacceptable input offset levels. All the results obtained with the previous analysis have been confirmed by AC and noise SPICE simulations of the voltage preamplifier followed by a band-pass filter with center frequency f_{op} and variable bandwidth.

Since, in QEPAS applications, several QTFs have been employed [1–14], the same study has also been carried out for a representative QTF characterized by an f_S = 12.484 kHz and a quality factor of 10^4 [9], and the same results were found.

The reported voltage amplifier, implementing the AD8067, will be employed to validate the results obtained both with our model and with the SPICE simulation.

Author Contributions: Conceptualization, M.D.G., G.M. (Giansergio Menduni), P.P. and V.S.; methodology, L.L., C.M. and G.M. (Giansergio Menduni); validation, M.D.G. and G.M. (Gianvito Matarrese); formal analysis, M.D.G., L.L. and C.M.; writing—original draft preparation, C.M.; writing—review and editing, M.D.G., L.L., Giansergio Menduni and P.P.; visualization, L.L., G.M. (Giansergio Menduni) and G.M. (Gianvito Matarrese); supervision, C.M., P.P. and V.S.; project administration, P.P. and V.S.; funding acquisition, V.S. All authors have read and agreed to the published version of the manuscript.

Funding: All authors acknowledge funding from the European Union's Horizon 2020 research and innovation program under grant agreement No. 101016956 PASSEPARTOUT, in the context of the "INDUSTRIAL LEADERSHIP—Leadership in enabling and industrial technologies - Information and Communication Technologies (ICT)".

Data Availability Statement: Not applicable.

Conflicts of Interest: The authors declare no conflict of interest.

References

1. Olivieri, M.; Menduni, G.; Giglio, M.; Sampaolo, A.; Patimisco, P.; Wu, H.; Dong, L.; Spagnolo, V. Characterization of H_2S QEPAS detection in methane-based gas leaks dispersed into environment. *Photoacoustics* **2023**, *29*, 100438. [CrossRef]
2. Lin, H.; Liu, Y.; Lin, L.; Zhu, W.; Zhou, X.; Zhong, Y.; Giglio, M.; Sampaolo, A.; Patimisco, P.; Tittel, F.K.; et al. Application of standard and custom quartz tuning forks for quartz-enhanced photoacoustic spectroscopy gas sensing. *Appl. Spectrosc. Rev.* **2022**, *57*, 1–23. [CrossRef]
3. Menduni, G.; Zifarelli, A.; Sampaolo, A.; Patimisco, P.; Giglio, M.; Amoroso, N.; Wu, H.; Dong, L.; Bellotti, R.; Spagnolo, V. High-concentration methane and ethane QEPAS detection employing partial least squares regression to filter out energy relaxation dependence on gas matrix composition. *Photoacoustics* **2022**, *26*, 100349. [CrossRef]
4. Yin, X.; Gao, M.; Miao, R.; Zhang, L.; Zhang, X.; Liu, L.; Shao, X.; Tittel, F.K. Near-infrared laser photoacoustic gas sensor for simultaneous detection of CO and H_2S. *Opt. Express.* **2021**, *29*, 34258. [CrossRef]
5. Ma, Y.; Lewicki, R.; Razeghi, M.; Tittel, F.K. QEPAS based ppb-level detection of CO and N_2O using a high power CW DFB-QCL. *Opt. Exp.* **2013**, *21*, 1008–1019. [CrossRef]
6. Ma, Y. Review of Recent Advances in QEPAS-Based Trace Gas Sensing. *Appl. Sci.* **2018**, *8*, 1822. [CrossRef]
7. Wu, H.; Dong, L.; Zheng, H.; Yu, Y.; Ma, W.; Zhang, L.; Yin, W.; Xiao, L.; Jia, S.; Tittel, F.K. Beat frequency quartz-enhanced photoacoustic spectroscopy for fast and calibration-free continuous trace-gas monitoring. *Nat. Commun.* **2017**, *8*, 15331. [CrossRef]
8. Liu, Y.; Lin, H.; Montano, B.A.Z.; Zhu, W.; Zhong, Y.; Kan, R.; Yuan, B.; Yu, J.; Shao, M.; Zheng, H. Integrated near-infrared QEPAS sensor based on a 28 kHz quartz tuning fork for online monitoring of CO_2 in the greenhouse. *Photoacoustics* **2022**, *25*, 100332. [CrossRef]
9. Menduni, G.; Zifarelli, A.; Kniazeva, E.; Dello Russo, S.; Ranieri, A.C.; Ranieri, E.; Patimisco, P.; Sampaolo, A.; Giglio, M.; Manassero, F.; et al. Measurement of methane, nitrous oxide and ammonia in atmosphere with a compact quartz-enhanced photoacoustic sensor. *Sens. Actuators B Chem.* **2023**, *375*, 132953. [CrossRef]
10. Zifarelli, A.; De Palo, R.; Patimisco, P.; Giglio, M.; Sampaolo, A.; Blaser, S.; Butet, J.; Landry, O.; Müller, A.; Spagnolo, V. Multi-gas quartz-enhanced photoacoustic sensor for environmental monitoring exploiting a Vernier effect-based quantum cascade laser. *Photoacoustics* **2022**, *28*, 100401. [CrossRef]
11. Liberatore, N.; Viola, R.; Mengali, S.; Masini, L.; Zardi, F.; Elmi, I.; Zampolli, S. Compact GC-QEPAS for On-Site Analysis of Chemical Threats. *Sensors* **2023**, *23*, 270. [CrossRef]
12. Zifarelli, A.; Menduni, G.; Giglio, M.; Elefante, A.; Sukhinets, A.; Sampaolo, A.; Patimisco, P.; Fangyuan, S.; Chongwu, W.; Wang, Q.J.; et al. Compact and versatile QEPAS-based sensor box for simultaneous detection of methane and infrared absorber gas molecules in ambient air. *Front. Environ. Chem.* **2022**, *3*, 926233. [CrossRef]
13. Li, B.; Feng, C.; Wu, H.; Jia, S.; Dong, L. Calibration-free mid-infrared exhaled breath sensor based on BF-QEPAS for real-time ammonia measurements at ppb level. *Sens. Actuators B Chem.* **2022**, *358*, 131510. [CrossRef]
14. Shang, Z.; Li, S.; Li, B.; Wu, H.; Sampaolo, A.; Patimisco, P.; Spagnolo, V.; Dong, L. Quartz-enhanced photoacoustic NH_3 sensor exploiting a large-prong-spacing quartz tuning fork and an optical fiber amplifier for biomedical applications. *Photoacoustics* **2022**, *26*, 100363. [CrossRef] [PubMed]
15. Dello Russo, S.; Sampaolo, A.; Patimisco, P.; Menduni, G.; Giglio, M.; Hoelzl, C.; Passaro, V.M.N.; Wu, H.; Dong, L.; Spagnolo, V. Quartz-enhanced photoacoustic spectroscopy exploiting low-frequency tuning forks as a tool to measure the vibrational relaxation rate in gas species. *Photoacoustics* **2021**, *21*, 100227. [CrossRef]
16. Kosterev, A.A.; Bakhirkin, Y.A.; Curl, R.F.; Tittel, F.K. Quartz-enhanced photoacoustic spectroscopy. *Opt. Lett.* **2002**, *27*, 1902–1904. [CrossRef]
17. Kosterev, A.A.; Tittel, F.K.; Serebryakov, D.; Malinovsky, A.; Morozov, A. Applications of quartz tuning fork in spectroscopic gas sensing. *Rev. Sci. Instrum.* **2005**, *76*, 043105. [CrossRef]
18. Menduni, G.; Sampaolo, A.; Patimisco, P.; Giglio, M.; Dello Russo, S.; Zifarelli, A.; Elefante, A.; Wieczorek, P.Z.; Starecki, T.; Passaro, V.M.N.; et al. Front-End Amplifiers for Tuning Forks in Quartz Enhanced PhotoAcoustic Spectroscopy. *Appl. Sci.* **2020**, *10*, 2947. [CrossRef]
19. Wieczorek, P.Z.; Starecki, T.; Tittel, F.K. Improving the signal to noise ratio of QTF preamplifiers dedicated for QEPAS applications. *Appl. Sci.* **2020**, *10*, 4105. [CrossRef]
20. Winkowski, M.; Stacewicz, T. Low noise, open-source QEPAS system with instrumentation amplifier. *Sci. Rep.* **2019**, *9*, 7–12. [CrossRef]
21. Starecki, T.; Wieczorek, P.Z. A High Sensitivity Preamplifier for Quartz Tuning Forks in QEPAS (Quartz Enhanced PhotoAcoustic Spectroscopy) Applications. *Sensors* **2017**, *17*, 2528. [CrossRef]

22. Principles of Lock-in Detection and the State of the Art, Zurich Instruments. Available online: https://www.zhinst.com/sites/default/files/documents/2020-06/zi_whitepaper_principles_of_lock-in_detection.pdf (accessed on 19 January 2023).
23. Kishore, K.; Akbar, S.A. Evolution of Lock-In Amplifier as Portable Sensor Interface Platform: A Review. *IEEE Sens. J.* **2020**, *20*, 10345–10354. [CrossRef]
24. Zhang, Z.; Li, C.; Huang, Z. Quadrature demodulation based on lock-in amplifier technique for self-mixing interferometry displacement sensor. *Appl. Opt.* **2019**, *58*, 6098–6104. [CrossRef]
25. Huang, K.; Geng, Y.; Zhang, X.; Chen, D.; Cai, Z.; Wang, M.; Zhu, Z.; Wang, Z. A Wide-Band Digital Lock-In Amplifier and Its Application in Microfluidic Impedance Measurement. *Sensors* **2019**, *19*, 3519. [CrossRef] [PubMed]
26. Patimisco, P.; Scamarcio, G.; Tittel, F.K.; Spagnolo, V. Quartz-Enhanced Photoacoustic Spectroscopy: A Review. *Sensors* **2014**, *14*, 6165–6206. [CrossRef]
27. Levy, R.; Duquesnoy, M.; Melkonian, J.-M.; Raybaut, M.; Aoust, G. New Signal Processing for Fast and Precise QEPAS Measurements. *IEEE Trans. Ultrason. Ferroelectr. Freq. Control* **2020**, *67*, 1230–1235. [CrossRef]
28. Serebryakov, D.V.; Cherkun, A.P.; Loginov, B.A.; Letokhov, V.S. Tuning-fork-based fast highly sensitive surface-contact sensor for atomic force microscopy/near-field scanning optical microscopy. *Rev. Sci. Instrum.* **2002**, *73*, 1795. [CrossRef]
29. Friedt, J.-M.; Carry, É. Introduction to the quartz tuning fork. *Am. J. Phys.* **2007**, *75*, 415–422. [CrossRef]
30. The Quartz Crystal Model and Its Frequencies. Technical Note 32, Rev. A. *Statek.* Available online: https://statek.com/wp-content/uploads/2018/03/tn32.pdf (accessed on 19 January 2023).
31. AD8067, High Gain Bandwidth Product, Precision Fast FET™ Op Amp. Analog Devices Inc. Available online: https://www.analog.com/en/products/ad8067.html (accessed on 19 January 2023).
32. Noise Analysis in Operational Amplifier Circuits, Application Report. Texas Instruments. Available online: https://www.ti.com/lit/an/slva043b/slva043b.pdf (accessed on 19 January 2023).
33. Current Noise in FET Input Amps. Analog Devices Inc. Available online: https://www.analog.com/en/analog-dialogue/raqs/raq-issue-174.html (accessed on 19 January 2023).
34. Sedra, A.S.; Smith, K.C. *Microelectronic Circuits*, 6th ed.; Oxford University Press: Oxford, UK, 2010; pp. 975–982.
35. Villa, T.F.; Salimi, F.; Morton, K.; Morawska, L.; Gonzalez, F. Development and Validation of a UAV Based System for Air Pollution Measurements. *Sensors* **2016**, *16*, 2202. [CrossRef] [PubMed]
36. Aasen, H.; Honkavaara, E.; Lucieer, A.; Zarco-Tejada, P.J. Quantitative Remote Sensing at Ultra-High Resolution with UAV Spectroscopy: A Review of Sensor Technology, Measurement Procedures, and Data Correction Workflows. *Remote Sens.* **2018**, *10*, 1091. [CrossRef]

MDPI
St. Alban-Anlage 66
4052 Basel
Switzerland
Tel. +41 61 683 77 34
Fax +41 61 302 89 18
www.mdpi.com

Micromachines Editorial Office
E-mail: micromachines@mdpi.com
www.mdpi.com/journal/micromachines

www.ingramcontent.com/pod-product-compliance
Lightning Source LLC
Chambersburg PA
CBHW061457210926
51850CB00008B/1744